STUDENT'S SOLUTIONS MANUAL

MARY WOLYNIAK
Broome Community College

INTRODUCTORY ALGEBRA THROUGH APPLICATIONS

Geoffrey Akst Sadie Bragg
Borough of Manhattan Community College, City University of New York

Boston San Francisco New York
London Toronto Sydney Tokyo Singapore Madrid
Mexico City Munich Paris Cape Town Hong Kong Montreal

Reproduced by Pearson Addison-Wesley from electronic files supplied by the author.

Publishing as Pearson Addison-Wesley, 75 Arlington Street, Boston, MA 02116.

ISBN 0-201-66229-9

3 4 5 6 BRG 08 07 06 05

Contents

Chapter R Prealgebra Review

Chapter R Pretest

1. 7^3
2. $16 \cdot 25 = 400$
3. $16 - 8 = 8$
4. $20 - 2(3 - 1^2) =$
$20 - 2(2) =$
$20 - 4 = 16$
5. 1, 2, 3, 4, 6, 12
6. $2^2 \cdot 5$
7. $\frac{2}{3}$
8. $\frac{3}{10} + \frac{1}{10} = \frac{4}{10} = \frac{2}{5}$
9.
$10\frac{1}{3=} - 2\frac{5}{6}$
$10\frac{2}{6} - 2\frac{5}{6} =$
$\frac{62}{6} - \frac{17}{6} =$
$\frac{45}{6} = 7\frac{3}{6} = 7\frac{1}{2}$
10. $\frac{2}{9} \cdot \frac{2}{3} = \frac{4}{27}$
11. $4 \div 1\frac{1}{2} = 4 \div \frac{3}{2} = \frac{4}{1} \cdot \frac{2}{3} = \frac{8}{3} = 2\frac{2}{3}$
12. nine and thirteen thousandths
13. 3.1
14. 20.3103
15. 5.66
16. 37.843
17. 23.5
18. 0.18
19. 0.0031
20. 0.07
21. $3 - 1.316 = 1.684$
$2 *million*
22. $\frac{1}{2} \bullet \frac{2}{3} = \frac{1}{3}$ *mile*
23. $0.1 \div 0.05 = 2$ *times*
24. $0.1 = 10\%$
25. $95\% = 0.95$

Practice Section R.1

1. $2 \cdot 2 \cdot 2 \cdot 2 \cdot 2 = 2^5$
2. $7^2 \cdot 2^4 = (7 \cdot 7)(2 \cdot 2 \cdot 2 \cdot 2) = 49 \cdot 16 = 784$
3. 10^9
4. $8 \div 2 + 4 \cdot 3 = 4 + 4 \cdot 3 = 4 + 12 = 16$
5.
$(4+1)^2 - 4 \cdot 6 = 5^2 - 4 \cdot 6 =$
$25 - 4 \cdot 6 = 25 - 24 = 1$
6. sum of this year's bills = $1200.
$A = \frac{\$1200}{12} = \$100$
No, the average this year was lower by $10.

Practice R.2

1.
$10 = 5 \cdot 2$
$25 = 5^2$
$LCM = 5^2 \cdot 2$
$LCM = 50$
2.
$20 = 5 \cdot 2^2$
$36 = 3^2 \cdot 2^2$
$60 = 5 \cdot 3 \cdot 2^2$
$LCM = 5 \cdot 3^2 \cdot 2^2$
$LCM = 180$
3.
$6 = 3 \cdot 2$
$8 = 2^3$
$LCM = 3 \cdot 2^3 = 24$
In 24 hours both medications will be given at the same time.

Practice R.3

1. $3\frac{2}{9} = \frac{(3 \cdot 9) + 2}{9} = \frac{29}{9}$

2. $\frac{8}{3}=2\frac{2}{3}$

3. $\frac{24}{30}=\frac{6\bullet 4}{6\bullet 5}=\frac{4}{5}$

4. $\frac{7}{15}+\frac{3}{15}=\frac{10}{15}=\frac{2}{3}$

5. $\frac{19}{20}-\frac{11}{20}=\frac{8}{20}=\frac{2}{5}$

6.

$\frac{11}{12}+\frac{3}{4}=\frac{11}{12}+\frac{9}{12}=$

$\frac{20}{12}=\frac{5}{3}$ or $1\frac{2}{3}$

7. $\frac{4}{5}-\frac{1}{2}=\frac{8}{10}-\frac{5}{10}=\frac{3}{10}$

8. $4\frac{5}{8}+3\frac{1}{2}=4\frac{5}{8}+3\frac{4}{8}=7\frac{9}{8}=8\frac{1}{8}$

9.

$15\frac{1}{12}=14+1+\frac{1}{12}=14+\frac{12}{12}+\frac{1}{12}=14\frac{13}{12}$

$15\frac{1}{12}-9\frac{11}{12}=14\frac{13}{12}-9\frac{11}{12}=5\frac{2}{12}=5\frac{1}{6}$

10.

$\frac{1}{2}\bullet\frac{3}{4}=\frac{1\bullet 3}{2\bullet 4}=\frac{3}{8}$

11.

$\frac{7}{10}\bullet\frac{5}{11}=\frac{7\bullet\overset{1}{\cancel{5}}}{\underset{2}{\cancel{10}}\bullet 11}=\frac{7}{22}$

12.

$3\frac{3}{4}\bullet 2\frac{1}{10}=$

$\frac{\overset{3}{\cancel{15}}}{4}\bullet\frac{21}{\underset{2}{\cancel{10}}}=\frac{63}{8}=7\frac{7}{8}$

13.

$\frac{3}{4}\div\frac{1}{8}=\frac{3}{\underset{1}{\cancel{4}}}\bullet\frac{\overset{2}{\cancel{8}}}{1}=\frac{6}{1}=6$

14.

$6\div 3\frac{3}{4}=\frac{6}{1}\div\frac{15}{4}=\frac{\overset{2}{\cancel{6}}}{1}\bullet\frac{4}{\underset{5}{\cancel{15}}}=\frac{8}{5}=1\frac{3}{5}$

15.

$1000\bullet\frac{3}{5}=6001000-600=400$ students

Practice R.4

1.

a. $\frac{5}{10}=\frac{1}{2}$

b. $2\frac{73}{1000}$

2. Four and three thousandths
3. 748.08
4. 42.092
5. 1.179
6. 9.835
7. 327,000
8. 18.04
9. 0.00086
10.
Drop 6.2 and 6.7.

$\frac{6.3+6.4+6.5}{3}=\frac{19.2}{3}=6.4$

Practice R.5

1. $\frac{7}{100}$
2. 0.05
3. 2.5%
4. 25%
5. 0.40 or 0.4

Chapter R Exercises

1. 6^5
3. $2^2\bullet 10^3$
5. 25,000
7. $6+9=15$

9.

$2+18\div 3(9-7)=$

$2+18\div 3(2)=$

$2+6(2)=$

$2+12=14$

11. $\frac{4^2+8}{9-3}=\frac{16+8}{6}=\frac{24}{6}=4$

13. 1, 2, 3, 5, 6, 10, 15, 25, 30, 50, 75, 150

15. Prime

17. Composite

19. $2\cdot 3\cdot 7$

21. $2^4\cdot 3$

23.

$6=3\cdot 2$

$8=2^3$

$LCM=2^3\cdot 3=24$

25.

$24=3\cdot 2^3$

$36=2^2\cdot 3^2$

$72=2^3\cdot 3^2$

$LCM=2^3\cdot 3^2=72$

27.

$3\frac{4}{5}=\frac{3\cdot 5+4}{5}=\frac{19}{5}$

29. $5\frac{3}{4}$

31.

$\frac{14}{28}=\frac{2\cdot 7}{2\cdot 2\cdot 7}=\frac{1}{2}$

33. $5\frac{1}{2}$

35.

$\frac{1}{9}+\frac{4}{9}=\frac{1+4}{9}=\frac{5}{9}$

37.

$\frac{3}{8}-\frac{1}{8}=\frac{3-1}{8}=\frac{2}{8}=\frac{1}{4}$

39.

$\frac{2}{5}+\frac{4}{7}=\frac{14}{35}+\frac{20}{35}=\frac{14+20}{35}=\frac{34}{35}$

41.

$\frac{3}{10}-\frac{1}{20}=\frac{6}{20}-\frac{1}{20}=\frac{6-1}{20}=\frac{5}{20}=\frac{1}{4}$

43.

$1\frac{1}{8}+5\frac{3}{8}=\frac{9}{8}+\frac{43}{6}=\frac{9+43}{8}=\frac{52}{8}=6\frac{4}{8}=6\frac{1}{2}$

45.

$8\frac{7}{10}+1\frac{9}{10}=9\frac{16}{10}=10\frac{6}{10}=10\frac{3}{5}$

47.

$9\frac{11}{12}-6\frac{7}{12}=3\frac{4}{12}=3\frac{1}{3}$

49.

$6\frac{1}{10}-4\frac{3}{10}=\frac{61}{10}-\frac{43}{10}=$

$\frac{61-43}{10}=\frac{18}{10}=1\frac{8}{10}=1\frac{4}{5}$

51.

$12-5\frac{1}{2}=\frac{24}{2}-\frac{11}{2}=\frac{24-11}{2}=\frac{13}{2}=6\frac{1}{2}$

53.

$7\frac{1}{2}-4\frac{5}{8}=\frac{15}{2}-\frac{37}{8}=\frac{60}{8}-\frac{37}{8}=\frac{23}{8}=2\frac{7}{8}$

55.

$\frac{2}{3}\cdot\frac{1}{5}=\frac{2\cdot 1}{3\cdot 5}=\frac{2}{15}$

57.

$1\frac{2}{5}\cdot 10=\frac{7}{5}\cdot\frac{10}{1}=\frac{7\cdot 10}{5\cdot 1}=\frac{70}{5}=14$

59

$3\frac{1}{4}\cdot 4\frac{2}{3}=\frac{13}{4}\cdot\frac{14}{3}=\frac{13\cdot 14}{4\cdot 3}=\frac{182}{12}=15\frac{1}{6}$

61.

$2\frac{5}{6}\div\frac{1}{2}=\frac{17}{6}\cdot\frac{2}{1}=\frac{34}{6}=5\frac{2}{3}$

63.

$\frac{2}{3}\div 6=\frac{2}{3}\cdot\frac{1}{6}=\frac{2\cdot 1}{3\cdot 6}=\frac{2}{18}=\frac{1}{9}$

65.

$8\div 2\frac{1}{3}=\frac{8}{1}\div\frac{7}{3}=\frac{8}{1}\cdot\frac{3}{7}=\frac{24}{7}=3\frac{3}{7}$

67.

$$(\frac{3}{4})^2-\frac{3}{8}\div 6=\frac{9}{16}-\frac{3}{8}\div\frac{6}{1}=$$

$$\frac{9}{16}-\frac{3}{8}\bullet\frac{1}{6}=\frac{9}{16}-\frac{1}{16}=\frac{8}{16}=\frac{1}{2}$$

69. $\frac{875}{1000}=\frac{7}{8}$

71. hundredths

73. seventy-two hundredths

75. Three and nine thousandths

77. 7.3

79. 4.39

81. 18.11

83. 1.873

85. 2.912

87. 2710

89.0.0015

91. 5

93. 7.35

95.

$$\frac{75}{100}=\frac{3}{4}$$

97.

$$1\frac{6}{100}=1\frac{3}{50}$$

99. 0.06

101. 1.5

103. 31%

105. 1.45%

107. 10%

109. 80%

111.

$$\frac{67+72+78+70+65+77+82}{7}=$$

$$\frac{511}{7}=73^o\ F$$

113.

$$\frac{1}{2}+\frac{1}{4}=\frac{2}{4}+\frac{1}{4}=\frac{3}{4}$$

115.

$$\frac{1}{3}\bullet 27=\frac{1}{3}\bullet\frac{27}{1}=\frac{27}{3}=\$9 \text{ text}$$

$\$27-\$9=\$18$

cost of roses = $18

117.

$$\frac{\$7.96}{4 \text{ lb}}=\$1.99 \text{ per pound}$$

$$\frac{\$9.87}{3 \text{ lb}}=\$3.29 \text{ per pound}$$

$\$3.29-\$1.99=\$1.30$

119.

$$\frac{15}{100}=\frac{3}{20}$$

Chapter R Posttest

1. $64\bullet 8=512$

2.

$11\bullet 2+5\bullet 3=22+15=37$

3. 1, 2, 4, 5, 10, 20

4.

$$3\frac{1}{4}=\frac{3\bullet 4+1}{4}=\frac{13}{4}$$

5.

$$\frac{10}{36}=\frac{2\bullet 5}{2\bullet 18}=\frac{5}{18}$$

6.

$$\frac{5}{8}+\frac{7}{8}=\frac{12}{8}=1\frac{4}{8}=1\frac{1}{2}$$

7.

$$7\frac{7}{8}+4\frac{1}{6}=7\frac{21}{24}+4\frac{4}{24}=11\frac{25}{24}=12\frac{1}{24}$$

8.

$$\frac{4}{9}-\frac{3}{10}=\frac{40}{90}-\frac{27}{90}=\frac{13}{90}$$

9.

$$12\frac{1}{4}-8\frac{3}{10}=12\frac{5}{20}-8\frac{6}{20}=$$

$$\frac{245}{20}-\frac{166}{20}=\frac{79}{20}=3\frac{19}{20}$$

10. $\frac{3}{4} \bullet \frac{4}{5} = \frac{3}{5}$

11.

$\frac{2}{3} \div \frac{1}{3} = \frac{2}{3} \bullet \frac{3}{1} = \frac{6}{3} = 2$

12.

$7 \div 3\frac{1}{5} = \frac{7}{1} \div \frac{16}{5} = \frac{7}{1} \bullet \frac{5}{16} = \frac{35}{16} = 2\frac{3}{16}$

13. two and three hundred ninety-six thousandths

14. 16.202

15. 6.99

16. 44.678

17. 2070

18. 0.0005

19. 0.125 = 12.5%

20. $70\% = \frac{7}{10}$

21. $3 \div \frac{1}{2} = \frac{3}{1} \bullet \frac{2}{1} = 6$ times

22. 1.8

23. $(30.5\text{m})(20.5\text{m}) = 625.25 \approx 625.3$ sq m

24. $9 \div 0.6 = 15$ times

25. 78% = 0.78

Chapter 1 Pretest

1. + \$2000.
2. Yes.
3.

4. 5
5. $\frac{2}{3}$
6. $-31 < -1$
7. $-6-7=-13$
8. $9+(-4)+2+(-9)=-2$
9.
$3(-7)-5=$
$-21-5=-26$
10. $\frac{4}{1}$ or 4
11. 9
12. ten less than the product of 3 and n (Answers may vary.)
13. -6^4
14.
$2(-2)-4(3)+8=$
$-4-12+8=-8$
15. $4n-7$
16.
$-5(2-x)+9x=$
$-10+5x+9x=$
$-10+14x=$
$14x-10$
17.
$5895-(-156)=$
$5895+156=6051m$
18. $\frac{L}{3}$ or $\frac{1}{3}L$
19.
$P=2(12cm)+2(3cm)=$
$24cm+6cm=30cm$
20.
total interest =
$0.06x+0.08(500-x)=$
$0.06x+40-0.08x=$
$40-0.02x$ dollars

1.1 Integers, Rational Numbers, and The Real-Number Line

Practice 1.1

1. -5° F
2. a. -1.7 **b.** $\frac{5}{4}$

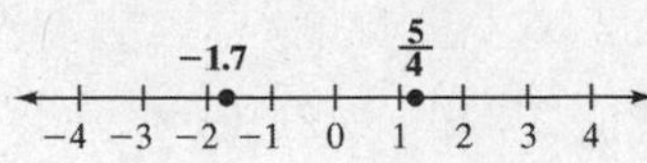

3.

Number	Opposite
a. -41	41
b. $-\frac{8}{9}$	$\frac{8}{9}$
c. 1.7	-1.7
d. $-\frac{2}{5}$	$\frac{2}{5}$

4.
a. $\frac{1}{2}$
b. 0
c. 9
d. -3
5.
a. True
b. False
c. True
d. True
e. False
6.

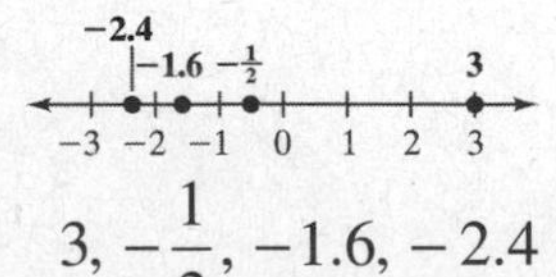

$3, -\frac{1}{2}, -1.6, -2.4$

7. Your house; -15 < -10.
8.
a.

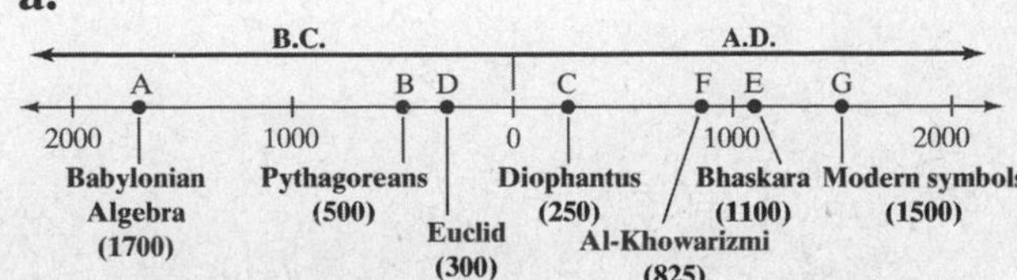

b. A, B, D, C, E, F, and G

Exercises 1.1

1. -5 km
3. -22.5^oC
5. $-\$160$
7.

9.

11.

13.

	Whole Numbers	Integers	Rational Numbers	Real Numbers
15. -7		✓	✓	✓
17. -1.9			✓	✓
19. 10	✓	✓	✓	✓

21. 3
23. 0
25. 3.5
27. 4
29. 0
31. 4.6
33. $-\frac{1}{2}$
35. 4 and -4
37. Impossible; absolute value is always positive or zero.
39. True
41. True
43. True
45. $0 > -1$
47. $-1.5 > -2$
49. $|-4| = 4$
51. $6.2 < |-7.1|$
53. $3\frac{1}{2}$, 0, $-\frac{1}{2}$, $-1\frac{1}{2}$

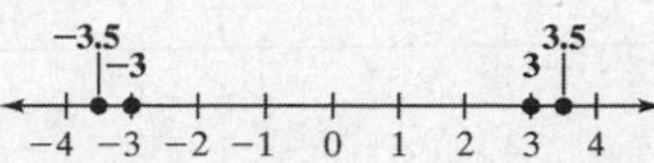

55. 3.5, 3, -3, -3.5

57. Today; - \$200 > - \$2000
59. $-64.8^o\,C$
61. a. Sirius
b. -13
c.

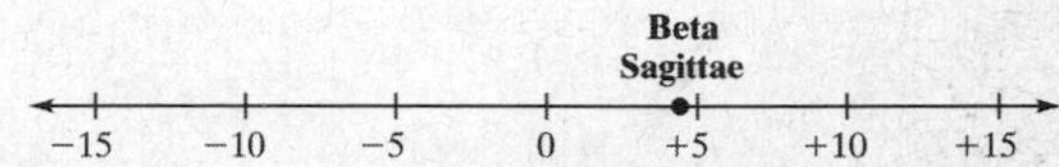

1.2 Addition of Real Numbers

Practice 1.2

1. -1

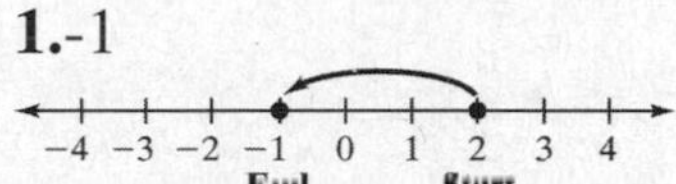

2. 0

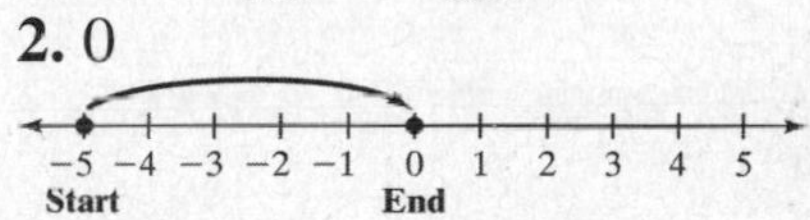

3. 0.5

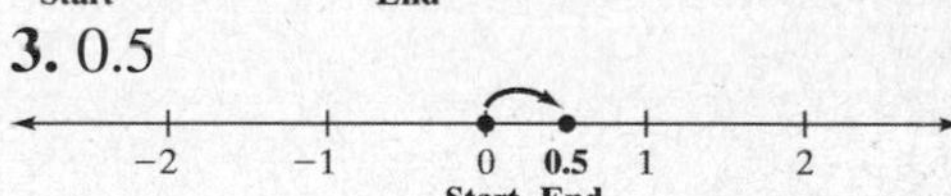

4. -31
5. 3.5
6. 0
7.
$-8 + (-4) + 7 + (-8) + 3 =$
$7 + 3 + (-8) + (-4) + (-8) =$
$10 + (-20) = -10$
8.
$37.50 + 2. + (-1) + (-2) =$
$39.50 + (-3) = \$36.50$
9. -15.592

Exercises 1.2

1. 1
3. 0
5. 5

7. Additive Inverse Property
9. Commutative Property of Addition
11. Additive Inverse Property
13. Commutative Property of Addition
15. Additive Identity Property
17. $24+(-1)=23$
19. $-10+5=-5$
21. $10+(-6)=4$
23. $-50+(-30)=-80$
25. $-10+2=-8$
27. $-18+18=0$
29. $5.2+(-0.9)=4.3$
31. $-0.2+0.8=0.6$
33. $0+(-0.3)=-0.3$
35. $-9.6+3.9=-5.7$
37. $(-9.8)+(-6.5)=-16.3$
39. $-\frac{1}{2}+(-\frac{1}{2})=-1$
41. $-1\frac{3}{5}+2=-1\frac{3}{5}+1\frac{5}{5}=\frac{2}{5}$
43.
$-24+(25)+(-89)=$
$-24+(-89)+(25)=$
$(-113)+(25)=-88$
45.
$15+(-9)+(-15)+9=$
$15+9+(-9)+(-15)=$
$24+(-24)=0$
47. $-0.4+(-2.6)+(-4)=-7$
49. $(-58)+10.48+58=10.48$
51.
$107+(-97)+(-45)+23=$
$107+23+(-97)+(-45)=$
$130+(-142)=-12$
53.
$-2.001+(0.59)+(-8.1)+10.756=$
$0.59+10.756+(-2.001)+(-8.1)=$
$11.346+(-10.101)=1.245$
55. $(-2)+7=5^o$
5° above 0°
57. $132{,}000+(-148{,}000)=-16{,}000$
Lost \$16,000
59. $-51+21=-30$
30 B.C.
61.
$371.25+(-71.33)+(-51.66)+35$
$371.25+35+(-71.33)+(-51.66)=$
$406.25+(-122.99)=283.26$
Yes, since you have \$283.26 in your account, you have enough money to cover a check of \$250.
63. $32{,}000+(-700)=31{,}300$ feet

1.3 Subtraction of Real Numbers

Practice 1.3

1. $4-(-1)=4+(+1)=5$
2. $-12-(-15)=-12+(+15)=3$
3. $8-12=8+(-12)=-4$
4. $-8.1-7.6=-8.1+(-7.6)=-15.7$
5.
$5-(-8)-(-15)=$
$5+(+8)-(-15)=$
$13+(+15)=28$
6.
$4+(-6)-(-11)+8=$
$-2+(+11)+8=$
$9+8=17$
7.
$770-(-100)=$
$770+100=870$ years older

Exercises 1.3

1. $25-8=25+(-8)=17$
3. $-24-7=-24+(-7)=-31$
5. $(-19)-25=(-19)+(-25)=-44$
7. $52-(-19)=52+(+19)=71$
9. $60-95=60+(-95)=-35$
11. $-34-(-2)=-34+(+2)=-32$
13. $16-(-16)=16+(+16)=32$
15. $0-45=0+(-45)=-45$

17. $-31-31=-31+(-31)=-62$

19. $22-(-22)=22+(+22)=44$

21. $200-(-800)=200+(+800)=1000$

23. $6-7.42=6+(-7.42)=-1.42$

25. $-7.3-(0.5)=-7.3+(-0.5)=-7.8$

27. $(-5.6)-(-5.6)=(-5.6)+(+5.6)=0$

29. $8.6-(-1.7)=8.6+(+1.7)=10.3$

31.

$-\frac{1}{3}-\frac{5}{6}=-\frac{1}{3}+(-\frac{5}{6})=$

$-\frac{2}{6}+(-\frac{5}{6})=-\frac{7}{6}=-1\frac{1}{6}$

33.

$-12-\frac{1}{4}=-12+(-\frac{1}{4})=$

$-\frac{48}{4}+(-\frac{1}{4})=-\frac{49}{4}=-12\frac{1}{4}$

35.

$4\frac{3}{5}-(-1\frac{1}{2})=4\frac{3}{5}+(+1\frac{1}{2})=$

$4\frac{6}{10}+(+1\frac{5}{10})=5\frac{11}{10}=6\frac{1}{10}$

37.

$3+(-6)-(-15)=$

$-3+(+15)=12$

39.

$8-10+(-5)=$

$8+(-10)+(-5)=$

$-2+(-5)=-7$

41.

$-9+(-4)-9+4=$

$-13+(-9)+4=$

$-22+4=-18$

43.

$9-12-18=9+(-12)+(-18)=$

$-3+(-18)=-21$

45.

$-10.722+(-3.913)-8.36-3.492=$

$-14.635+(-8.36)+(-3.492)=$

$-22.995+(-3.492)=-26.487$

Applications

47.

$2002-(-776)=2002+(+776)=2778\ years$

49. $400-(-500)=400+(+500)=900\,mi$

51.

$-281{,}330-(-5291)=$

$-281{,}330+(+5291)=$

$-276{,}039$

The losses were reduced by $276,039.

53.

$10{,}152-(-184)=$

$10{,}152+(+184)=$

$10{,}336\,\text{ft}$

55.

a. radon:

$-61.8^o-(-71^o)=-61.8^o+(+71^o)=9.2^o$

neon:

$-246^o-(-248.7^o)=-246^o+(+248.7^o)=2.7^o$

bromine:

$58.8^o-(-7.2^o)=58.8^o+(+7.2^o)=66^o$

b. bromine

c. bromine

1.4 Multiplication of Real Numbers

Practice 1.4

1.

$|-1|=1 \quad |-100|=100$

$1 \cdot 100=100$

$(-1)(-100)=100$

2.

$|-5|=5 \quad |3|=3$

$5 \cdot 3=15$

$(-5)(3)=-15$

3.

a. $(-\frac{2}{3})(-12) = \frac{-2}{3} \cdot \frac{-12}{1} = \frac{24}{3} = 8$

b. $(-\frac{1}{3})(\frac{5}{9}) = \frac{-1}{3} \cdot \frac{5}{9} = \frac{-5}{27}$

c. $(-0.4)(-0.3) = 0.12$

d. $2.5(-1.9) = -4.75$

e. $0 \cdot (-2.8) = 0$

f. $1 \cdot \frac{2}{3} = \frac{2}{3}$

4. $-8(4)(-2) = (-32)(-2) = 64$

5.

$(-6)(-1)(4)(2)(-5) =$

$(4)(2)(-6)(-1)(-5) =$

$(8)(-30) =$

-240

6.

$4(-25) - (-2)(36) =$

$-100 - (-72) =$

$-100 + (+72) = -28$

7.

$-5(-9 + 15) =$

$-5(6) = -30$

8.

$5 + (-32)(2) =$

$5 + (-64) = -59$

The object is moving downward at a velocity of 59 ft/sec.

Exercises 1.4

1. Commutative Property of Multiplication

3. Associative Property of Multiplication

5. Multiplicative Identity Property

7. Multiplication Property of Zero

9. $6(-2) = -12$

11. $-7(-3) = 21$

13. $-12(\frac{1}{4}) = (\frac{-12}{1})(\frac{1}{4}) = \frac{-12}{4} = -3$

15. $(-\frac{5}{6})(-\frac{2}{7}) = \frac{10}{42} = \frac{5}{21}$

17. $-1.5(-0.6) = 0.9$

19. $1.2(-50) = -60$

21. $3(-2)(-20) = (-6)(-20) = 120$

23. $-15(-3)(0) = (45)(0) = 0$

25.

$-6(1)(-2)(-3)(-4) =$

$(-6)(-2)(-3)(-4) =$

$(12)(-3)(-4) =$

$(-36)(-4) = 144$

27.

$-4(5)(-6)(1) =$

$-4(-6)(5)(1) =$

$(24)(5) = 120$

29.

$(-\frac{1}{3})(-\frac{1}{3})(-\frac{1}{3}) =$

$(\frac{1}{9})(-\frac{1}{3}) = -\frac{1}{27}$

31.

$-6.24(0.08)(-1.97) =$

$-6.24(-1.97)(0.08) \approx$

$12.29(0.08) \approx 0.98$

33.

$-7 + 3(-2) - 10 =$

$-7 + (-6) - 10 =$

$-13 + (-10) = -23$

35.

$4(-3) + (-2)(-6) = -12 + 12 = 0$

37. $-3 - 5(-6) = -3 + 30 = 27$

39.

$(\frac{3}{5})(-15) - 6 =$

$(\frac{3}{5})(\frac{-15}{1}) - 6 =$

$(\frac{-45}{5}) - 6 =$

$(-9) - 6 = (-9) + (-6) = -15$

41.

$-5 \cdot (-3 + 4) = -5 \cdot 1 = -5$

43.

Input	Output
a. – 2	5
b. -1	1
c. 0	-3
d. 1	-7
e. 2	-11

Applications

45. $-\$5+(-\$5)=-\$10$. Lost $10

47.

$3(4)+2(1)+4(-3)+1(0)=$

$12+2+(-12)+0=2$

The team scored 2 more points than its opponents.

49. $5(-3)=-15$; dropped 15 inches

51.a. $3(30)+\frac{1}{2}(-450)=$

$90+(-225)=-135$ *calories*

b. $\frac{1}{2}(-288)+2(125)=$

$(-144)+250=+106$ *calories*

c. $3(80)+2(210)+2(-612)+1(-288)=$

$240+420+(-1224)+(-288)=$

$660+(-1512)=-852$ *calories*

1.5 Division of Real Numbers

Practice 1.5

1.a.

$|40|=40 \quad |-5|=5$

$40\div 5=8$

$40\div(-5)=-8$

b.

$|-42|=42 \quad |-6|=6$

$\frac{42}{6}=7 \qquad \frac{-42}{-6}=7$

c.

$|-6.3|=6.3 \quad |9|=9$

$\frac{6.3}{9}=0.7 \qquad \frac{-6.3}{9}=-0.7$

d.

$|-24|=24 \quad |-0.4|=0.4$

$\frac{24}{0.4}=60 \qquad \frac{-24}{-0.4}=60$

2.

a. $-\frac{1}{5}$ because $-5(-\frac{1}{5})=1$

b. $-\frac{8}{1}$ or -8 because $-8(\frac{1}{-8})=1$

c. $\frac{3}{4}$ because $\frac{4}{3}(\frac{3}{4})=1$

d. $-\frac{5}{8}$ because $-\frac{8}{5}(-\frac{5}{8})=1$

e. $-\frac{9}{2}$ because $-\frac{2}{9}(-\frac{9}{2})=1$

3.a.

$-\frac{8}{9}\div\frac{2}{3}=-\frac{8}{9}\bullet\frac{3}{2}=-\frac{24}{18}=-\frac{4}{3}$ or $-1\frac{1}{3}$

b.

$-10\div(-\frac{2}{5})=\frac{-10}{1}\bullet\frac{-5}{2}=\frac{50}{2}=25$

4. a.

$(-3)(-4)\div(2)(-2)=$

$12\div(2)(-2)=$

$6(-2)=-12$

b.

$\frac{-9-(-3)}{2}=\frac{-6}{2}=-3$

5.

$\frac{-300+200+-500+100}{4}=\frac{-500}{4}=-125$

The average weekly change was -$125.

Exercises 1.5

1.

Number	Reciprocal
a. $-\frac{1}{2}$	$-\frac{2}{1}$ or -2
b. 5	$\frac{1}{5}$
c. $-\frac{3}{4}$	$-\frac{4}{3}$ or $-1\frac{1}{3}$
d. $3\frac{1}{5}$	$\frac{5}{16}$
e. -1	$\frac{1}{-1}$ or -1

3. $-8 \div (-1) = 8$

5. $-63 \div 7 = -9$

7. $\frac{0}{-9} = 0$

9. $-2{,}500 \div 100 = -25$

11. $-200 \div (-8) = 25$

13. $-64 \div (-16) = 4$

15. $\frac{-25}{-5} = 5$

17. $\frac{-2}{16} = -\frac{1}{8}$ or -0.125

19. $\frac{10}{-20} = -\frac{1}{2}$ or -0.5

21.

$\frac{4}{5} \div (-\frac{2}{3}) = \frac{4}{5} \bullet (-\frac{3}{2}) =$

$-\frac{12}{10} = -\frac{6}{5}$ or $-1\frac{1}{5}$

23.

$8 \div (-\frac{1}{4}) = 8 \bullet (-\frac{4}{1}) =$

$-\frac{32}{1} = -32$

25.

$2\frac{1}{2} \div (-20) = \frac{5}{2} \bullet (-\frac{1}{20}) =$

$-\frac{5}{40} = -\frac{1}{8}$

27. $(-3.5) \div 7 = -0.5$

29. $10 \div (-0.5) = -20$

31. $\frac{-7.2}{0.9} = -8$

33. $\frac{-3}{-0.3} = 10$

35. $(-15.5484) \div (-6.13) \approx 2.54$

37. $-0.8385 \div 0.715 \approx -1.17$

39.

$-16 \div (-2)(-2) =$

$8(-2) = -16$

41.

$(3-7) \div (-4) =$

$(-4) \div (-4) = 1$

43.

$\frac{2+(-6)}{-2} = \frac{-4}{-2} = 2$

45.

$(4-6) \div (1-5) = (-2) \div (-4) = \frac{1}{2}$

47.

$-56 \div 7 - 4 \bullet (-3) =$

$-8 - (-12) =$

$-8 + (+12) = 4$

49. $(-130) \div 5 = -26$

51.

$-47{,}355 \div 10 = -4735.5$

A decrease of about 4,736 people per year

53. $-24 \div 6 = -4$

An average loss of 4 yards

55. $\frac{-72{,}000}{12} = -6{,}000$

Expenses averaged $6,000 per month.

57. $\frac{2+0+(-7)+(-11)+1}{5} = \frac{-15}{5} = -3$

Yes, -3°F was correct.

1.6 Algebraic Expressions, Translations and Exponents

Practice 1.6

1.
a. 3 terms
b. 1 term
2.
a. one-third of p
b. the difference between 9 and x
c. s divided by – 8
d. n plus -6
e. the product of $\frac{3}{8}$ and m.
3.
a. twice x minus the product of 3 and y
b. four plus the product of 3 and m.
c. five times the difference between a and b
d. the difference between r and s divided by the sum of r and s
4.
a. $\frac{1}{6}n$
b. $n+(-5)$
c. $m-(-4)$
d. $\frac{100}{x}$
e. $-2y$
5.
a. $m+(-n)$
b. $5y-11$
c. $\frac{m+n}{mn}$
d. $-6(x+y)$
6. $60(m+1)$ words
7.
a. $-6^2=-(6)(6)=-36$
b. $(-6)^2 \bullet (-3)=36 \bullet (-3)=324$
8. $2^4(-5)^2$
9.
a. $-x^5$
b. $2m^3n^4$
10. $10 \div 2=5$ There are 5 two-hour periods.
$3^5 \bullet x=243x$
The population after 10 hours was 243x.

Exercises 1.6

1. 1 term
3. 3 terms
5. 2 terms
7. three plus t
9. four less than x
11. seven times r
13. the quotient of *a* and four
15. the product of four-fifths and w
17. the sum of negative three and z
19. twice n plus one
21. four times the quantity x minus y
23. one minus three times
25. the product of a and b divided by the sum of a and b
27. twice x minus five times y
29. $x+5$
31. $d-4$
33. $-6a$
35. $y+(-15)$
37. $\frac{1}{8}k$
39. $\frac{m}{n}$
41. $a-2b$
43. $4z+5$
45. $12(x-y)$
47. $\frac{b}{a-b}$
49. $-3^2=-(3)(3)=-9$
51.
$(-3)^3 \bullet (-4)^2=$
$(-27)(16)=-432$
53. $(-2)^3 \bullet (4)^2$
55. $6^2 \bullet (-3)^3$
57. $3n^3$
59. $-4a^3b^2$
61. $-y^3$
63. $10a^3b^2c$

65. $-x^2y^3$

67. $90^o + x^o + y^o$

69. $\frac{30,000}{p}$ dollars

71. $1+x$ dollars

73. $2^3 \cdot 5000$ dollars

75. s^2

77.

$10,000(\frac{1}{20})(20-n)$ dollars or

$500(20-n)$ dollars

79. $(ab-cd)$ sq ft

1.7 Evaluating Algebraic Expressions and Formulas

Practice 1.7

1.

a. $25+m=25+(-10)=15$

b. $-3xy=(-3)(-2)(5)=(6)(5)=30$

2.

a. $5a-2c=5(2)-2(-4)=10+8=18$

b. $2(5+3)=2(8)=16$

c. $2(d-b)=2(5-(-3))=$

$3cd^2=3(-4)(5)^2=(-12)(25)=-300$

d. $2a^3+4b^2=2(2)^3+4(-3)^2=$

$2(8)+4(9)=16+36=52$

3.

a. $\frac{x-2z}{y}=\frac{-5-2(1)}{-3}=\frac{-7}{-3}=$

$\frac{7}{3}$ or $2\frac{1}{3}$

b. $\frac{x-z}{x+y}=\frac{-5-1}{-5+(-3)}=\frac{-6}{-8}=\frac{3}{4}$

c. $(-y)^4=(-[-3])^4=$

$(3)^4=81$

d. $-y^4=-(-3)^4=-81$

4.

$\frac{a^2+b^2+c^2}{3}=\frac{(-0.5)^2+(0.3)^2+(0.2)^2}{3}=$

$\frac{0.38}{3}=0.12\bar{6}\approx 0.1$

5. $F=\frac{9}{5}C+32$

6.

$d=rt$

$d=50 \cdot 1.6=80\,mi$

7.

a. $K=C+273$

b. $K=-6+273=267$

Exercises 1.7

1. $b-5=3-5=-2$

3. $-2ac=-2(4)(-2)=16$

5. $-2a^2=-2(4)^2=-32$

7. $2a-15=2(4)-15=$

$8-15=-7$

9. $a+2c=4+2(-2)=$

$4+(-4)=0$

11. $2(a-c)=2(4-(-2))=2(6)=12$

13. $-a+b^2=-4+3^2=-4+9=5$

15. $3a^2-c^3=3(4)^2-(-2)^3=48+8=56$

17. $\frac{a+b}{b-a}=\frac{4+3}{3-4}=\frac{7}{-1}=-7$

19.

$\frac{3}{5}(a+b+c)^2=\frac{3}{5}(4+3+(-2))^2=$

$\frac{3}{4}(5)^2=\frac{3}{5}(25)=15$

21.

$2w^2-3x+y-4z=$

$2(-0.5)^2-3(2)+(-3)-4(1.5)=$

$0.5-6-3-6=-14.5$

23.

$$w-7z-\frac{1}{4}(x-6y)=$$

$$-0.5-7(1.5)-\frac{1}{4}(2-6(-3))=$$

$$-0.5-10.5-\frac{1}{4}(2+18)=$$

$$-0.5-10.5-5=-16$$

25.

$$\frac{-10xy}{(w-z)^2}=\frac{-10(2)(-3)}{(-0.5-1.5)^2}=$$

$$\frac{60}{4}=15$$

27.

x	**2x+5**
0	5
1	7
2	9
-1	3
-2	1

29.

y	**y – 0.5**
0	-0.5
1	0.5
2	1.5
3	2.5
4	3.5

31.

x	$-\frac{1}{2}x$
0	0
2	-1
4	-2
-2	1
-4	2

33.

n	$\frac{n}{2}$
2	1
4	2
6	3
-2	-1
-4	-2

35.

g	$-g^2$
0	0
1	-1
2	-4
-1	-1
-2	-4

37.

a	a^2+2a-2
0	-2
1	1
2	6
-1	-3
-2	-2

39.

$$C=\frac{5}{9}(-4-32)=$$

$$\frac{5}{9}(-36)=-20^o C$$

41.

$$P=2(2\frac{1}{2})+2(1\frac{1}{4})=$$

$$2(\frac{5}{2})+2(\frac{5}{4})=\frac{10}{2}+\frac{10}{4}=$$

$$5+2\frac{1}{2}=7\frac{1}{2}\text{ ft}$$

43. $C\approx(3.14)(100)\approx 314m$

45. $A=6(1.5)^2=6(2.25)=13.5\text{cm}^2$

Applications

47. $A = \dfrac{a+b+c}{3}$

49. $P = 2(l+w)$

51. $E = mc^2$

53. $I = 0.4w + 25$

55.

$S = \frac{1}{2}(32)(2)^2 =$

$\frac{1}{2}(32)(4) = 64\,\text{ft}$

The object falls 64 feet.

57.

a. $m = \dfrac{100(s-c)}{c}$

b.

$m = \dfrac{100(8.75-6.25)}{6.25} =$

$\dfrac{100(2.50)}{6.25} = \dfrac{250}{6.25} = 40\%$

1.8 Simplifying Algebraic Expressions

Practice 1.8

1.

a. Terms: m and -3m; like terms

b. Terms: 5x and 7; unlike terms

c. Terms: $2x^2y$ and $-3xy^2$; unlike terms

d. Terms: m, 2m, and -4m; like terms

2.

a. $(-10)(4r+s) = -10 \cdot 4r + (-10) \cdot s =$

$-40r - 10s$

b. $(5+1)w = 5 \cdot w + 1 \cdot w = 5w + w$

c. $3(g-3h) = 3 \cdot g - 3 \cdot 3h = 3g - 9h$

d. $1.5(y+2) = 1.5 \cdot y + 1.5 \cdot 2 =$

$1.5y + 3$

3.

a. $5x + x = 6x$

b. $-5y - y = (-5-1)y = -6y$

c. $a - 3a + b = (1-3)a + b = -2a + b$

d. $-9t + 3t + 6t = (-9+3+6)t = 0 \cdot t = 0$

4.

a. Cannot be simplified.

b. Cannot be simplified

c. $4xy^2 - xy^2 = (4-1)xy^2 = 3xy^2$

5. $3(y-4) + 2 = 3y - 12 + 2 = 3y - 10$

6.

$-(2a-3b) = -1(2a) + (-1)(-3b) =$

$-2a + 3b$

7.

$5y - 6 - (y-5) = 5y - 6 - 1(y-5) =$

$5y - 6 - y + 5 = 4y - 1$

8.

$(y+3) - 3(y+7) = y + 3 - 3y - 21 = -2y - 18$

9.

$10 - [4y + 3(2y-1)] =$

$10 - [4y + 6y - 3] =$

$10 - [10y - 3] =$

$10 - 10y + 3 = -10y + 13$

10.

$5c + 12(c-40) =$

$5c + 12c - 480 =$

$17c - 480$ dollars

Exercises 1.8

1. 7

3. 1

5. -0.1

7. Terms: $2a$ and $-a$; like terms

9. Terms: $5p$ and 3; unlike terms

11. Terms: $4x^2$ and $-6x^2$ like terms

13. Terms: x^2 and $7x^3$; unlike terms

15. $-7(x-y) = -7 \cdot x - 7(-y) = -7x + 7y$

17. $(1-10)a = 1 \cdot a + (-10)a = a - 10a$

19.

$-0.5(r+3) = -0.5 \cdot r + (-0.5)3 =$

$-0.5r - 1.5$

21. $3x + 7x = (3+7)x = 10x$

23. $-10n - n = (-10 + (-1))n = -11n$

25. $20a - 10a + 4a = 14a$

27. $3y - y + 2 = (3-1)y + 2 = 2y + 2$

29. $8b^3 + b^3 - 9b^3 = (8+1-9)b^3 = 0$

31. Cannot be simplified.
33. $3r^2t^2 + r^2t^2 = (3+1)r^2t^2 = 4r^2t^2$
35. Cannot be simplified..
37. $2(x+3)-4 = 2x+6-4 = 2x+2$
39. $(7x+1)+(2x-1) = 7x+1+2x-1 = 9x$
41. $-(3y-10) = -1(3y)-1(-10) = -3y+10$
43.
$5x-3-(x+6) =$
$5x-3-1(x)-1(6) =$
$5x-3-x-6 = 4x-9$
45.
$-4(n-9)+3(n+1) =$
$-4n-4(-9)+3n+3(1) =$
$-4n+36+3n+3 =$
$-n+39$
47.
$x-4-2(x-1)+3(2x+1) =$
$x-4-2x-2(-1)+3(2x)+3(1) =$
$x-4-2x+2+6x+3 =$
$5x+1$
49.
$7+3[x-2(x-1)] =$
$7+3[x-2x+2] =$
$7+3[-x+2] =$
$7+3(-x)+3(2) =$
$7-3x+6 = -3x+13$
51.
$10-3[4(a+2)-3a] =$
$10-3[4a+4(2)-3a] =$
$10-3[4a+8-3a] =$
$10-3[a+8] =$
$10-3a-24 = -3a-14$

Applications

53.
$x^o + x^o + 40^o =$
$2x^o + 40^o$

55.
$d+2(d+4) =$
$d+2d+8 =$
$3d+8$ dollars
57.
$n+(n+1)+(n+2) =$
$3n+3$
59.
$0.05x+0.04(1000-x) =$
$0.05x+40-0.04x =$
$0.01x+40$ dollars

Chapter 1 Review Exercises

[1.1]
1. +3 miles
2. -$160.
3.
−4 −3 −2 −1 0 1 2 3 4
4.
−4 −3 −2 −1 0 1 2 3 4
5.
−4 −3 −2 −1 0 1 2 3 4
6.
−4 −3 −2 −1 0 1 2 3 4
7. 4
8. -6.5
9. $-\frac{2}{3}$
10. 0.7
11. 4
12. 0
13. 2.6
14. $\frac{5}{9}$
15. True
16. False

[1.2]
17. $-4+(-1) = -5$
18. $3+(-7) = -4$
19. Commutative property of addition

20. Additive Identity Property
21. Associative Property of Addition
22. Additive Inverse Property
23. 0
24. 2
25. 2
26. -15
27. -5
28.
$-3+7+(-89)=$
$4+(-89)=-85$
29.
$-0.5+(-3.6)+(-4)=$
$-4.1+(-4)=-8.1$
30.
$-2+5.3+12=$
$3.3+12=15.3$

[1.3]
31. $12-3=12+(-3)=9$
32. $36-47=36+(-47)=-11$
33. $-52-3=-52+(-3)=-55$
34. $2-5=2+(-5)=-3$
35. $-19-8=-19+(-8)=-27$
36. $24-(-3)=24+(+3)=27$
37. $8-(-8)=8+(+8)=16$
38. $0-5=0+(+5)=5$
39. $6-7.42=-1.42$
40. $-9-(-\frac{3}{8})=-9+(+\frac{3}{8})=-8\frac{5}{8}$
41.
$2+(-4)-(-7)=$
$-2+(+7)=5$
42.
$-3-(-1)+12=$
$-3+(+1)+12=$
$-2+12=10$

[1.4]
43. Commutative Property of Multiplication
44. Associative Property of Multiplication
45. Multiplicative Identity Property
46. Multiplication Property of Zero
47. -10
48. -21
49. -5400
50. 2400
51. 27
52. $-\frac{1}{4}$
53.
$5(-4)(-300)=$
$-20(-300)=6000$
54.
$(-1)(-12)(3)=$
$12(3)=36$
55.
$-8+3(-2)-9=$
$-8+(-6)+(-9)=$
-23
56.
$3-2(-3)-(-5)=$
$3+6+5=14$
57.
$-9-5(-7)=$
$-9+35=26$
58.
$20-3(-6)=$
$20+18=38$
59.
$-4(-2+5)=$
$-4(3)=-12$
60.
$(-12+6)(-1)=$
$(-6)(-1)=6$

[1.5]
61. $-\frac{3}{2}$
62. $\frac{1}{8}$
63. 3
64. -6
65. $-2\frac{1}{5}$

66.
$\frac{4}{5} \div (-\frac{2}{3}) = \frac{4}{5} \bullet \frac{-3}{2} =$
$\frac{-12}{10} = -\frac{6}{5} = -1\frac{1}{5}$

67. $-16 \div 2(-4) = -8(-4) = 32$

68.
$(9-23) \div (-13+6) =$
$-14 \div (-7) = 2$

69. $\frac{3+(-1)}{-2} = \frac{2}{-2} = -1$

70. $\frac{5(7-3)}{-8-2} = \frac{5(4)}{-10} = \frac{20}{-10} = -2$

71.
$(-3)+8-2\bullet(-4) =$
$(-3)+8-(-8) =$
$(-3)+8+(+8) = 13$

72.
$10 \div (-2) + (-3) \bullet 5 =$
$-5 + (-15) = -20$

[1.6]

73. 3 terms
74. 2 terms
75. 1 terms
76. 4 terms
77. the sum of – 6 and w
78. the product of $-\frac{1}{3}$ and x
79. six more than -3 times n
80. five times the quantity p minus q
81. $x-10$
82. $\frac{1}{2}s$
83. $\frac{p}{q}$
84. $R-2V$
85. $6(4n-2)$
86. $\frac{-4a}{5b+c}$
87. $(-3)^4$
88. $(-5)^3 3^2$
89. $4x^3$
90. $-5a^2b^3c$

[1.7]

91. $30 + (-1) = 29$

92. $-\frac{4}{9}(5) = \frac{-4}{9} \bullet \frac{5}{1} = \frac{-20}{9} = -2\frac{2}{9}$

93. $-5(2)^2 = -5(4) = -20$

94. $10(5-(-1)) = 10(6) = 60$

95. $\frac{1-2}{-1} = \frac{-1}{-1} = 1$

96.
$4(2)^2 - 4(2)(5) + (5)^2 =$
$4(4) - 4(2)(5) + 25 =$
$16 - 40 + 25 = 1$

[1.8]

97.
$-5(x-y) =$
$-5(x) + (-5)(-y) =$
$-5x + 5y$

98.
$4x + 10x - 2y =$
$(4+10)x - 2y =$
$14x - 2y$

99.
$3x^2 - x^2 - 4x^2 =$
$(3-1-4)x^2 =$
$-2x^2$

100.
$2r^2t^2 - r^2t^2 =$
$(2-1)r^2t^2 =$
r^2t^2

101.
$2(a-5)+1 =$
$2(a) + 2(-5) + 1 =$
$2a + (-10) + 1 =$
$2a - 9$

102.
$-(3x+2) =$
$(-1)(3x) + (-1)(2) =$
$-3x - 2$

103.
$-3x-5-(x+10)=$
$-3x-5+(-1)x+(-1)10=$
$-3x-5-x-10=$
$-4x-15$

104.
$(2a-4)+2(a-5)-3(a+1)=$
$2a-4+2a-10-3a-3=$
$a-17$

Mixed Applications

105. $+\$700$
106. $-\$7000$
107. $-\frac{1}{2}\bullet 4=-2;\quad -2$ dollars
108.
$-4-(-7)=-4+(+7)=3$
$+3^o C$ Exothermic
109. $I-0.5h$ degrees
110.
$F=\frac{9}{5}(-10)+32$
$F=-18+32$
$F=14^o F$
111. $2(2w)+2w=6w$
112. -4°
113. $-61.8-(-71)=-61.8+(+71)=9.2$
The boiling point is 9.2^o higher.
114. $\$12.34+(-\$0.43)=\$11.91$
The price per share on Tuesday was $11.91.
115. $3^3\bullet 10$
116.
$S=0.5(90)+26$
$S=45+26=71\ ft$
117. $-496+90=406BC$
118. $\frac{-60{,}000}{-20{,}000}=3$
The first loss is 3 times the second loss.
119. $I=\frac{2300}{115}=20$ amperes
120.
$410+900+(-720)+2(-300)=$
$410+900+(-720)+(-600)=$
$1310+(-1320)=-10$
The account is overdrawn by $10.
121.
$x+(32-20)y=$
$(x+12y)$ dollars
122. $(f+4s)$ students

Chapter 1 Posttest

1. $-10{,}000$
2. yes
3.
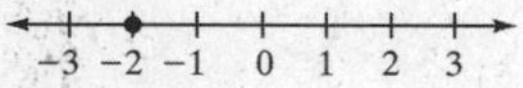

4. -7
5. 3.5
6. True
7. $10+(-3)=7$
8. $2+(-3)+(-1)+5=-1+(-1)+5=-2+5=3$
9. $4+(-1)(-6)=4+6=10$
10. $\frac{1}{12}$
11. $-15\div 3=-5$
12. $x+2y$
13. $(-6)^3$
14. $3(-1)+0-2=-3+0-2=-5$
15. $4y+3-7y+10y+1=7y+4$
16. $8t+1-2(3t-1)=8t+1-6t+2=2t+3$
17. 2°
18. $d+0.05d=1.05d$ dollars
19.
$50{,}000-(-20{,}000)=$
$50{,}000+20{,}000=\$70{,}000$
An improvement of $70,000
20. $4(-\$200)=-\$800$
The net debt is $800.

Chapter 2 Solving Linear Equations and Inequalities

Pretest

1. no

$$7-2x=3x-11$$

$$7-2(4)\stackrel{?}{=}3(4)-11$$

$$7-8\stackrel{?}{=}12-11$$

$$-1\neq 1$$

2.

$$n+2=-6$$

$$n+2+(-2)=-6+(-2)$$

$$n=-8$$

check:

$$-8+2=-6$$

$$-6=-6$$

3.

$$\frac{y}{-5}=1$$

$$(-5)\frac{y}{-5}=1(-5)$$

$$y=-5$$

check :

$$\frac{-5}{-5}\stackrel{?}{=}1$$

$$1=1$$

4.

$$-n=8$$

$$(-1)(-n)=8(-1)$$

$$n=-8$$

check :

$$-(-8)\stackrel{?}{=}8$$

$$8=8$$

5.

$$\frac{2}{3}x-3=-9$$

$$\frac{2}{3}x-3+3=-9+3$$

$$\frac{2}{3}x=-6$$

$$(\frac{3}{2})(\frac{2}{3}x)=-6(\frac{3}{2})$$

$$x=-9$$

check :

$$\frac{2}{3}(-9)-3\stackrel{?}{=}-9$$

$$-6-3\stackrel{?}{=}-9$$

$$-9=-9$$

6.

$$4x-8=-10$$

$$4x-8+8=-10+8$$

$$4x=-2$$

$$(\frac{1}{4})4x=-2(\frac{1}{4})$$

$$x=-\frac{1}{2}$$

check :

$$4(-\frac{1}{2})-8\stackrel{?}{=}-10$$

$$-2-8\stackrel{?}{=}-10$$

$$-10=-10$$

7.

$$6-y=-5$$

$$6+(-6)-y=-5+(-6)$$

$$-y=-11$$

$$(-1)(-y)=(-11)(-1)$$

$$y=11$$

check :

$$6-11\stackrel{?}{=}-5$$

$$-5=-5$$

8.

$9x+13=7x+19$

$9x+13-13=7x+19-13$

$9x-7x=7x-7x+6$

$2x=6$

$(\frac{1}{2})(2x)=6(\frac{1}{2})$

$x=3$

check :

$9(3)+13\stackrel{?}{=}7(3)+19$

$27+13\stackrel{?}{=}21+19$

$40=40$

9.

$-2(3n-1)=-7n$

$-6n+2=-7n$

$-6n+6n+2=-7n+6n$

$2=-n$

$(-1)2=(-n)(-1)$

$-2=n$

check :

$-2(3(-2)-1)\stackrel{?}{=}-7(-2)$

$-2(-7)\stackrel{?}{=}14$

$14=14$

10.

$14x-(8x-13)=12x+3$

$14x-8x+13=12x+3$

$6x+13=12x+3$

$6x-6x+13=12x-6x+3$

$13-3=6x+3-3$

$10=6x$

$(\frac{1}{6})(10)=6x(\frac{1}{6})$

$\frac{10}{6}=x$

$x=\frac{5}{3} \quad or \quad 1\frac{2}{3}$

check :

$14(\frac{5}{3})-(8(\frac{5}{3})-13)=12(\frac{5}{3})+3$

$\frac{70}{3}-(\frac{40}{3}-\frac{39}{3})\stackrel{?}{=}\frac{60}{3}+3$

$\frac{69}{3}=\frac{69}{3}$

11.

$v-5u=w$

$v-5u+5u=w+5u$

$v=w+5u$

12.

x =percent (expressed as decimal)

$36x=9$

$\frac{1}{36}(36x)=9(\frac{1}{36})$

$x=\frac{9}{36}$

$x=.25$

$x=25\%$

13.

$x = a\ number$

$.60x=12$

$\frac{.60x}{.60}=\frac{12}{.60}$

$x=20$

14.
$x \le 2$

−3 −2 −1 0 1 2 3

15.
$x + 3 > 3$

$x > 0$

−3 −2 −1 0 1 2 3

16.
(pages per minute) • (minutes)=number pages
x = number of minutes

$30x = 360$

$\frac{30x}{30} = \frac{360}{30}$

$x = 12$

x = 12 minutes

17. x = number of centerpieces

$100 - 100 + 70x = 1500 - 100$

$70x = 1400$

$\frac{70x}{70} = \frac{1400}{70}$

$x = 20$ centerpieces

18.

$E = \frac{1}{2}mv^2$

$2E = mv^2$

$\frac{2E}{v^2} = m$

19. x = amount invested at 5%
2x = amount invested at 8%

$.05x + .08(2x) = 420$

$.05x + .16x = 420$

$.21x = 420$

$\frac{.21x}{.21} = \frac{420}{.21}$

$x = \$2000$ was invested at 5%

$2x = \$4000$ was invested at 8%

20. h = number of hours

$10 + 3h > 55$

$3h > 45$

$\frac{3h}{3} < \frac{45}{3}$

$h > 15$

Option A is a better deal if you use the gym more than 15 hours per month.

2.1 Solving Linear Equations: The Addition Property

Practice 2.1

1.

$5x - 4 = 2x + 5$

$5(4) - 4 \stackrel{?}{=} 2(4) + 5$

$20 - 4 \stackrel{?}{=} 8 + 5$

$16 \ne 13$

No, 4 is not a solution.

2.

$5(x + 3) = 3x - 1$

$5((-8) + 3) \stackrel{?}{=} 3(-8) - 1$

$5(-5) \stackrel{?}{=} (-24) - 1$

$-25 = -25$

Yes, -8 is a solution.

3.

$y - 12 = -7$

$y - 12 + 12 = -7 + 12$

$y = 5$

check :

$5 - 12 \stackrel{?}{=} -7$

$-7 = -7$

4.

$-2 = n + 15$

$-2 + (-15) = n + 15 + (-15)$

$-17 = n$

or

$n = -17$

check :

$-2 \stackrel{?}{=} (-17) + 15$

$-2 = -2$

5.

$5 = 4.9 - (-x)$

$5 = 4.9 + x$

$5 + (-4.9) = 4.9 + (-4.9) + x$

$0.1 = x$

or

$x = 0.1$

check :

$5 \stackrel{?}{=} 4.9 - (-0.1)$

$5 \stackrel{?}{=} 4.9 + 0.1$

$5 = 5$

6.

x = mass of the empty bottle

24.56 *is* 9.68 *plus x*

$24.56 = 9.68 + x$

$24.56 + (-9.68) = 9.68 + (-9.68) + x$

$14.88 = x$ *or* $x = 14.88$ *grams*

check :

$24.56 \stackrel{?}{=} 9.68 + 14.88$

$24.56 = 24.56$

Exercises 2.1

1.

Value of x	Equation	True or False
a. -8	3x +13 = -11	True
b. 7	28 – x = 7 – 4x	False
c. 9	2(x-3) = 12	True
d. $\frac{2}{3}$	$12x - 2 = 6x + 2$	True

a.

$3x + 13 = -11$

$3(-8) + 13 \stackrel{?}{=} -11$

$-24 + 13 \stackrel{?}{=} -11$

$-11 = -11$

b.

$28 - x = 7 - 4x$

$28 - 7 \stackrel{?}{=} 7 - 4(7)$

$21 \stackrel{?}{=} 7 - 28$

$21 \neq -21$

c.

$2(x - 3) = 12$

$2(9 - 3) \stackrel{?}{=} 12$

$2(6) \stackrel{?}{=} 12$

$12 = 12$

d.

$12x - 2 = 6x + 2$

$12(\frac{2}{3}) - 2 \stackrel{?}{=} 6(\frac{2}{3}) + 2$

$8 - 2 \stackrel{?}{=} 4 + 2$

$6 = 6$

3. $x + 4 = -6$ Subtract 4 or add (-4) to each side.

5. $-2 = -1 + x$ Add 1 to each side.

7. $z - (-3.5) = 5$ Subtract 3.5 or add (-3.5) to each side.

9. $9 = x - 2\frac{1}{5}$ Add $2\frac{1}{5}$ to each side.

11.

$y+9=-14$

$y+9+(-9)=-14+(-9)$

$y=-23$

check :

$-23+9\overset{?}{=}-14$

$-14=-14$

13.

$t-4=-4$

$t-4+4=-4+4$

$t=0$

check :

$0-4\overset{?}{=}-4$

$-4=-4$

15.

$9+a=-3$

$9+(-9)+a=-3+(-9)$

$a=-12$

check :

$9+(-12)\overset{?}{=}-3$

$-3=-3$

17.

$z-4=-10$

$z-4+4=-10+4$

$z=-6$

check :

$-6-4\overset{?}{=}-10$

$-10=-10$

19.

$12=x+12$

$12+(-12)=x+12+(-12)$

$0=x \;\; or \;\; x=0$

check :

$12\overset{?}{=}0+12$

$12=12$

21.

$-19=r-19$

$-19+19=r-19+19$

$0=r$ or $r=0$

check :

$-19\overset{?}{=}0-19$

$-19=-19$

23.

$-4=-10+r$

$-4+10=-10+10+r$

$6=r$ or $r=6$

check :

$-4\overset{?}{=}-10+6$

$-4=-4$

25.

$-15+n=-2$

$-15+n+15=-2+15$

$-15+15+n=-2+15$

$n=13$

check :

$-15+(13)\overset{?}{=}-2$

$-2=-2$

27.

$x+\frac{2}{3}=-\frac{1}{3}$

$x+\frac{2}{3}+(-\frac{2}{3})=-\frac{1}{3}+(-\frac{2}{3})$

$x=-1$

check :

$-1+\frac{2}{3}\overset{?}{=}-\frac{1}{3}$

$-\frac{1}{3}=-\frac{1}{3}$

29.

$8+y=4\frac{1}{2}$

$8+(-8)+y=4\frac{1}{2}+(-8)$

$y=-3\frac{1}{2}$

$check: \quad 8+(-3\frac{1}{2})\stackrel{?}{=}4\frac{1}{2}$

$4\frac{1}{2}=4\frac{1}{2}$

31.

$m+2.4=5.3$

$m+2.4+(-2.4)=5.3+(-2.4)$

$m=2.9$

$check:$

$2.9+2.4\stackrel{?}{=}5.3$

$5.3=5.3$

33.

$-2.3+t=-5.9$

$-2.3+t+2.3=-5.9+2.3$

$-2.3+2.3+t=-3.6$

$t=-3.6$

$check:$

$-2.3+(-3.6)\stackrel{?}{=}-5.9$

$-5.9=-5.9$

35.

$a-(-35)=30$

$a+35=30$

$a+35+(-35)=30+(-35)$

$a=-5$

$check:$

$-5-(-35)\stackrel{?}{=}30$

$-5+35\stackrel{?}{=}30$

$30=30$

37.

$m-(-\frac{1}{4})=-\frac{1}{4}$

$m+\frac{1}{4}=-\frac{1}{4}$

$m+\frac{1}{4}+(-\frac{1}{4})=-\frac{1}{4}+(-\frac{1}{4})$

$m=-\frac{2}{4}$

$m=-\frac{1}{2}$

$check:$

$-\frac{1}{2}-(-\frac{1}{4})\stackrel{?}{=}-\frac{1}{4}$

$-\frac{1}{2}+\frac{1}{4}\stackrel{?}{=}-\frac{1}{4}$

$-\frac{1}{4}=-\frac{1}{4}$

39.

$y+2.932=4.811$

$y+2.932+(-2.932)=4.811+(-2.932)$

$y=1.879$

$y=1.88$ (rounded to nearest hundredth)

$check:$

$1.88+2.932\stackrel{?}{=}4.811$

$4.81=4.81$ (rounded to nearest hundredth)

41.

$x+2=12$

$x+2+(-2)=12+(-2)$

$x=10$

$check:$

$10+2\stackrel{?}{=}12$

$12=12$

43.

$n-4=21$

$n-4+4=21+4$

$n=25$

check :

$25-4\overset{?}{=}21$

$21=21$

45.

$x+(-3)=-1$

$x+(-3)+3=-1+3$

$x=2$

check :

$2+(-3)\overset{?}{=}-1$

$-1=-1$

47.

$n+7=11$

$n+7+(-7)=11+(-7)$

$n=4$

check :

$4+7\overset{?}{=}11$

$11=11$

49.

d. $x-6=127$

51.

a. $x-4.7=250$ megabytes

53.

x = the speed of the car now

$x+10=44$

$x+10+(-10)=44+(-10)$

$x=34$

check :

$34+10\overset{?}{=}44$

$44=44$

55.

x = number of calories used in one hour of roller-skating

$x-130=220$

$x-130+130=220+130$

$x=350$ *calories*

check :

$350-130\overset{?}{=}220$

$220=220$

57.

h = original altitude of the helicopter

$h-170=215$

$h-170+170=215+170$

$h=385$ meters

check :

$365-170\overset{?}{=}215$

$215=215$

59.

x = the measure of $\angle$ x

$x+118.5^o=180^o$

$x+118.5^o+(-118.5^o)=180^o+(-118.5^o)$

$x=61.5^o$

check :

$61.5^o+118.5^o\overset{?}{=}180^o$

$180^o=180^o$

2.2 Solving Linear Equations: The Multiplication Property

Practice 2.2

1.

$\frac{y}{3} = 21$

$(3)\frac{y}{3} = 21(3)$

$y = 63$

check :

$\frac{63}{3} \stackrel{?}{=} 21$

$21 = 21$

2.

$7y = 63$

$(\frac{1}{7})7y = 63(\frac{1}{7})$

$y = 9$

check :

$7(9) \stackrel{?}{=} 63$

$63 = 63$

3.

$-x = 10$

$\frac{-x}{-1} = \frac{10}{-1}$

$x = -10$

check :

$-(-10) \stackrel{?}{=} 10$

$10 = 10$

4.

$-11.7 = -0.9z$

$\frac{-11.7}{-0.9} = \frac{-0.9}{-0.9}z$

$13 = z$

check :

$-11.7 \stackrel{?}{=} -0.9(13)$

$-11.7 = -11.7$

5.

$\frac{6}{7}y = -12$

$\frac{7}{6} \bullet \frac{6}{7}y = -12 \bullet \frac{7}{6}$

$y = -14$

check :

$\frac{6}{7} \bullet -14 \stackrel{?}{=} -12$

$-12 = -12$

6.

T = total bill

One fourth of the total bill was labor.

$\frac{1}{4}T = \$189.50$

$4 \bullet \frac{1}{4}T = 189.50 \bullet 4$

$T = \$758.$

7.

$d = rt$

$130 = 60t$

$\frac{130}{60} = \frac{60t}{60}$

$2.2 \approx t$

$t \approx 2.2$ hours

It will take about 2.2 hours.

Exercises 2.2

1. Multiply each side of the equation by 3.
3. Divide each side of the equation by -5.
5. Divide each side of the equation by -2.2
7. Multiply each side of the equation by $\frac{4}{3}$.

9.

$6x = -30$

$\frac{6x}{6} = \frac{-30}{6}$

$x = -5$

check :

$6(-5) \stackrel{?}{=} -30$

$-30 = -30$

11.

$\frac{n}{2} = 9$

$2 \bullet \frac{n}{2} = 9 \bullet 2$

$n = 18$

check :

$\frac{18}{2} \stackrel{?}{=} 9$

$9 = 9$

13.

$\frac{a}{4} = 1.2$

$4 \bullet \frac{a}{4} = 1.2 \bullet 4$

$a = 4.8$

check :

$\frac{4.8}{4} \stackrel{?}{=} 1.2$

$1.2 = 1.2$

15.

$-5x = 2.5$

$\frac{-5x}{-5} = \frac{2.5}{-5}$

$x = -0.5$

check :

$-5(-0.5) \stackrel{?}{=} 2.5$

$2.5 = 2.5$

17.

$42 = -6c$

$\frac{42}{-6} = \frac{-6c}{-6}$

$-7 = c$

$c = -7$

check :

$42 \stackrel{?}{=} -6(-7)$

$42 = 42$

19.

$11 = -\frac{r}{2}$

$(-2)11 = -\frac{r}{2}(-2)$

$-22 = r$

$r = -22$

check :

$11 \stackrel{?}{=} -\frac{-22}{2}$

$11 \stackrel{?}{=} -(-11)$

$11 = 11$

21.

$\frac{5}{6}x = 10$

$\frac{6}{5} \bullet \frac{5}{6}x = 10 \bullet \frac{6}{5}$

$x = 12$

check :

$\frac{5}{6} \bullet 12 \stackrel{?}{=} 10$

$10 = 10$

23.

$-\frac{2}{5}y = 1$

$(-\frac{5}{2})(-\frac{2}{5}y) = (1)(-\frac{5}{2})$

$y = -\frac{5}{2}\ or\ -2\frac{1}{2}$

check :

$(-\frac{2}{5})(-\frac{5}{2}) \stackrel{?}{=} 1$

$1 = 1$

25.

$\frac{3n}{4} = 6$

$\frac{3}{4}n = 6$

$\frac{4}{3} \bullet \frac{3}{4}n = 6 \bullet \frac{4}{3}$

$n = 8$

check :

$\frac{3 \bullet 8}{4} \stackrel{?}{=} 6$

$\frac{24}{4} \stackrel{?}{=} 6$

$6 = 6$

27.

$\frac{4c}{3} = -4$

$\frac{4}{3}c = -4$

$\frac{3}{4} \bullet \frac{4}{3}c = -4 \bullet \frac{3}{4}$

$c = -3$

check :

$\frac{(4)(-3)}{3} \stackrel{?}{=} -4$

$(4)(-1) \stackrel{?}{=} -4$

$-4 = -4$

29.

$\frac{x}{2.4} = -1.2$

$2.4 \bullet \frac{x}{2.4} = (-1.2) \bullet 2.4$

$x = -2.88$

check :

$\frac{-2.88}{2.4} \stackrel{?}{=} -1.2$

$-1.2 = -1.2$

31.

$-2.5a = 5$

$\frac{-2.5a}{-2.5} = \frac{5}{-2.5}$

$a = -2$

check :

$(-2.5)(-2) \stackrel{?}{=} 5$

$5 = 5$

33.

$\frac{2}{3}y = \frac{4}{9}$

$\frac{3}{2} \bullet \frac{2}{3}y = \frac{4}{9} \bullet \frac{3}{2}$

$y = \frac{2}{3}$

check :

$\frac{2}{3} \bullet \frac{2}{3} \stackrel{?}{=} \frac{4}{9}$

$\frac{4}{9} = \frac{4}{9}$

35.

$\frac{x}{-1.515} = 1.515$

$(-1.515)\frac{x}{-1.515} = 1.515(-1.515)$

$x \approx -2.295$

$x = -2.30$

37.

$-3.14x = 21.4148$

$\frac{-3.14x}{-3.14} = \frac{21.4148}{-3.14}$

$x = -6.82$

39.

$-4x = 56$

$\frac{-4x}{-4} = \frac{56}{-4}$

$x = -14$

check :

$(-4)(-14) \stackrel{?}{=} 56$

$56 = 56$

41.

$\frac{n}{0.2} = 1.1$

$0.2 \bullet \frac{n}{0.2} = 1.1 \bullet 0.2$

$n = 0.22$

check :

$\frac{0.22}{0.2} \stackrel{?}{=} 1.1$

$1.1 - 1.1$

43.

$\frac{x}{-3} = 20$

$(-3) \bullet \frac{x}{-3} = 20 \bullet (-3)$

$x = -60$

check :

$\frac{-60}{-3} \stackrel{?}{=} 20$

$20 = 20$

45.

$\frac{1}{6}x = 2\frac{4}{5}$

$6 \bullet \frac{1}{6}x = 2\frac{4}{5} \bullet 6$

$x = \frac{84}{5}$ or $16\frac{4}{5}$

check;

$\frac{1}{6} \bullet \frac{84}{5} \stackrel{?}{=} 2\frac{4}{5}$

$\frac{14}{5} \stackrel{?}{=} 2\frac{4}{5}$

$2\frac{4}{5} = 2\frac{4}{5}$

47. c. $20 = \frac{m}{2}$

49. a. $6p = 150$

51. x = number of years

$0.02x = 10.5$

$\frac{0.02x}{0.02} = \frac{10.5}{0.02}$

$x = 525$ *years*

check :

$0.02 \bullet 525 \stackrel{?}{=} 10.5$

$10.5 = 10.5$

53. r = average speed of bus

$rt = d$

$70r = 3348$

$\frac{70r}{70} = \frac{3348}{70}$

$r \approx 47.83$ miles per hour

$r \approx 48$ miles per hour

check :

$70 \bullet 47.83 \stackrel{?}{=} 3348$

$3348.1 \approx 3348$

$3348 = 3348$

55. x = number of copies

$0.05c = 20$

$\frac{0.05c}{0.05} = \frac{20}{0.05}$

$c = 400$ copies

check :

$0.05 \bullet 400 \stackrel{?}{=} 20$

$20 = 20$

57. x = projected income

$\frac{2}{3}x = 800,000$

$\frac{3}{2} \bullet \frac{2}{3}x = 800,000 \bullet \frac{3}{2}$

$x = \$1,200,000$

check :

$\frac{2}{3} \bullet \$1,200,000 \stackrel{?}{=} 800,000$

$800,000 = 800,000$

59. d =depth of Mediterranean Sea

$\frac{1}{5}d = 1000$

$5 \bullet \frac{1}{5}d = 1000 \bullet 5$

$d = 5000$ meters

check :

$\frac{1}{5} \bullet 5000 \stackrel{?}{=} 1000$

$1000 = 1000$

61. h = number of hours

$7.50h = 187.50$

$\frac{7.50h}{7.50} = \frac{187.50}{7.50}$

$h = 25$ hours

check :

$7.50 \bullet 25 \stackrel{?}{=} 187.50$

$187.50 = 187.50$

63. x = monthly rent

$12x = 10,020$

$\frac{12x}{12} = \frac{10,020}{12}$

$x = \$835$ per month

check :

$12 \bullet 835 \stackrel{?}{=} 10,020$

$10,020 = 10,020$

2.3 Solving Equations by Combining Properties

Practice 2.3

1.

$2y + 1 = 9$

$2y + 1 - 1 = 9 - 1$

$2y = 8$

$\frac{2y}{2} = \frac{8}{2}$

$y = 4$

check :

$2(4) + 1 \stackrel{?}{=} 9$

$8 + 1 \stackrel{?}{=} 9$

$9 = 9$

2.

$$\frac{c}{5}-1=8$$
$$\frac{c}{5}-1+1=8+1$$
$$\frac{c}{5}=9$$
$$5\bullet\frac{c}{5}=9\bullet5$$
$$c=45$$
check:
$$\frac{45}{5}-1\stackrel{?}{=}8$$
$$9-1\stackrel{?}{=}8$$
$$8=8$$

3.

$$-6b-5=13$$
$$-6b-5+5=13+5$$
$$-6b=18$$
$$\frac{-6b}{-6}=\frac{18}{-6}$$
$$b=-3$$
check:
$$(-6)(-3)-5\stackrel{?}{=}13$$
$$18-5\stackrel{?}{=}13$$
$$13=13$$

4.

$$5-t-t=-1$$
$$5-2t=-1$$
$$5-2t-5=-1-5$$
$$-2t=-6$$
$$\frac{-2t}{-2}=\frac{-6}{-2}$$
$$t=3$$
check:
$$5-3-3\stackrel{?}{=}-1$$
$$2-3\stackrel{?}{=}-1$$
$$-1=-1$$

5.

$$3f-12=-f-15$$
$$f+3f-12=f-f-15$$
$$4f-12=-15$$
$$4f-12+12=-15+12$$
$$4f=\ 3$$
$$\frac{4f}{4}=\frac{-3}{4}$$
$$f=-\frac{3}{4}$$
check:
$$(3)(-\frac{3}{4})-12\stackrel{?}{=}-(-\frac{3}{4})-15$$
$$-\frac{9}{4}-12\stackrel{?}{=}\frac{3}{4}-15$$
$$-14\frac{1}{4}=-14\frac{1}{4}$$

6.

$-5(z+6)=z$

$-5z-30=z$

$5z-5z-30=z+5z$

$-30=6z$

$\frac{-30}{6}=\frac{6z}{6}$

$-5=z$

$z=-5$

check :

$-5(-5+6)\overset{?}{=}-5$

$25-30\overset{?}{=}-5$

$-5=-5$

7.

$2(t-3)-3(t-2)=t+8$

$2t-6-3t+6=t+8$

$-t=t+8$

$-t-t=-t+t+8$

$-2t=8$

$\frac{-2t}{-2}=\frac{8}{-2}$

$t=-4$

check :

$2(-4-3)-3(-4-2)\overset{?}{=}-4+8$

$2(-7)-3(-6)\overset{?}{=}4$

$-14+18\overset{?}{=}4$

$4=4$

8.

$4[5y-(y-1)]=7(y-2)$

$4[5y-y+1]=7(y-2)$

$4[4y+1]=7(y-2)$

$16y+4=7y-14$

$16y+4-7y=7y-14-7y$

$9y+4=-14$

$9y+4-4=-14-4$

$9y=-18$

$\frac{9y}{9}=\frac{-18}{9}$

$y=-2$

check :

$4[5(-2)-(-2-1)]\overset{?}{=}7(-2-2)$

$4[-10-(-3)]\overset{?}{=}7(-4)$

$4[-7]\overset{?}{=}7(-4)$

$-28=-28$

9.

$y = \textit{number of years}$

$12{,}000-1100y=6500$

$12{,}000-12{,}000-1100y=6500-12{,}000$

$-1100y=-5500$

$\frac{-1100y}{-1100}=\frac{-5500}{-1100}$

$y=5 \textit{ years}$

check :

$12.000-1100(5)\overset{?}{=}6500$

$12{,}000-5500\overset{?}{=}6500$

$6500=6500$

10. t = time of express train

	Rate	**Time**	**Distance**
Express	**60**	**t**	**60t**
Local	**50**	**t+1/2**	**50(t+1/2)**

$60t = 50(t + \frac{1}{2})$

$60t = 50t + 25$

$60t - 50t = 50t + 25 - 50t$

$10t = 25$

$\frac{10t}{10} = \frac{25}{10}$

$t = 2.5$ hours or $2\frac{1}{2}$ hours

check :

$60(2.5) \stackrel{?}{=} 50(2.5 + \frac{1}{2})$

$60(2.5) \stackrel{?}{=} 50(3)$

$150 = 150$

11. d = distance

	rate	time	distance
Going	25	d/25	d
Return	30	d/30	d

$\frac{d}{25} + \frac{d}{30} = 5.5$

$30d + 25d = (5.5)(30)(25)$

$55d = 4125$

$\frac{55d}{55} = \frac{4125}{55}$

$d = 75$ miles

check :

$\frac{75}{25} + \frac{75}{30} \stackrel{?}{=} 5.5$

$3 + 2.5 \stackrel{?}{=} 5.5$

$5.5 = 5.5$

12. x = speed uphill

	rate	time	distance
Up	x	2	2x
Down	x+10	1	x+10

$2x + x + 10 = 40$

$3x + 10 = 40$

$3x + 10 - 10 = 40 - 10$

$3x = 30$

$\frac{3x}{3} = \frac{30}{3}$

$x = 10$ mph

$x + 10 = 20$ mph

check :

$2 \cdot 10 + 10 + 10 \stackrel{?}{=} 40$

$40 = 40$

Exercises 2.3

1.

$3x-1=8$

$3x-1+1=8+1$

$3x=9$

$\frac{3x}{9}=\frac{9}{9}$

$x=3$

check :

$3 \cdot 3-1\overset{?}{=}8$

$9-1\overset{?}{=}8$

$8=8$

3.

$9t+17=-1$

$9t+17-17=-1-17$

$9t=-18$

$\frac{9t}{9}=\frac{-18}{9}$

$t=-2$

check :

$9(-2)+17\overset{?}{=}-1$

$-18+17\overset{?}{=}-1$

$-1=-1$

5.

$20-5m=45$

$20-20-5m=45-20$

$-5m=25$

$\frac{-5m}{-5}=\frac{25}{-5}$

$m=-5$

check :

$20-5(-5)=45$

$20+25=45$

$45=45$

7.

$\frac{n}{2}-1=5$

$\frac{n}{2}-1+1=5+1$

$\frac{n}{2}=6$

$2 \cdot \frac{n}{2}=6 \cdot 2$

$n=12$

check :

$\frac{12}{2}-1\overset{?}{=}5$

$6-1\overset{?}{=}5$

$5=5$

9.

$\frac{x}{5}+15=0$

$\frac{x}{5}+15-15=0-15$

$\frac{x}{5}=-15$

$(5)\frac{x}{5}=-15(5)$

$x=-75$

check :

$\frac{-75}{5}+15\overset{?}{=}0$

$-15+15\overset{?}{=}0$

$0=0$

11.

$3-t=1$

$3-3-t=1-3$

$-t=-2$

$t=2$

check :

$3-2\overset{?}{=}1$

$1=1$

13.

$-8-b=11$

$8-8-b=11+8$

$-b=19$

$b=-19$

check :

$-8-(-19)\overset{?}{=}11$

$11=11$

15.

$\frac{2}{3}x-9=17$

$\frac{2}{3}x=26$

$(\frac{3}{2})(\frac{2}{3})x=26(\frac{3}{2})$

$x=39$

check :

$\frac{2}{3}(39)-9\overset{?}{=}17$

$26-9\overset{?}{=}17$

$17=17$

17.

$\frac{4}{5}r+20=-20$

$\frac{4}{5}r+20-20=-20-20$

$\frac{4}{5}r=-40$

$\frac{4}{5}r(\frac{5}{4})=-40(\frac{5}{4})$

$r=-50$

check :

$\frac{4}{5}(-50)+20\overset{?}{=}-20$

$-40+20\overset{?}{=}-20$

$-20=-20$

19.

$3y+y=-8$

$4y=-8$

$(\frac{1}{4})4y=-8(\frac{1}{4})$

$y=-2$

check :

$3(-2)+(-2)\overset{?}{=}-8$

$-6+(-2)\overset{?}{=}-8$

$-8=-8$

21.

$7z-2z=-30$

$5z=-30$

$(\frac{1}{5})5z=-30(\frac{1}{5})$

$z=-6$

check :

$7(-6)-2(-6)\overset{?}{=}-30$

$-42+12\overset{?}{=}-30$

$-30=-30$

23.

$28-a+4a=7$

$28+3a=7$

$28-28+3a=7-28$

$3a=-21$

$(\frac{1}{3})3a=-21(\frac{1}{3})$

$a=-7$

check :

$28-(-7)+4(-7)\overset{?}{=}7$

$28+7-28\overset{?}{=}7$

$7=7$

25.

$1 = 1 - 6t - 4t$

$1 = 1 - 10t$

$1 - 1 = 1 - 1 - 10t$

$0 = -10t$

$(-\frac{1}{10})(0) = (-\frac{1}{10})(-10t)$

$0 = t \quad or \quad t = 0$

check :

$1 = 1 - 6(0) - 4(0)$

$1 = 1$

27.

$3y + 2 = -y - 2$

$3y + 2 - 2 = -y - 2 - 2$

$3y = -y - 4$

$3y + y = -y + y - 4$

$4y = -4$

$(\frac{1}{4})4y = (\frac{1}{4})(-4)$

$y = -1$

check :

$3(-1) + 2 \stackrel{?}{=} -(-1) - 2$

$-3 + 2 \stackrel{?}{=} 1 - 2$

$-1 = -1$

29.

$5r - 4 = 2r + 6$

$5r - 4 + 4 = 2r + 6 + 4$

$5r = 2r + 10$

$5r - 2r = 2r - 2r + 10$

$3r = 10$

$(\frac{1}{3})3r = (\frac{1}{3})10$

$r = \frac{10}{3} \quad or \quad r = 3\frac{1}{3}$

check :

$5(\frac{10}{3}) - 4 \stackrel{?}{=} 2(\frac{10}{3}) + 6$

$\frac{50}{3} - 4 \stackrel{?}{=} \frac{20}{3} + 6$

$\frac{38}{3} = \frac{38}{3}$

31.

$4(x + 7) = 7 + x$

$4x + 28 = 7 + x$

$4x - x + 28 = 7 + x - x$

$3x + 28 = 7$

$3x + 28 - 28 = 7 - 28$

$3x = -21$

$(\frac{1}{3})3x = (\frac{1}{3})(-21)$

$x = -7$

33.

$5(y-1)=2y+1$

$5y-5=2y+1$

$5y-5+5=2y+1+5$

$5y=2y+6$

$5y-2y=2y-2y+6$

$3y=6$

$(\frac{1}{3})3y=(\frac{1}{3})6$

$y=2$

check :

$5(2-1)\overset{?}{=}2(2)+1$

$5(1)\overset{?}{=}4+1$

$5=5$

35.

$3a-2(a-9)=4+2a$

$3a-2a+18=4+2a$

$a+18=4+2a$

$a+18-18=4-18+2a$

$a=-14+2a$

$a-2a=-14+2a-2a$

$-a=-14$

$(-1)(-a)=(-1)(-14)$

$a=14$

check :

$3(14)-2(14-9)\overset{?}{=}4+2(14)$

$42-2(5)\overset{?}{=}4+28$

$42-10\overset{?}{=}4+28$

$32=32$

37.

$5(2-t)-(1-3t)=6$

$10-5t-1+3t=6$

$9-2t=6$

$9-9-2t=6-9$

$-2t=-3$

$(-\frac{1}{2})(-2t)=(-\frac{1}{2})(-3)$

$t=\frac{3}{2} \quad or \quad t=1\frac{1}{2}$

check :

$5(2-\frac{3}{2})-(1-3(\frac{3}{2}))\overset{?}{=}6$

$5(\frac{1}{2})-(1-\frac{9}{2})\overset{?}{=}6$

$\frac{5}{2}-(-\frac{7}{2})\overset{?}{=}6$

$6=6$

39.

$2y-3(y+1)=-(5y+3)+y$

$2y-3y-3=-5y-3+y$

$-y-3=-4y-3$

$-y-3+3=-4y-3+3$

$-y=-4y$

$-y+4y=-4y+4y$

$-3y=0$

$(-\frac{1}{3})(-3y)=(-\frac{1}{3})(0)$

$y=0$

check :

$2(0)-3((0)+1)\overset{?}{=}-(5(0)+3)+0$

$0-3\overset{?}{=}-3+0$

$0=0$

41.

$2[3z-5(2z-3)]=3z-4$

$2[3z-10z+15]=3z-4$

$2[-7z+15]=3z-4$

$-14z+30=3z-4$

$-14z=3z-34$

$-17z=-34$

$(-\frac{1}{17})(-17z)=(-\frac{1}{17})(-34)$

$z=2$

check :

$2[3(2)-5(2(2)-3)]\stackrel{?}{=}3(2)-4$

$2[6-5(4-3)]\stackrel{?}{=}6-4$

$2[6-5]\stackrel{?}{=}6-4$

$2=2$

43.

$-8m-[2(11-2m)+4]=9m$

$-8m-[22-4m+4]=9m$

$-8m-[26-4m]=9m$

$-8m-26+4m=9m$

$-4m-26=9m$

$-4m=9m+26$

$-13m=26$

$(-\frac{1}{13})(-13m)=(-\frac{1}{13})(26)$

$m=-2$

check :

$-8(-2)-[2(11-2(-2))+4]\stackrel{?}{=}9(-2)$

$16-[2(11+4)+4]\stackrel{?}{=}-18$

$16-[30+4]\stackrel{?}{=}-18$

$16-34\stackrel{?}{=}-18$

$-18=-18$

45.

$\frac{y}{0.87}+2.51=4.03$

$\frac{y}{0.87}+2.51-2.51=4.03-2.51$

$\frac{y}{0.87}=1.52$

$(0.87)\frac{y}{0.87}=1.52(0.87)$

$y\approx 1.3224$

$y=1.32$

47.

$7.37n+4.06=-1.98n+6.55$

$7.37n+4.06-4.06=-1.98n+6.55-4.06$

$7.37n=-1.98n+2.49$

$7.37n+1.98n=-1.98n+1.98n+2.49$

$9.35n=2.49$

$n\approx 0.2663$

$n=0.27$

Applications

49. t = travel time of first car
t – 1 = travel time of second car

	r	**t**	**D**
Car1	45	t	45t
Car2	54	(t-1)	54(t-1)

a. $54(t-1)=45t$

51. x = number of miles traveled

d. $1.50+1.25(x-1)=4.25$

53.
x = number of credits

$50+120x=1010$

$50-50+120x=1010-50$

$120x=960$

$(\frac{1}{120})120x=(\frac{1}{120})(960)$

$x=8$ credits

check :

$50+120(8)\stackrel{?}{=}1010$

$50+960\stackrel{?}{=}1010$

$1010=1010$

55. x = number of votes received by one candidate
2x = number of votes received by the other candidate

$x+2x=3690$

$3x=3690$

$(\frac{1}{3})3x=(\frac{1}{3})3690$

$x=1230$ votes

$2x=2460$ votes

57. t = number of hours car was parked in garage

$3+2(t-1)=9$

$3+2t-2=9$

$1+2t=9$

$1-1+2t=9-1$

$2t=8$

$(\frac{1}{2})2t=(\frac{1}{2})(8)$

$t=4$ hours

check :

$3+2(4-1)\stackrel{?}{=}9$

$3+2(3)\stackrel{?}{=}9$

$3+6\stackrel{?}{=}9$

$9=9$

59. x = number of large postcards
(5000 – x) = number of small postcards

$0.02x+0.01(5000-x)=85$

$0.02x+50-0.01x=85$

$0.01x+50=85$

$0.01x+50-50=85-50$

$0.01x=35$

$\frac{0.01x}{0.01}=\frac{35}{0.01}$

$x=3500$

x = 3500 large postcards
5000 – x = 1500 small postcards

check :

$0.02(3500)+0.01(5000-3500)\stackrel{?}{=}85$

$0.02(3500)+0.01(1500)\stackrel{?}{=}85$

$70+15\stackrel{?}{=}85$

$85=85$

61. t = time it takes to catch bus

	r	**t**	**d**
bus	24	t + 1/3	24(t+1/3)
car	36	t	36t

$24(t+\frac{1}{3})=36t$

$24t+8=36t$

$24t+8-8=36t-8$

$24t=36t-8$

$24t-36t=36t-36t-8$

$-12t=-8$

$(-\frac{1}{12})(-12t)=(-\frac{1}{12})(-8)$

$t=\frac{-8}{-12}=\frac{2}{3}$

$t=\frac{2}{3}$ hour

t = 40 minutes

63. r = rate of one snail
r+2 = rate of the other snail

	r	t	d
Snail 1	R	27	27r
Snail 2	r + 2	27	27(r+2)

$27r+27(r+2)=432$

$27r+27r+54=432$

$54r+54=432$

$54r+54-54=432-54$

$54r=378$

$(\frac{1}{54})54r=(\frac{1}{54})378$

$r=7$ cm/min

$r+2=9$ cm/min

65. r = speed of slower truck
r + 4 = speed of faster truck

	r	t	d
slower truck	r	2 hr	2r
faster truck	r + 4	2 hr	2(r + 4)

$2r+2(r+4)=212$

$2r+2r+8=212$

$4r+8=212$

$4r+8-8=212-8$

$4r=204$

$(\frac{1}{4})4r=(\frac{1}{4})(204)$

$r=51$ mph

2.4 Solving Literal Equations and Formulas

Practice 2.4

1.

$3r-s=t$

$3r-s+s=t+s$

$3r=t+s$

$\frac{3r}{3}=\frac{t+s}{3}$

$r=\frac{t+s}{3}$

2.

$\frac{4x}{5a}=c$

$(5a)\frac{4x}{5a}=c(5a)$

$4x=5ac$

$\frac{4x}{4}=\frac{5ac}{4}$

$x=\frac{5ac}{4}$

3.a.

$y=mx+b$

$y-b=mx+b-b$

$y-b=mx$

$\frac{y-b}{m}=\frac{mx}{m}$

$\frac{y-b}{m}=x$

$x=\frac{y-b}{m}$

b.

$x=\frac{10-7}{-3}$

$x=\frac{3}{-3}$

$x=-1$

4.a.

$A = P(1+rt)$

$A = P + \Pr t$

$A - P = P - P + \Pr t$

$A - P = \Pr t$

$\frac{A-P}{Pt} = \frac{\Pr t}{Pt}$

$\frac{A-P}{Pt} = r$

$r = \frac{A-P}{Pt}$

b.

$r = \frac{2100-2000}{(2000)(2)}$

$r = \frac{100}{4000}$

$r = 0.025$ or 2.5%

5. a.

$A = lw$

$\frac{A}{l} = w$

$w = \frac{A}{l}$

b.

$w = \frac{63}{9}$

$w = 7$ in

6.a.

$A = \frac{1}{2}h(b+B)$

b

$A = \frac{1}{2}h(b+B)$

$2A = hb + hB$

$2A - hB = hb + hB - hB$

$2A - hB = hb$

$\frac{2A-hB}{h} = \frac{hb}{h}$

$\frac{2A-hB}{h} = b$

$b = \frac{2A-hB}{h}$

c.

$b = \frac{2A-hB}{h}$

$b = \frac{2(32)-(4)(11)}{4}$

$b = \frac{20}{4}$

$b = 5\,cm$

Exercises 2.4

1.

$y + 10 = x$

$y + 10 - 10 = x - 10$

$y = x - 10$

3.

$d - c = 4$

$d - c + c = 4 + c$

$d = c + 4$

5.

$-3y = da$

$\frac{-3y}{a} = \frac{da}{a}$

$\frac{-3y}{a} = d$

$d = \frac{-3y}{a}$

7.

$\frac{1}{2}n = 2p$

$(2)(\frac{1}{2}n) = 2p(2)$

$n = 4p$

9.

$a = \frac{1}{2}xyz$

$(2)a = (2)\frac{1}{2}xyz$

$2a = xyz$

$\frac{2a}{xy} = \frac{xyz}{xy}$

$\frac{2a}{xy} = z$ *or* $z = \frac{2a}{xy}$

11.

$3x + y = 7$

$3x + y - y = 7 - y$

$3x = 7 - y$

$\frac{3x}{3} = \frac{7-y}{3}$

$x = \frac{7-y}{3}$

13.

$3x + 4y = 12$

$3x - 3x + 4y = 12 - 3x$

$4y = 12 - 3x$

$\frac{4y}{4} = \frac{12-3x}{4}$

$y = \frac{12-3x}{4}$

15.

$y - 4t = 0$

$y - 4t + 4t = 0 + 4t$

$y = 4t$

17.

$-5b + p = r$

$-5b + p - p = r - p$

$-5b = r - p$

$\frac{-5b}{-5} = \frac{r-p}{-5}$

$b = \frac{r-p}{-5}$ or $b = \frac{p-r}{5}$

19.

$h = 2(m - 2l)$

$h = 2m - 4l$

$h - 2m = 2m - 2m - 4l$

$h - 2m = -4l$

$\frac{h-2m}{-4} = \frac{-4l}{-4}$

$\frac{h-2m}{-4} = l$ or $l = \frac{2m-h}{4}$

21.

$I = prt$

$\frac{I}{pt} = \frac{prt}{pt}$

$\frac{I}{pt} = r$

$r = \frac{I}{pt}$

23.

$d = rt$

$\frac{d}{t} = \frac{rt}{t}$

$\frac{d}{t} = r$

$r = \frac{d}{t}$

25.

$P = a + b + c$

$P - a - c = a + b + c - a - c$

$P - a - c = b$

$b = P - a - c$

27.

$C = \pi d$

$\frac{C}{\pi} = \frac{\pi d}{\pi}$

$\frac{C}{\pi} = d$

$d = \frac{C}{\pi}$

29.

$P = I^2 R$

$\frac{P}{I^2} = \frac{I^2 R}{I^2}$

$\frac{P}{I^2} = R$ or $R = \frac{P}{I^2}$

31.

$A = \frac{a+b+c}{3}$

$3A = \frac{a+b+c}{3}(3)$

$3A = a + b + c$

$3A - b - c = a + b + c - b - c$

$3A - b - c = a$ or $a = 3A - b - c$

33.

$S = a + (n-1)d$

$S - (n-1)d = a + (n-1)d - (n-1)d$

$S - (n-1)d = a$

$S - dn + d = a$ or $a = S - dn + d$

35.

$3x + y = 6$

$3x - 3x + y = 6 - 3x$

$y = 6 - 3x$

when $x = -12$

$y = 6 - 3(-12)$

$y = 6 + 36$

$y = 42$

37.

$3x - 7 = y$

$3x - 7 + 7 = y + 7$

$3x = y + 7$

$\frac{3x}{3} = \frac{y+7}{3}$

$x = \frac{y+7}{3}$

when $y = 5$

$x = \frac{5+7}{3}$

$x = \frac{12}{3}$

$x = 4$

39.

$-\frac{1}{3}y = x$

$(-3)(-\frac{1}{3})y = -3x$

$y = -3x$

when $x = \frac{1}{2}$

$y = -3(\frac{1}{2})$

$y = -\frac{3}{2}$ or $-1\frac{1}{2}$

41.

$ax + by = c$

$ax + by - by = c - by$

$ax = c - by$

$\frac{ax}{a} = \frac{c-by}{a}$

$x = \frac{c-by}{a}$

when $a = 1, b = 3, y = -4, c = 2$

$x = \frac{2-(3)(-4)}{1}$

$x = 14$

Applications

43.

a. $K = \frac{V}{T}$

b.

$K = \frac{V}{T}$

$TK = \frac{V}{T}(T)$

$TK = V \quad or \quad V = KT$

45.

a. $C = \frac{w}{150} \bullet A$

b.

$C = \frac{w}{150} \bullet A$

$\frac{150}{w} \bullet C = \frac{150}{w} \bullet \frac{w}{150} \bullet A$

$\frac{150C}{w} = A$

$A = \frac{150C}{w}$

47.

a. $m = \frac{t}{5}$

b.

$m = \frac{t}{5}$

$5m = \frac{t}{5}(5)$

$5m = t \quad \text{or} \quad t = 5m$

c.

$t = 5m$

$t = (5)(2.5)$

t = 12.5 seconds

49.

a. $C = 2\pi r$

b.

$C = 2\pi r$

$\frac{C}{2\pi} = \frac{2\pi r}{2\pi}$

$\frac{C}{2\pi} = r \quad \text{or} \quad r = \frac{C}{2\pi}$

c.

$r = \frac{C}{2\pi}$

$r = \frac{5}{2(3.14)}$

$r = 0.8 \text{ ft}$

2.5 Solving Equations Involving Percent

Practice 2.5

1.

$0.4x = 8$

$\frac{0.4x}{0.4} = \frac{8}{0.4}$

$x = 20$

2.

$0.12s = 3{,}000{,}000$

$\frac{0.12s}{0.12} = \frac{3{,}000{,}000}{0.12}$

$s = \$25{,}000{,}000$

3.

$x = 0.23 \bullet 45$

$x = 10.35 \text{ m}$

4. x = tax

$x = 0.065 \bullet 35$

$x = \$2.28$

total cost = x + \$35 = \$37.28

5.

$x \cdot 16 = 14$

$\frac{x \cdot 16}{16} = \frac{14}{16}$

$x = .875$

$x = 87.5\%$

6.

$x \cdot 42 = 13$

$\frac{x \cdot 42}{42} = \frac{13}{42}$

$x \approx 0.3095$

$x = 31\%$ of the Presidents had been Vice Presidents

7.

$361 - 300 = 61$

$x \cdot 300 = 61$

$\frac{x \cdot 300}{300} = \frac{61}{300}$

$x = 0.20\bar{3}$

$x = 20.3\%$

The number of nursing homes increased by 20.39%.

8.

$300 - 230 = 70$

$x \cdot 300 = 70$

$\frac{x \cdot 300}{300} = \frac{70}{300}$

$x = 0.23\bar{3}$

$x = 23.3\%$ drop in stocks on October 29, 1929

$2250 - 1750 = 500$

$x \cdot 2250 = 500$

$\frac{x \cdot 2250}{2250} = \frac{500}{2250}$

$x = 0.22\bar{2}$

$x = 22.2\%$ drop in stocks on October 19, 1987

The stock dropped more in 1929.

9.

$I = P \cdot r \cdot t$

$\$130 = \$2000 \cdot r \cdot 1$

$130 = 2000r$

$\frac{130}{2000} = \frac{2000r}{2000}$

$0.065 = r$

$r = 6.5\%$

10.

	P	**r**	**t**	**I**
Mutual fund	x	-0.1	1	-0.1x
Bonds	2x	0.1	1	0.2x
Total				$700

$0.2x + (-0.1x) = 700$

$0.1x = 700$

$\frac{0.1x}{0.1} = \frac{700}{0.1}$

$x = \$7000$ was invested in mutual fund

$2x = \$14,000$ was invested in bonds

11.

Action	**Substance**	**Amount**	**Amount of salt**
Start with	20% salt solution	30g	.20(30)
Add	100% Salt	x	1x
Finish with	25% salt solution	30 +x	.25(30+x)

$0.20(30) + x = 0.25(30 + x)$

$6 + x = 7.5 + 0.25x$

$6 - 6 + x = 7.5 + 0.25x - 6$

$x = 1.5 + 0.25x$

$x - 0.25x = 1.5 + 0.25x - 0.25x$

$0.75x = 1.5$

$\frac{0.75x}{0.75} = \frac{1.5}{0.75}$

$x = 2$ g of salt

Exercises 2.5

1.
$x = 0.75 \cdot 8$

$x = 6$

3.
$x = 1.00 \cdot 23$

$x = 23$

5.
$x = 0.41 \cdot 7$

$x = 2.87\text{kg}$

7.
$x = 0.08 \cdot \$500$

$x = \$40$

9.
$x = 0.125 \cdot 32$

$x = 4$

11.
$.25x = 8$

$\frac{.25x}{.25} = \frac{8}{.25}$

$x = 32 \text{ sq in}$

13.
$0.10x = \$12$

$\frac{0.10x}{0.10} = \frac{12}{0.10}$

$x = \$120$

15.
$0.10x = \$20$

$\frac{0.10x}{0.10} = \frac{120}{0.10}$

$x = \$200$

17.
$2.00x = 3.5$

$\frac{2.00x}{2.00} = \frac{3.5}{2.00}$

$x = 1.75$

19.
$0.005x = 23$

$\frac{0.005x}{0.005} = \frac{23}{0.005}$

$x = 4600 \text{ m}$

21.
$x \cdot 80 = 50$

$\frac{x \cdot 80}{80} = \frac{50}{80}$

$x = 0.625$

$x = 62.5\% \text{ or } 62\frac{1}{2}\%$

23.
$x \cdot 15 = 5$

$\frac{x \cdot 15}{15} = \frac{5}{15}$

$x = 0.3\bar{3}$

$x = 33.3\% \text{ or } 33\frac{1}{3}\%$

25.
$x \cdot 8 = 10$

$\frac{x \cdot 8}{8} = \frac{10}{8}$

$x = 1.25$

$x = 125\%$

27.
$x \cdot 5 = \frac{1}{2}$

$\frac{x \cdot 5}{5} = \frac{\frac{1}{2}}{5}$

$x = 0.1$

$x = 10\%$

29.
$x \cdot 4 = 2.5$

$\frac{x \cdot 4}{4} = \frac{2.5}{4}$

$x = 0.625$

$x = 62.5\% \text{ or } 62\frac{1}{2}\%$

31.
$x = 0.35 \cdot \$400$

$x = \$140$

33.

$x \cdot 50 \text{ mi} = 20 \text{ mi}$

$\frac{x \cdot 50}{50} = \frac{20}{50}$

$x = 0.40$

$x = 40\%$

35.

$0.70x = 14$

$\frac{0.70x}{0.70} = \frac{14}{0.70}$

$x = 20$

37.

$x = 0.001 \cdot 35$

$x = 0.035$

39.

$x \cdot \$240 = \$8$

$\frac{x \cdot 240}{240} = \frac{8}{240}$

$x = 0.03\overline{3}$

$x = 3.3\% \text{ or } 3\frac{1}{3}\%$

41.

$0.20x = 3oz.$

$\frac{0.20x}{0.20} = \frac{3}{0.20}$

$x = 15 \; oz.$

Applications

43.

$w = 0.55 \cdot 160$

$w = 88$

$m = 0.45 \cdot 160$

$m = 72$

$88 - 72 = 16$

There are 16 more women than men.

45.

$x \cdot 4 = \frac{9}{12}$

$x \cdot 4 = 0.75$

$\frac{x \cdot 4}{4} = \frac{0.75}{4}$

$x = 0.1875$

$x = 18.75\% \text{ or } 18\frac{3}{4}\%$

47.

$0.25x = 8$

$\frac{0.25x}{0.25} = \frac{8}{0.25}$

$x = 32$ employees

49.

$x \cdot \$250,000 = \$50,000$

$\frac{x \cdot 250,000}{250,000} = \frac{50,000}{250,000}$

$x = 0.20$

$x = 20\%$

51.

$0.08x = 1.5$ million

$\frac{0.08x}{0.08} = \frac{1.5}{0.08}$

$x = 18.75$ million people

53.

$x = 0.60 \cdot 90$

$x = 54$ tables

55.

$4000 - 1800 = 2200$

$x \cdot 4000 = 2200$

$\frac{x \cdot 4000}{4000} = \frac{2200}{4000}$

$x = 0.55$

$x = 55\%$

No, the animal is not endangered.

57.

$I = P \cdot r \cdot t$

$I = \$2000 \cdot 0.05 \cdot 1$

$I = \$100$

59.

$I = P \cdot r \cdot t$

$I = \$5000 \cdot 0.05 \cdot 1$

$I = \$250$

61.

P	r	t	I
x	0.08	1	0.08x
34000-x	0.1	1	0.1(34000-x)
			$3000

$0.08x + 0.1(34000 - x) = \$3000$

$0.08x + 3400 - 0.1x = \$3000$

$-0.02x + 3400 = 3000$

$-0.02x + 3400 - 3400 = 3000 - 3400$

$-0.02x = -400$

$\frac{-0.02x}{-0.02} = \frac{-400}{-0.02}$

$x = \$20,000$ invested at 8%

$34000 - x = \$14,000$ invested at 10%

63.

P	r	t	I
$20,000	0.08	1	$1600
x	0.05	1	0.05x
			$2100

$\$1600. + 0.05x = \$2100.$

$1600 - 1600 + 0.05x = 2100 - 1600$

$0.05x = 500$

$x = \$10,000$

65.

Action	**Substance**	**Amount**	**Amount of vinegar**
Start with	20% vinegar solution	2 cups	.20(2)
Add	10% vinegar solution	x	.10x
Finish with	12% vinegar solution	2 +x	.12(2+x)

$0.20(2) + 0.10x = 0.12(2 + x)$

$0.4 + 0.10x = 0.24 + 0.12x$

$0.4 - 0.4 + 0.10x = 0.24 - 0.4 + 0.12x$

$0.10x = -0.16 + 0.12x$

$0.10x - 0.12x = -0.16 + 0.12x - 0.12x$

$-0.02x = -0.16$

$\frac{-0.02x}{-0.02} = \frac{-0.16}{-0.02}$

$x = 8$ cups

67.

Action	**Substance**	**Amount**	**Amount of acid**
Start with	4% acid solution	30 oz	0.04(30)
Add	40% acid solution	x	0.4x
Finish with	10% acid solution	30 +x	0.1(30+x)

$0.04(30) + 0.4x = 0.1(30 + x)$

$1.2 + 0.4x = 3 + 0.1x$

$0.4x = 1.8 + 0.1x$

$0.3x = 1.8$

$x = 6$ oz

2.6 Solving Inequalities

Practice 2.6

1.

$\frac{1}{2}x - 2 < -1$

$\frac{1}{2}(4) - 2 \overset{?}{<} -1$

$2 - 2 \overset{?}{<} -1$

$0 > -1$

No, 4 is not a solution.

2.

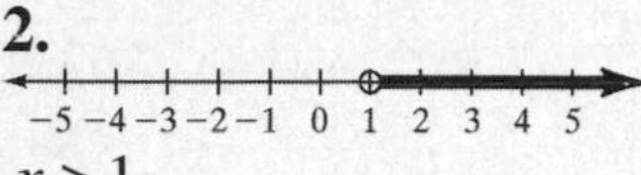

$x > 1$

3.

$x \leq -1\frac{1}{2}$

4.

$-3 < x < 4$

5.

$n + 5 > 4$

$n + 5 - 5 > 4 - 5$

$n > -1$

6.

$x - 4 \leq 1\frac{1}{2}$

$x - 4 + 4 \leq 1\frac{1}{2} + 4$

$x \leq 5\frac{1}{2}$

7.

$4x + 5 \geq 3x - 2$

$4x + 5 - 5 \geq 3x - 2 - 5$

$4x \geq 3x - 7$

$4x - 3x \geq 3x - 3x - 7$

$x \geq -7$

8.

$\frac{x}{3} \leq 1$

$(3)\frac{x}{3} \leq 1(3)$

$x \leq 3$

9.

$-3x > 15$

$\frac{-3x}{-3} < \frac{15}{-3}$

$x < -5$

10.

$10 > 5x - 7x$

$10 > -2x$

$\frac{10}{-2} < \frac{-2x}{-2}$

$-5 < x$

$x > -5$

11.

$-6 \geq 3z + 4 - z$

$-6 \geq 2z + 4$

$-6 - 4 \geq 2z + 4 - 4$

$-10 \geq 2z$

$\frac{-10}{2} \geq \frac{2z}{2}$

$-5 \geq x$

$x \leq -5$

12.

$7x - (9x + 1) > 5$

$7x - 9x - 1 > 5$

$-2x - 1 > 5$

$-2x - 1 + 1 > 5 + 1$

$-2x > 6$

$\frac{-2x}{-2} < \frac{6}{-2}$

$x < -3$

13.

$x+(x+2)+(x+3)\geq 14$

$3x+5\geq 14$

$3x+5-5\geq 14-5$

$3x\geq 9$

$\frac{3x}{3}\geq\frac{9}{3}$

$x\geq 3$

14.

$15(\$8.50)+\$7.50t\geq \$300$

$127.50+7.50t\geq 300$

$127.50-127.50+7.50t\geq 300-127.50$

$7.50t\geq 172.5$

$t\geq 23$ hours

Exercises 2.6

1.

Value of x	Inequality	True or False
a. 1	$8-3x>5$	False
b. 4	$4x-7\leq 2x+1$	True
c. -7	$6(x+6)<-9$	False
d. -3/4	$8x+10\geq 12x+15$	False

3. $x>3$

5. $x\geq -5$

7. $x>0$

9. $x\leq 0.5$

11. $x>-2\frac{1}{2}$

13. $-3<x<1$

15. $-\frac{1}{2}\leq x<2$

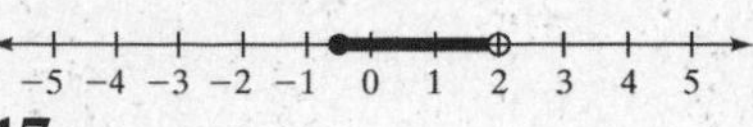

17.

$v+2<-5$

$v+2-2<-5-2$

$v<-7$

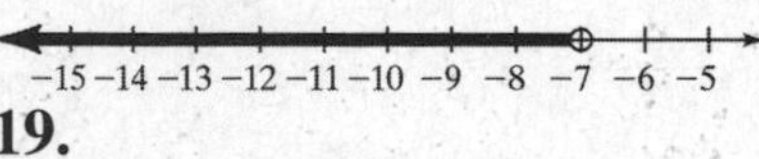

19.

$y-5>-5$

$y-5+5>-5+5$

$y>0$

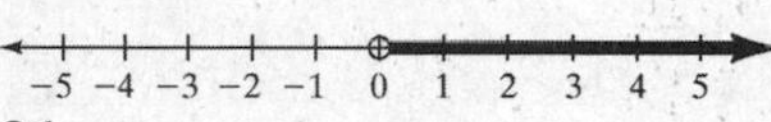

21.

$y+2\leq 5.5$

$y+2-2\leq 5.5-2$

$y\leq 3.5$

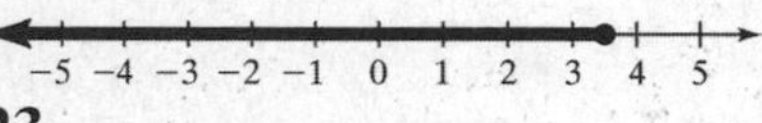

23.

$v-17\leq -15$

$v-17+17\leq -15+17$

$v\leq 2$

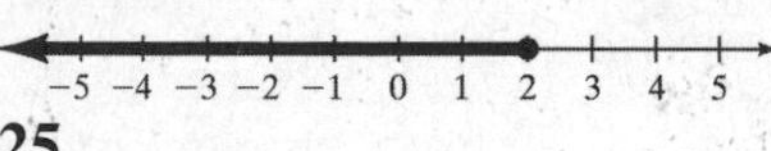

25.

$-2\geq x-4$

$-2+4\geq x-4+4$

$2\geq x$

$x\leq 2$

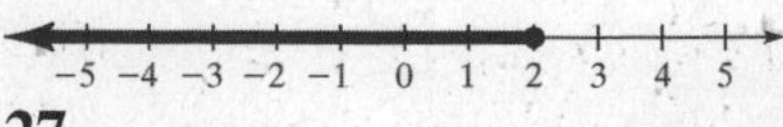

27.

$\frac{1}{3}a<-1$

$(3)\frac{1}{3}a<(-1)(3)$

$a<-3$

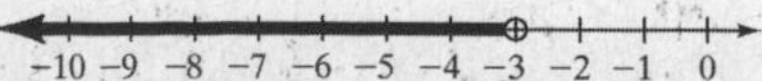

29.

$-5y > 10$

$\frac{-5y}{-5} < \frac{10}{-5}$

$y < -2$

−10 −9 −8 −7 −6 −5 −4 −3 −2 −1 0

31.

$2x \geq 0$

$\frac{2x}{2} \geq \frac{0}{2}$

$x \geq 0$

−5 −4 −3 −2 −1 0 1 2 3 4 5

33.

$-\frac{3}{4}a \geq 3$

$(-\frac{4}{3})(-\frac{3}{4})a \leq 3(-\frac{4}{3})$

$a \leq -4$

−12 −11 −10 −9 −8 −7 −6 −5 −4 −3 −2

35.

$6 \leq \frac{-2}{3}n$

$(-\frac{3}{2})(6) \geq (-\frac{3}{2})\frac{-2}{3}n$

$-9 \geq n$

$n \leq -9$

−18 −17 −16 −15 −14 −13 −12 −11 −10 −9 −8

37.

$\frac{n}{3} + 2 > 3$

$\frac{n}{3} + 2 - 2 > 3 - 2$

$\frac{n}{3} > 1$

$(3)\frac{n}{3} > 1(3)$

$n > 3$

39.

$3x - 12 \leq 6$

$3x - 12 + 12 \leq 6 + 12$

$3x \leq 18$

$\frac{3x}{3} \leq \frac{18}{3}$

$x \leq 6$

41.

$-21 - 3y > 0$

$-21 + 21 - 3y > 0 + 21$

$-3y > 21$

$\frac{-3y}{-3} < \frac{21}{-3}$

$y < -7$

43.

$5n - 11 \geq 2n + 28$

$5n - 11 + 11 \geq 2n + 28 + 11$

$5n \geq 2n + 39$

$5n - 2n \geq 2n - 2n + 39$

$3n \geq 39$

$\frac{3n}{3} \geq \frac{39}{3}$

$n \geq 13$

45.

$-4m + 8 \leq -3m + 1$

$-4m + 8 - 8 \leq -3m + 1 - 8$

$-4m \leq -3m - 7$

$-4m + 3m \leq -3m + 3m - 7$

$-m \leq -7$

$m \geq 7$

47.

$-7x + 4x + 23 < 2$

$-3x + 23 < 2$

$-3x + 23 - 23 < 2 - 23$

$-3x < -21$

$\frac{-3x}{-3} > \frac{-21}{-3}$

$x > 7$

49.

$-3(z+5) > -15$

$-3z - 15 > -15$

$-3z - 15 + 15 > -15 + 15$

$-3z > 0$

$z < 0$

51.

$0.5(2x+1) \geq 3x$

$x + 0.5 \geq 3x$

$x - x + 0.5 \geq 3x - x$

$0.5 \geq 2x$

$\frac{0.5}{2} \geq \frac{2x}{2}$

$0.25 \geq x$

$x \leq 0.25$

53.

$2(x-2) - 3x \geq -1$

$2x - 4 - 3x \geq -1$

$-x - 4 \geq -1$

$-x - 4 + 4 \geq -1 + 4$

$-x \geq 3$

$x \leq -3$

55.

$7y - (9y+1) < 5$

$7y - 9y - 1 < 5$

$-2y - 1 < 5$

$-2y - 1 + 1 < 5 + 1$

$-2y < 6$

$\frac{-2y}{-2} > \frac{6}{-2}$

$y > -3$

57.

$0.4(5x+1) \geq 3x$

$2x + 0.4 \geq 3x$

$2x - 2x + 0.4 \geq 3x - 2x$

$0.4 \geq x$

$x \leq 0.4$

59.

$5x + 1 < 3x - 2(4x-3)$

$5x + 1 < 3x - 8x + 6$

$5x + 1 - 1 < -5x + 6 - 1$

$5x < -5x + 5$

$5x + 5x < -5x + 5x + 5$

$10x < 5$

$\frac{10x}{10} < \frac{5}{10}$

$x < \frac{1}{2}$

61.

$3 + 5n \leq 6(n-1) + n$

$3 + 5n \leq 6n - 6 + n$

$3 + 5n \leq 7n - 6$

$3 - 3 + 5n \leq 7n - 6 - 3$

$5n \leq 7n - 9$

$5n - 7n \leq 7n - 7n - 9$

$-2n \leq -9$

$n \geq \frac{9}{2}$ or $4\frac{1}{2}$

63.

$-\frac{4}{3}x - 16 > x + \frac{1}{3}x$

$-\frac{4}{3}x - 16 > \frac{4}{3}x$

$-\frac{4}{3}x + \frac{4}{3}x - 16 > \frac{4}{3}x + \frac{4}{3}x$

$-16 > \frac{8}{3}x$

$(\frac{3}{8})(-16) > (\frac{3}{8})(\frac{8}{3}x)$

$-6 > x$

$x < -6$

65.
$0.2y > 1500 + 2.6y$
$0.2y - 2.6y > 1500 + 2.6y - 2.6y$
$-2.4y > 1500$
$\frac{-2.4y}{-2.4} < \frac{1500}{-2.4}$
$y < -625$

Applications

67. d. $a \geq 18$

69. d. $\frac{9}{5}C + 32 > 98.6$

71.
$\frac{81+85+91+x}{4} > 85$
$\frac{257+x}{4} > 85$
$(4)(\frac{257+x}{4}) > (85)(4)$
$257 + x > 340$
$257 - 257 + x > 340 - 257$
$x > 83$

73.
$\frac{250+250+150+130+180+x}{6} \geq 200$
$\frac{960+x}{6} \geq 200$
$(6)(\frac{960+x}{6}) \geq (200)(6)$
$960 + x \geq 1200$
$960 - 960 + x \geq 1200 - 960$
$x \geq \$240$

75.
$0.50 + 0.10x \geq 2$
$0.50 - 0.50 + 0.10x \geq 2 - 0.50$
$0.10x \geq 1.5$
$x \geq 15 \text{ min}$

77.
$1000 + 1500h > 1500 + 1200h$
$1000 - 1000 + 1500h > 1500 + 1200h - 1000$
$1500h > 500 + 1200h$
$1500h - 1200h > 500 + 1200h - 1200h$
$300h > 500$
$\frac{300h}{300} > \frac{500}{300}$
$h > \frac{5}{3} \text{ or } h > 1\frac{2}{3}$
He should accept the deal if he sells two or more houses each month.

79.
$200 - 2.5x < 180$
$200 - 200 - 2.5x < 180 - 200$
$-2.5x < 20$
$\frac{-2.5x}{-2.5} > \frac{20}{-2.5}$
$x > -8$
He will weigh less than 180 lb after 8 months.

Chapter 2 Review Exercises

[2.1]

1. No
$5x + 3 = 7 - 4x$
$5(2) + 3 \stackrel{?}{=} 7 - 4(2)$
$13 \neq -1$

2. 0 is a solution.
$4x - 15 = 5(x - 3)$
$4(0) - 15 \stackrel{?}{=} 5(0 - 3)$
$-15 = -15$

3.

$x-3=-12$

$x-3+3=-12+3$

$x=-9$

check :

$-9-3\stackrel{?}{=}-12$

$-12=-12$

4.

$t+10=8$

$t+10-10=8-10$

$t=-2$

check :

$-2+10\stackrel{?}{=}8$

$8=8$

5.

$-9=a+5$

$-9-5=a+5-5$

$a=-14$

check :

$-9\stackrel{?}{=}-14+5$

$-9=-9$

6.

$4=n-7$

$4+7=n-7+7$

$11=n \text{ or } n=11$

check :

$4\stackrel{?}{=}11-7$

$4=4$

7.

$y-(-3.1)=11$

$y-(-3.1)+(-3.1)=11+(-3.1)$

$y=7.9$

check :

$7.9-(-3.1)\stackrel{?}{=}11$

$11=11$

8.

$r+4.8=20$

$r+4.8-4.8=20-4.8$

$r=15.2$

check :

$15.2+4.8\stackrel{?}{=}20$

$20=20$

[2.2]

9.

$\frac{x}{3}=-2$

$(3)\frac{x}{3}=(-2)(3)$

$x=-6$

check :

$\frac{-6}{3}\stackrel{?}{=}-2$

$-2=-2$

10.

$\frac{z}{2}=-5$

$(2)\frac{z}{2}=(-5)(2)$

$z=-10$

check :

$\frac{-10}{2}\stackrel{?}{=}-5$

$-5=-5$

11.

$2x=-20$

$\frac{2x}{2}=\frac{-20}{2}$

$x=-10$

check :

$2(-10)\stackrel{?}{=}-20$

$-20=-20$

12.

$-5d = 15$

$\frac{-5d}{-5} = \frac{15}{-5}$

$d = -3$

check :

$-5(-3) \stackrel{?}{=} 15$

$15 = 15$

13.

$-y = -4$

$(-1)(-y) = (-1)(-4)$

$y = 4$

check :

$-(4) \stackrel{?}{=} -4$

$-4 = -4$

14.

$-x = 3$

$(-1)(-x) = (-1)(3)$

$x = -3$

check :

$-(-3) \stackrel{?}{=} 3$

$3 = 3$

15.

$20.5 = 0.5n$

$\frac{20.5}{0.5} = \frac{0.5n}{0.5}$

$41 = n$

$n = 41$

check :

$20.5 \stackrel{?}{=} 0.5(41)$

$20.5 = 20.5$

16.

$30 = -0.2r$

$\frac{30}{-0.2} = \frac{-0.2r}{-0.2}$

$-150 = r$

$r = -150$

check :

$30 \stackrel{?}{=} -0.2(-150)$

$30 = 30$

17.

$\frac{2t}{3} = -6$

$(\frac{3}{2})(\frac{2t}{3}) = (-6)(\frac{3}{2})$

$t = -9$

check :

$\frac{2(-9)}{3} \stackrel{?}{=} -6$

$6 = 6$

18.

$\frac{5y}{6} = -10$

$(\frac{6}{5})(\frac{5y}{6}) = (\frac{6}{5})(-10)$

$y = -12$

check :

$\frac{5(-12)}{6} \stackrel{?}{=} -10$

$-10 = -10$

[2.3]

19.

$2x + 1 = 7$

$2x = 6$

$\frac{2x}{2} = \frac{6}{2}$

$x = 3$

check :

$2(3) + 1 \stackrel{?}{=} 7$

$7 = 7$

20.

$-t-4=5$

$-t-4+4=5+4$

$-t=9$

$t=-9$

check :

$-(-9)-4\stackrel{?}{=}5$

$5-5$

21.

$\frac{a}{2}-3=-10$

$\frac{a}{2}-3+3=-10+3$

$\frac{a}{2}=-7$

$a=-14$

check :

$\frac{-14}{2}-3\stackrel{?}{=}-10$

$-10=-10$

22.

$\frac{r}{3}-6=12$

$\frac{r}{3}-6+6=12+6$

$\frac{r}{3}=18$

$(3)\frac{r}{3}=18(3)$

$r=54$

check :

$\frac{54}{3}-6\stackrel{?}{=}12$

$12=12$

23.

$-y+7=-2$

$-y+7-7=-2-7$

$-y=-9$

$y=9$

check;

$-9+7\stackrel{?}{=}-2$

$-2=-2$

24.

$-2t+3=1$

$-2t+3-3=1-3$

$-2t=-2$

$\frac{-2t}{-2}=\frac{-2}{-2}$

$t=1$

check :

$-2(1)+3\stackrel{?}{=}1$

$1=1$

25.

$4x-2x-5=7$

$2x-5=7$

$2x-5+5=7+5$

$2x=12$

$\frac{2x}{2}=\frac{12}{2}$

$x=6$

check :

$4(6)-2(6)-5\stackrel{?}{=}7$

$24-12-5\stackrel{?}{=}7$

$7=7$

26.

$3y - y + 12 = 6$

$2y + 12 = 6$

$2y + 12 - 12 = 6 - 12$

$2y = -6$

$\frac{2y}{2} = \frac{-6}{2}$

$y = -3$

check :

$3(-3) - (-3) + 12 \stackrel{?}{=} 6$

$-9 + 3 + 12 \stackrel{?}{=} 6$

$6 = 6$

27.

$z + 1 = -2z + 10$

$z + 1 - 1 = -2z + 10 - 1$

$z = -2z + 9$

$z + 2z = -2z + 2z + 9$

$3z = 9$

$z = 3$

check :

$4 \stackrel{?}{=} -6 + 10$

$4 = 4$

28.

$n - 3 = -n + 7$

$n - 3 + 3 = -n + 7 + 3$

$n = -n + 10$

$n + n = -n + n + 10$

$2n = 10$

$n = 5$

check :

$5 - 3 \stackrel{?}{=} -5 + 7$

$2 = 2$

29.

$c = -2(c + 1)$

$c = -2c - 2$

$c + 2c = -2c + 2c - 2$

$3c = -2$

$c = -\frac{2}{3}$

check :

$-\frac{2}{3} \stackrel{?}{=} -2(-\frac{2}{3} + 1)$

$-\frac{2}{3} \stackrel{?}{=} \frac{4}{3} - 2$

$-\frac{2}{3} = -\frac{2}{3}$

30.

$p = -(p - 5)$

$p = -p + 5$

$p + p = -p + p + 5$

$2p = 5$

$p = \frac{5}{2} \text{ or } 2\frac{1}{2}$

check :

$\frac{5}{2} \stackrel{?}{=} -(\frac{5}{2} - 5)$

$\frac{5}{2} = \frac{5}{2}$

31.

$2(x+1)-(x-8)=-x$

$2x+2-x+8=-x$

$x+10=-x$

$x+x+10=-x+x$

$2x+10=0$

$2x+10-10=0-10$

$2x=-10$

$x=-5$

check :

$2(-5+1)-(-5-8)\overset{?}{=}-(-5)$

$2(-4)-(-13)\overset{?}{=}5$

$-8+13\overset{?}{=}5$

$5=5$

32.

$-(x+2)-(x-4)=-5x$

$-x-2-x+4=-5x$

$-2x+2=-5x$

$-2x+2-2=-5x-2$

$-2x=-5x-2$

$-2x+5x=-5x+5x-2$

$3x=-2$

$x=-\frac{2}{3}$

check :

$-(-\frac{2}{3}+2)-(-\frac{2}{3}-4)\overset{?}{=}-5(-\frac{2}{3})$

$-\frac{4}{3}-(-\frac{14}{3})\overset{?}{=}\frac{10}{3}$

$\frac{10}{3}=\frac{10}{3}$

33.

$10-[3+(2x-1)]=3x$

$10-[2+2x]=3x$

$8-2x=3x$

$8-8-2x=3x-8$

$-2x=3x-8$

$-2x-3x=3x-3x-8$

$-5x=-8$

$x=\frac{8}{5}\ or\ 1\frac{3}{5}$

check :

$10-[3+(2(\frac{8}{5})-1)]\overset{?}{=}3(\frac{8}{5})$

$10-[3+\frac{11}{5}]\overset{?}{=}\frac{24}{5}$

$10-\frac{26}{5}\overset{?}{=}\frac{24}{5}$

$\frac{24}{5}=\frac{24}{5}$

34.

$x-[5+(3x-4)]=-x$

$x-[1+3x]=-x$

$-2x-1=-x$

$-2x-1+1=-x+1$

$-2x=-x+1$

$-2x+x=-x+x+1$

$-x=1$

$x=-1$

check :

$-1-[5+((3)(-1)-4)]\overset{?}{=}-(-1)$

$-1-[5+(-7)]\overset{?}{=}1$

$-1-[-2]\overset{?}{=}1$

$1=1$

[2.5]

35.

$a-5b=2c$

$a-5b+5b=2c+5b$

$a=2c+5b$

36.

$\frac{2a}{b}=n$

$(b)(\frac{2a}{b})=n(b)$

$2a=bn$

$\frac{2a}{2}=\frac{bn}{2}$

$a=\frac{bn}{2}$

37.

a.

$Ax+By=C$

$Ax+By-By=C-By$

$Ax=C-By$

$x=\frac{C-By}{A}$

b.

$x=\frac{c-By}{A}$

$x=\frac{0-(-1)(5)}{2}$

$x=\frac{5}{2}\ or\ 2\frac{1}{2}$

38.

a.

$A=\frac{bh}{2}$

$2A=bh$

$\frac{2A}{b}=h\quad or\quad h=\frac{2A}{b}$

b.

$h=\frac{2A}{b}$

$h=\frac{2(12)}{4}$

$h=\frac{24}{4}$

$h=6cm$

[2.5]

39.

$0.30x=12$

$\frac{0.30x}{0.30}=\frac{12}{.030}$

$x=40$

40.

$1.25x=5$

$\frac{1.25x}{1.25}=\frac{5}{1.25}$

$x=4$

41.

$x\bullet 5=8$

$\frac{x\bullet 5}{5}=\frac{8}{5}$

$x=1.6$

160%

42.

$x\bullet 8=5$

$\frac{x\bullet 8}{8}=\frac{5}{8}$

$x=0.625$

$x=62.5\%$

43.

$x=0.085\bullet\$300$

$x=25.5$

44.

$x=0.035\bullet\$2000$

$x=70$

[2.6]

45. $x < 2$

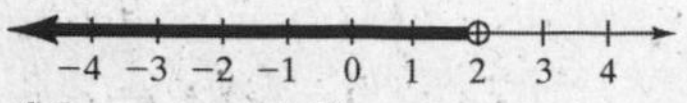

46. $x \geq -4.5$

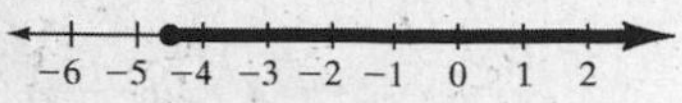

47. $3\frac{1}{2} \geq x$

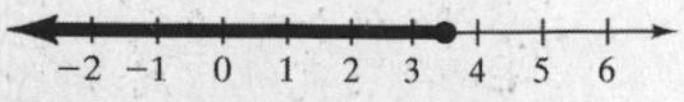

48. $-0.5 < x < 5$

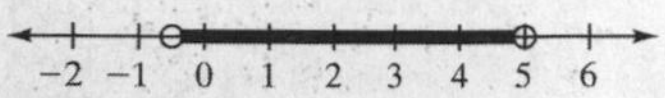

49.

$y + 1 > 6$

$y + 1 - 1 > 6 - 1$

$y > 5$

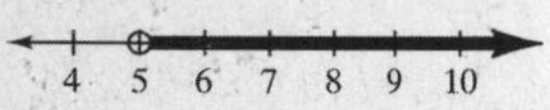

50.

$-\frac{1}{2}t + 3 \leq 3$

$-\frac{1}{2}t + 3 - 3 \leq 3 - 3$

$-\frac{1}{2}t \leq 0$

$(-\frac{2}{1})(-\frac{1}{2}t) \geq (-\frac{2}{1})(0)$

$t \geq 0$

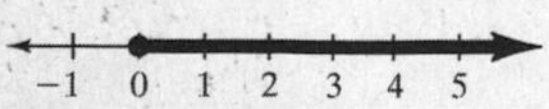

51.

$8y - 2 \leq 6y + 2$

$8y - 2 + 2 \leq 6y + 2 + 2$

$8y \leq 6y + 4$

$8y - 6y \leq 6y - 6y + 4$

$2y \leq 4$

$y \leq 2$

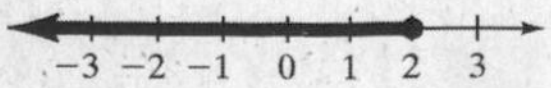

52.

$\frac{1}{2}(8 - 12x) \leq x - 10$

$4 - 6x \leq x - 10$

$4 - 4 - 6x \leq x - 10 - 4$

$-6x \leq x - 14$

$-6x - x \leq x - x - 14$

$-7x \leq -14$

$x \geq 2$

53.

$0.5n - 0.3 < 0.2(2n + 1)$

$0.5n - 0.3 < 0.4n + 0.2$

$0.5n - 0.3 + 0.3 < 0.4n + 0.2 + 0.3$

$0.5n < 0.4n + 0.5$

$0.5n - 0.4n < 0.4n - 0.4n + 0.5$

$0.1n < 0.5$

$n < 5$

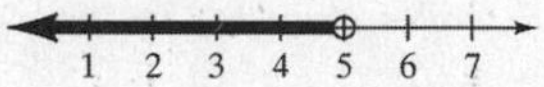

Mixed Applications

54.

$E = \frac{B}{W}$

$8 = \frac{B}{2000}$

$2000 \cdot 8 = \frac{B}{2000} \cdot 2000$

$16000 = B$

$B = 16,000$ Btu

55.

$x + x + x = 180^0$

$3x = 180^0$

$x = 60^o$

56.

$\$5000+\$50g=\$12,000$

$5000-5000+50g=12,000-5000$

$50g=7000$

$\frac{50g}{50}=\frac{7000}{50}$

$g=140$ guests

57.

$180(n-2)=540$

$180n-360=540$

$180n-360+360=540+360$

$180n=900$

$\frac{180n}{180}=\frac{900}{180}$

$n=5$ sides

58.

$x+(x+15,360)=39,210$

$2x+15,360=39,210$

$2x+15,360-15,360=39,210-15,360$

$2x=23,850$

$\frac{2x}{2}=\frac{23,850}{2}$

$x=11,925$ votes

$x+15,360=27,285$ votes

59.

a. $C=2+16y$

b. $C=2+16y$

$C-2=2-2+16y$

$C-2=16y$

$\frac{C-2}{16}=\frac{16y}{16}$

$y=\frac{C-2}{16}$

60.

	r	**t**	**D**
Truck 1	**45**	**t**	**45t**
Truck 2	**50**	**t**	**50t**

$45t+50t=380$

$95t=380$

$\frac{95t}{95}=\frac{380}{95}$

$t=4$ hours

61.

	r	**t**	**d**
Friend 1	40	t	40t
Friend 2	32	t	32t

$40t+32t=18$

$72t=18$

$\frac{72t}{72}=\frac{18}{72}$

$t=\frac{1}{4}$

$t=15$ min

10:15PM

62.

	r	**t**	**d**
Late Plane	400	t+0.5	400t+200
On-time Plane	500	t	500t

$400t + 200 = 500t$

$400t - 400t + 200 = 500t - 400t$

$200 = 100t$

$\frac{200}{100} = \frac{100t}{100}$

$2 = t$

$t = 2$ hours

$d = 500t$

$d = 500 \cdot 2$

$d = 1000$ miles

63.

$x \cdot 430 = 425$

$\frac{x \cdot 430}{430} = \frac{425}{430}$

$x \approx 0.988$

$x = 99\%$

64.

$8.8 - 5.7 = 3.1$

$x \cdot 8.8 = 3.1$

$\frac{x \cdot 8.8}{8.8} = \frac{3.1}{8.8}$

$x \approx 0.352$

$x = 35\%$

65.

$x = 0.007 \cdot 4000$

$x = 28$ students

66.

$I = P \cdot r \cdot t$

$I = \$500 \cdot 0.06 \cdot 1$

$I = \$30$

67.

$170 - 60 = 110$

$x \cdot 170 = 110$

$\frac{x \cdot 170}{170} = \frac{110}{170}$

$x \approx 0.647$

$x = 65\%$

Van Buren's electoral vote count dropped 65%.

68.

Action	**Substance**	**Amount**	**Amount of acid**
Start with	60% alcohol solution	6 L	3.6 L
Add	100% alcohol	x	x
Finish with	70% alcohol solution	6 +x	.7(6+x)

$3.6 + x = 0.7(6 + x)$

$3.6 + x = 4.2 + 0.7x$

$3.6 - 3.6 + x = 4.2 - 3.6 + 0.7x$

$x = 0.6 + 0.7x$

$x - 0.7x = 0.6 + 0.7x - 0.7x$

$0.3x = 0.6$

$\frac{0.3x}{0.3} = \frac{0.6}{0.3}$

$x = 2$ L of pure alcohol must be mixed with the solution

69.

Action	Substance	Amount	Amount of acid
Start with	10% disinfectant solution	4 pt	0.4
Add	1% disinfectant solution	x	0.01x
Finish with	5% disinfectant solution	4 +x	0.05(4+x)

$0.4+0.01x=0.05(4+x)$

$0.4+0.01x=0.2+0.05x$

$0.4-0.4+0.01x=0.2-0.4+0.05x$

$0.01x=-0.2+0.05x$

$0.01x-0.05x=-0.2+0.05x-0.05x$

$-0.04x=-0.2$

$\frac{-0.04x}{-0.04}=\frac{-0.2}{-0.04}$

$x=5$ pt of disinfectant must be added to the solution

70.

a. $p=2.2k$

b. $p=2.2k$

$\frac{p}{2.2}=\frac{2.2k}{2.2}$

$\frac{p}{2.2}=k$

$k=\frac{p}{2.2}$

71.

$100,000+25b=50b$

$100,000-100,000+25b=50b-100,000$

$25b=50b-100,000$

$25b-50b=50b-50b-100,000$

$-25b=-100,000$

$\frac{-25b}{-25}=\frac{-100,000}{-25}$

$b=4000$ books

72.

$5+0.50h\leq 16$

$5-5+0.50h\leq 16-5$

$0.50h\leq 11$

$\frac{0.50h}{0.50}\leq\frac{11}{0.50}$

$h\leq 22$ hours

You can surf the web 22 hours or less.

Chapter 2 Posttest

1.

$3x-4=6(x+2)$

$3(-2)-4\overset{?}{=}6(-2)+12$

$-6-4\overset{?}{=}-12+12$

$-10\neq 0$

-2 is not a solution.

2.

$x-1=-10$

$x-1+1=-10+1$

$x=-9$

check :

$-9-1\overset{?}{=}-10$

$-10=-10$

3.

$\frac{n}{2}=-3$

$(2)\frac{n}{2}=-3(2)$

$n=-6$

check :

$\frac{-6}{2}\overset{?}{=}-3$

$-3=-3$

4.

$-y=-11$

$y=11$

check :

$-(11)\stackrel{?}{=}-11$

$-11=-11$

5.

$\frac{3y}{4}=6$

$(\frac{4}{3})(\frac{3y}{4})=(6)(\frac{4}{3})$

$y=8$

check :

$\frac{3(8)}{4}\stackrel{?}{=}6$

$6=6$

6.

$2x+5=11$

$2x+5-5=11-5$

$2x=6$

$\frac{2x}{2}=\frac{6}{2}$

$x=3$

check :

$2(3)+5\stackrel{?}{=}11$

$11=11$

7.

$-s+4=2$

$-s+4-4=2-4$

$-s=-2$

$s=2$

check :

$-2+4\stackrel{?}{=}2$

$2=2$

8.

$10x+1=-x+23$

$10x+1-1=-x+23-1$

$10x=-x+22$

$10x+x=-x+x+22$

$11x=22$

$\frac{11x}{11}=\frac{22}{11}$

$x=2$

check :

$10(2)+1\stackrel{?}{=}-2+23$

$21=21$

9.

$16a=-4(a-5)$

$16a=-4a+20$

$16a+4a=-4a+4a+20$

$20a=20$

$\frac{20a}{20}=\frac{20}{20}$

$a=1$

check :

$16(1)\stackrel{?}{=}-4(1-5)$

$16\stackrel{?}{=}-4(-4)$

$16=16$

10.

$2(x+5)-(x+4)=7x+1$

$2x+10-x-4=7x+1$

$x+6=7x+1$

$x+6-6=7x+1-6$

$x=7x-5$

$x-7x=7x-7x-5$

$-6x=-5$

$x=\frac{5}{6}$

check :

$2(\frac{5}{6}+5)-(\frac{5}{6}+4)\overset{?}{=}7(\frac{5}{6})+1$

$2(\frac{35}{6})-\frac{29}{6}\overset{?}{=}\frac{41}{6}$

$\frac{41}{6}=\frac{41}{6}$

11.

$5n+p=t$

$5n-5n+p=t-5n$

$p=t-5n$

12.

$0.40x=8$

$\frac{0.40x}{0.40}=\frac{8}{0.40}$

$x=20$

13.

$x\cdot5=10$

$\frac{x\cdot5}{5}=\frac{10}{5}$

$x=2$

200%

14.

$-1\le x<3$

−3 −2 −1 0 1 2 3

15.

$-2z\le6$

$\frac{-2z}{-2}\ge\frac{6}{-2}$

$z\ge-3$

−3 −2 −1 0 1 2 3

16.

$\$2.00+\$1.25m=\$13.25$

$2.00-2.00+1.25m=13.25-2.00$

$1.25m=11.25$

$\frac{1.25m}{1.25}=\frac{11.25}{1.25}$

$m=9$

The ride was 9 miles.

17.

$S=3L-21$

$S+21=3L-21+21$

$S+21=3L$

$\frac{S+21}{3}=\frac{3L}{3}$

$\frac{S+21}{3}=\text{L or L}=\frac{S+21}{3}$

18.

$0.37x=8{,}400{,}000$

$\frac{0.37x}{0.37}=\frac{8{,}400{,}000}{0.37}$

$x\approx23{,}000{,}000$

19.

	r	T	d
Friend 1	R	1.5	1.5R
Friend 2	R+2	1.5	1.5(R+2)

$1.5R+1.5(R+2)=33$

$1.5R+1.5R+3=33$

$3R+3=33$

$3R+3-3=33-3$

$3R=30$

$\frac{3R}{3}=\frac{30}{3}$

$R=10$ mph

$R+2=12$ mph

20.

$39.99+0.79m>54.99+0.59m$

$39.99-39.99+0.79m>54.99-39.99+0.59m$

$0.79m>15+0.59m$

$0.79m-0.59m>15+0.59m-0.59m$

$0.2m>15$

$\frac{0.2m}{0.2}>\frac{15}{0.2}$

$m>75$

The monthly cost of Plan A exceeds the monthly cost of Plan B if more than 75 minutes of calls are made outside the network.

Cumulative Review Exercises

1. -6 yd

2.

−3 −2 −1 0 1 2 3

3. $|-2|=2$

4. 1 > -2, true

5. $2+(-1)+(-4)+4=1$

6. $(-10)\div(-2)=5$

7.

$3x+1-7(2x-5)=$

$3x+1-14x+35=$

$-11x+36$

8. 3 – 5 = – 2 lb

9. 3^4•\$1000.

10.

a. $A=50+25(t-1)$

b. $A=50+25(t-1)$

$A=50+25t-25$

$A=25+25t$

$A-25=25-25+25t$

$A-25=25t$

$\frac{A-25}{25}=\frac{25t}{25}$

$\frac{A-25}{25}=t$

$t=\frac{A-25}{25}$

c. $t=\frac{125-25}{25}$

$t=\frac{100}{25}$

$t=4$ hours

Chapter 3 Graphing Linear Equations and Inequalities

Chapter 3 Pretest

1.

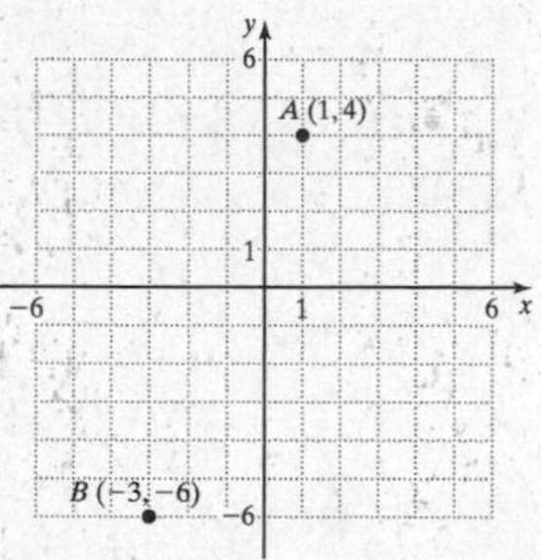

2. Quadrant IV

3. $m = \dfrac{2-5}{1-7} = \dfrac{-3}{-6} = \dfrac{1}{2}$

4 $\overleftrightarrow{AB}$ is parallel to $\overleftrightarrow{CD}$ since they have the same slope.

$\overleftrightarrow{AB} \quad m = \dfrac{5-1}{0-2} = \dfrac{4}{-2} = -2$

$\overleftrightarrow{CD} \quad m = \dfrac{1-7}{4-1} = \dfrac{-6}{3} = -2$

5. $\overleftrightarrow{PQ}$ is perpendicular to $\overleftrightarrow{RS}$, since the product of their slopes is -1.

$\overleftrightarrow{PQ} \quad m = \dfrac{(-1)-3}{1-(-3)} = \dfrac{-4}{4} = -1$

$\overleftrightarrow{RS} \quad m = \dfrac{4-(-2)}{4-(-2)} = \dfrac{6}{6} = 1$

$(-1)(1) = -1$

6. x-intercept: (3,0); y-intercept: (0,4)

7. The slope of the line is positive. As the population of a state increases, the number of representatives in Congress from that state increases.

8. Variety A grows faster.

9.

x	Y
4	3
7	9
$\frac{5}{2}$	0
2	-1

10.

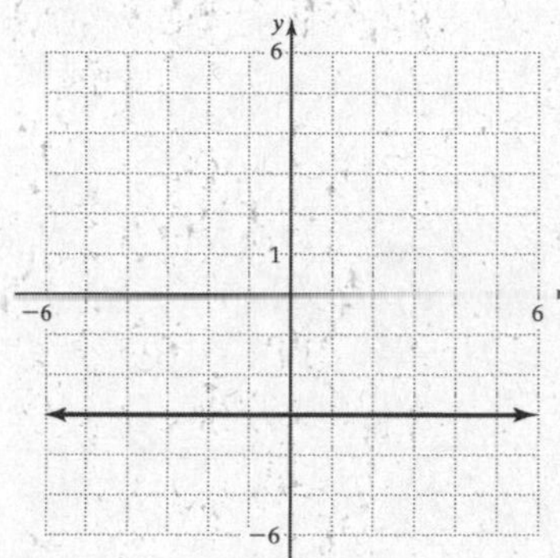

11.

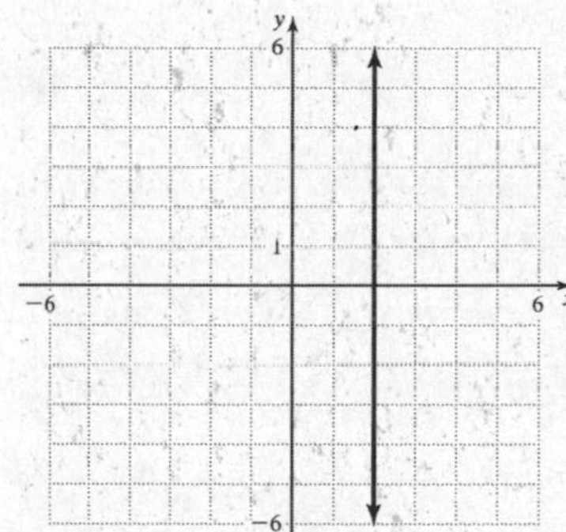

12.

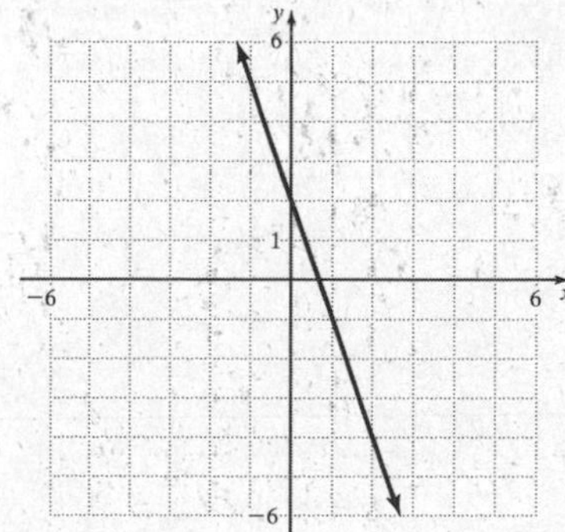

13.

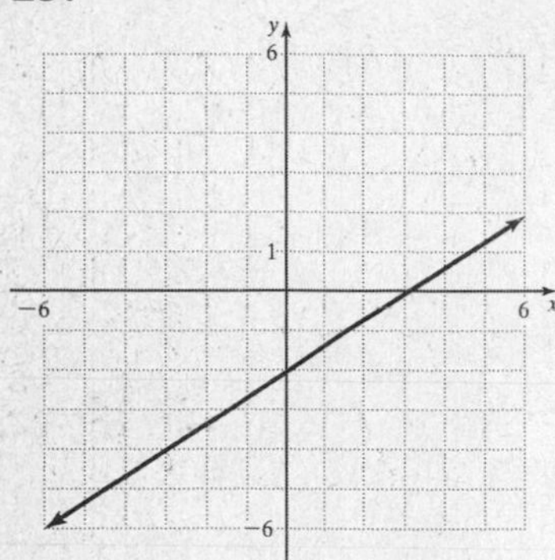

14. The slope is 2; the y-intercept is (0, -5).

15.

$y = 5x - 8$

16.

Slope-intercept form: $y = 2x + 8$;

Point-slope form: $y - 8 = 2(x - 0)$

17.

$m = \frac{-1-1}{2-4} = \frac{-2}{-2} = 1$

$y - 1 = 1(x - 4)$

$y - 1 = x - 4$

$y = x - 3$

Slope-intercept form: $y = x - 3$; Point-slope form: $y - 1 = (x - 4)$

18.

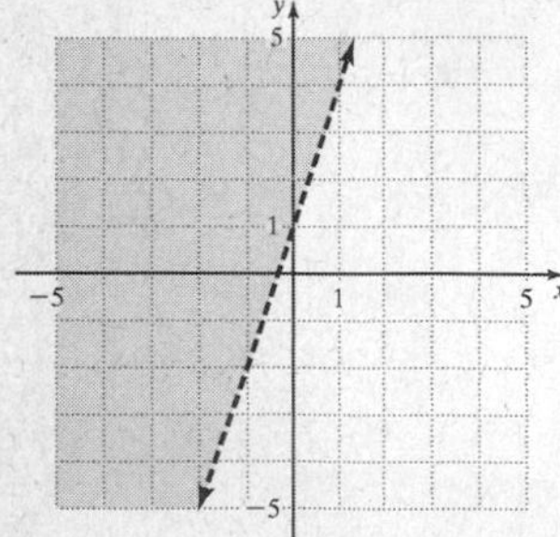

19.

a. $c = 2.5x$

b.

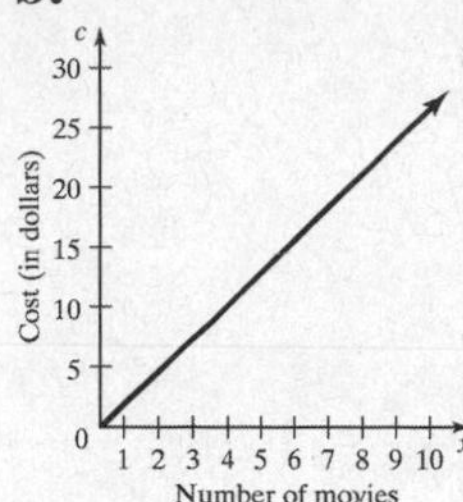

c. (0,0) (2,5)

$m = \frac{5-0}{2-0} = \frac{5}{2} = 2.5$

The slope of the line is 2.5. It represents the cost of renting a movie.

20.

a. $d = 50t$

b.

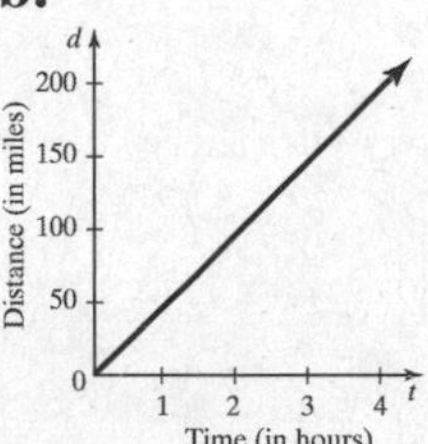

c. The slope of the graph is 50. It represents the speed the sales representative is driving.

3.1 Introduction to Graphing

Practice 3.1

1.

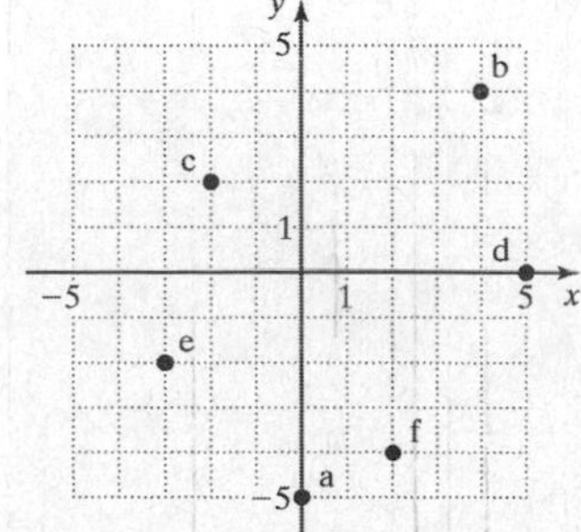

2.

Point	Quadrant
$\left(-\dfrac{1}{2},3\right)$	II
$(6,-7)$	IV
$(-1,-4)$	III
$(2,9)$	I

3.

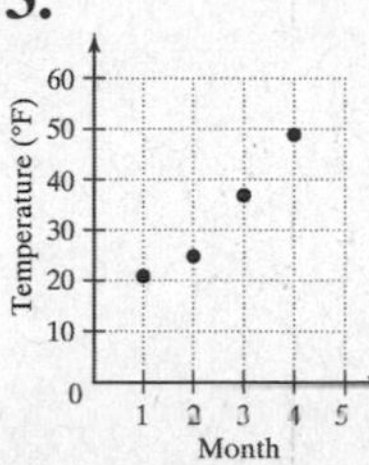

4. The value of the car decreases as the number of years increases.

5. From A to B and B to C the line segment slants up and to the right, indicating that the runner's heart beats per minute increase over this period of time. From C to D, the line segment slants down to the right, indicating that the runner's heartbeats per minute decrease.

Possible scenario: The runner starts out warming up by jogging slowly for a certain length of time (A to B), then the runner jogs more quickly for some time (B to C), and finally, the runner jogs more slowly (C to D), coming to a stop at D.

Exercises 3.1

1.

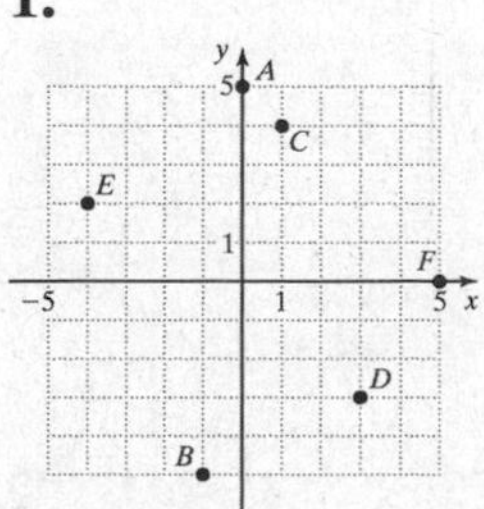

3.

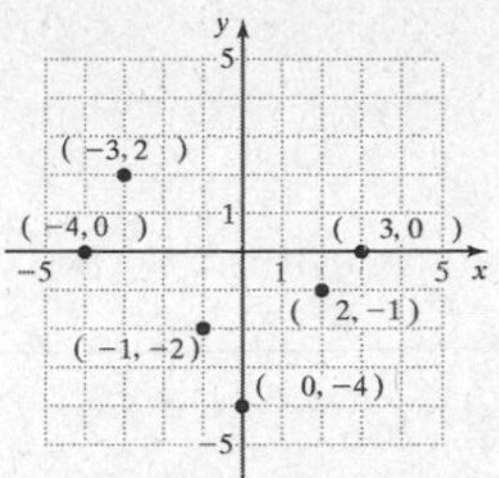

5. III

7. IV

9. I

11. II

13.

a. A(20, 40), B(52, 90), C(76, 80), D(90,28)

b. Students A, B, and C scored higher in English than in mathematics.

15.

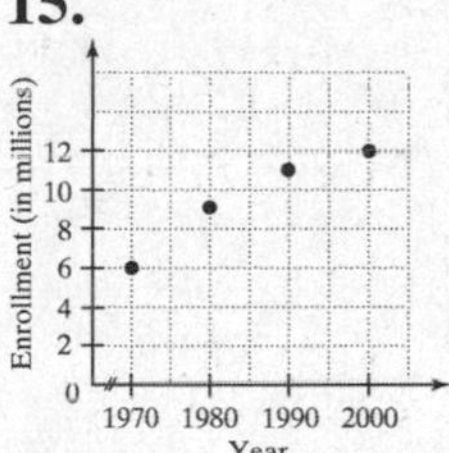

17.

a.

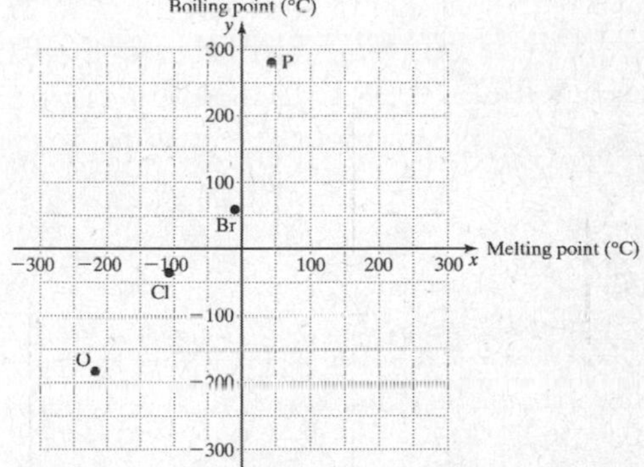

b. The y-coordinate is larger. The pattern shows that for each substance its boiling point is higher than its melting point.

19. The number of senators from a state (2) is the same regardless of the size of the state's population.

21. The graph in (a) could describe this motion. As the child moves away from the wall the distance from the wall increases (line segments slants up to the right). When the child stands still, the distance from the wall does not change (horizontal line segment). Finally, as the child moves toward the wall, the child's distance from the wall decreases (line segment slants down to the right).

3.2 Slope

Practice 3.2

1. $m=\dfrac{3-2}{4-1}=\dfrac{1}{3}$

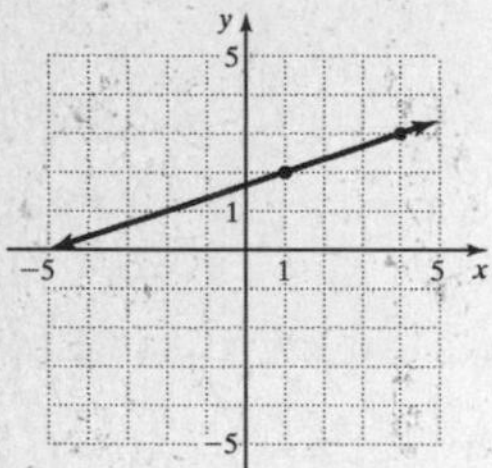

2. $m=\dfrac{-5-1}{3-(-2)}=\dfrac{-6}{5}=-\dfrac{6}{5}$

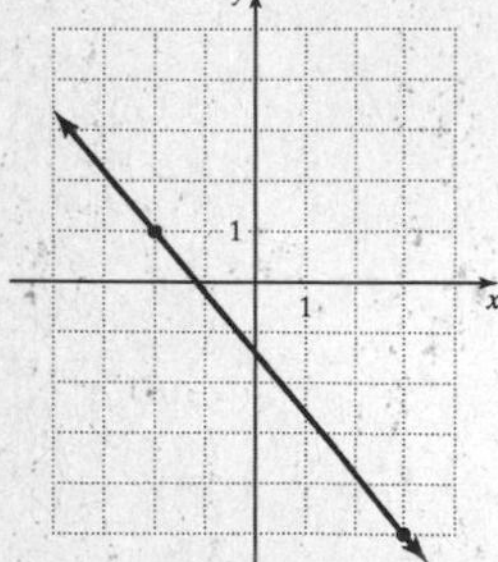

3. $m=\dfrac{-1-(-1)}{6-2}=\dfrac{0}{4}=0$

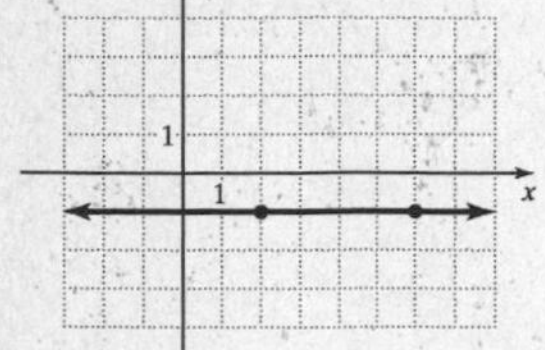

4. $m=\dfrac{0-7}{-2-(-2)}=\dfrac{-7}{0}=\text{undefined}$

The slope is undefined.

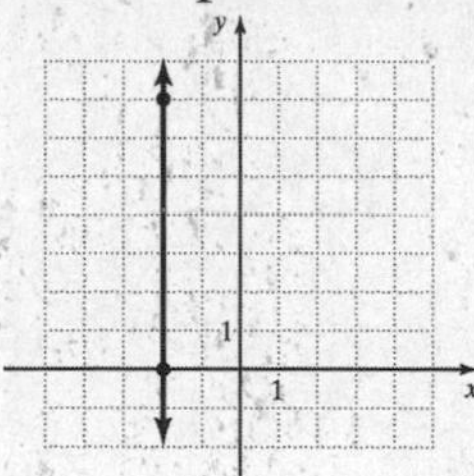

5.

$\overleftrightarrow{PQ}$ $m=\dfrac{0-4}{6-0}=\dfrac{-4}{6}=-\dfrac{2}{3}$

$\overleftrightarrow{RS}$ $m=\dfrac{-3-(-1)}{4-0}=\dfrac{-2}{4}=-\dfrac{1}{2}$

6. Scenario A is the most desirable. The slope of the line is negative, which indicates a decrease in the number of people ill over time.

7.

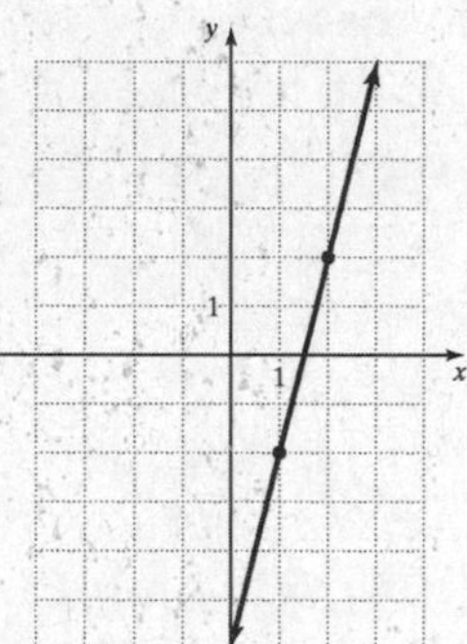

8.

a.

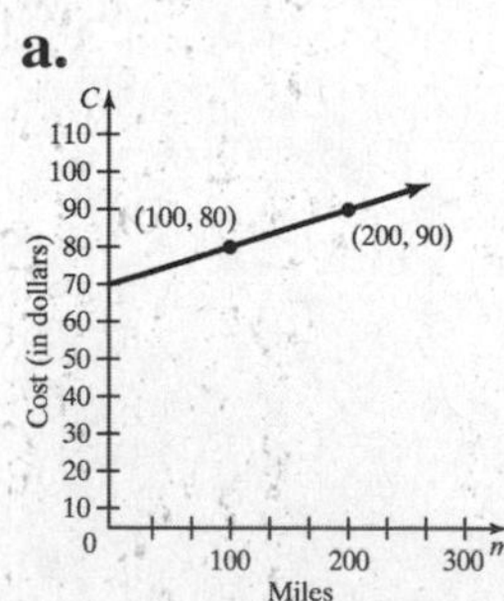

b. $m=\dfrac{90-80}{200-100}=\dfrac{10}{100}=0.1$

c. The slope represents the cost per mile for renting a car.

9. $\overleftrightarrow{EF}$ is parallel to $\overleftrightarrow{GH}$ since they have the same slope.

$\overleftrightarrow{EF}$ $m=\frac{-1-4}{4-0}=\frac{-5}{4}=-\frac{5}{4}$

$\overleftrightarrow{GH}$ $m=\frac{-2-8}{8-0}=\frac{-10}{8}=-\frac{5}{4}$

10.a. The lines are parallel since they have the same slope.

$\overleftrightarrow{\text{Technician}}$ $m=\frac{80-65}{6-2}=\frac{15}{4}$

$\overleftrightarrow{\text{Designer}}$ $m=\frac{45-30}{6-2}=\frac{15}{4}$

b. Yes, the lines on the graph appear to be parallel.
c. The salaries increased at the same rate.
d. The starting salary of the computer lab technician was about $57,000.

11. $\overleftrightarrow{AB}$ is not perpendicular to $\overleftrightarrow{AC}$, since the product of their slopes is not equal to -1.

$\overleftrightarrow{AB}$ $m=\frac{5-3}{2-1}=\frac{2}{1}=2$

$\overleftrightarrow{AC}$ $m=\frac{2-3}{-1-1}=\frac{-1}{-2}=\frac{1}{2}$

$(2)\left(\frac{1}{2}\right)=1\neq-1$

12. The diagonals are perpendicular to each other, since the product of their slopes is -1.

$m_1=\frac{0-6}{6-0}=\frac{-6}{6}=-1$

$m_2=\frac{0-6}{0-6}=\frac{-6}{-6}=1$

$(-1)(1)=-1$

The slope of the diagonal from (0,0) to (6,6) is +1. The slope of the diagonal from (0,6) to (6,0) is -1. Since the product of the slopes is -1, the diagonals of the square are perpendicular.

Exercises 3.2

1.

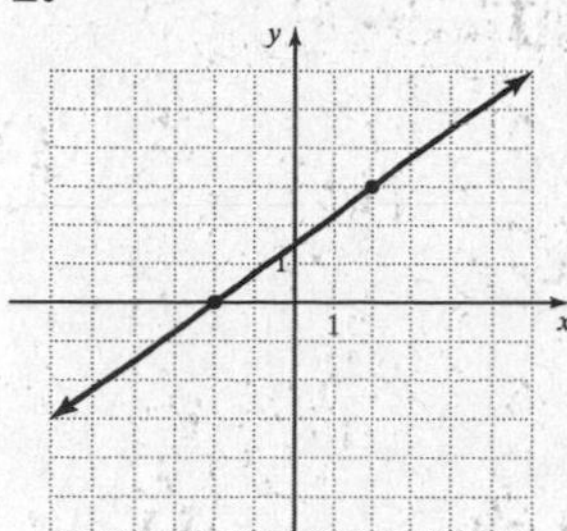

$m=\frac{y_2-y_1}{x_2-x_1}=\frac{6-3}{6-2}=\frac{3}{4}$

3.

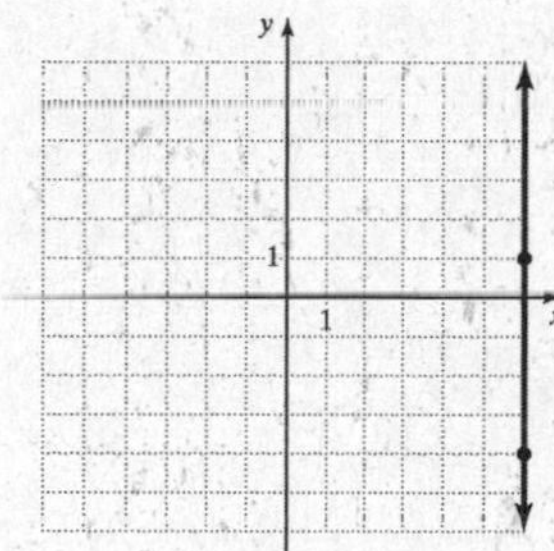

$m=\frac{y_2-y_1}{x_2-x_1}=\frac{1-(-4)}{6-6}=\frac{5}{0}=\text{undefined}$

5.

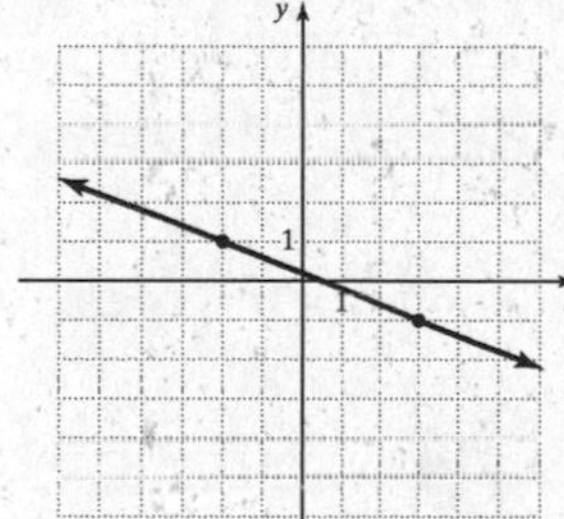

$m=\frac{y_2-y_1}{x_2-x_1}=\frac{-1-1}{3-(-2)}=\frac{-2}{5}=-\frac{2}{5}$

7.

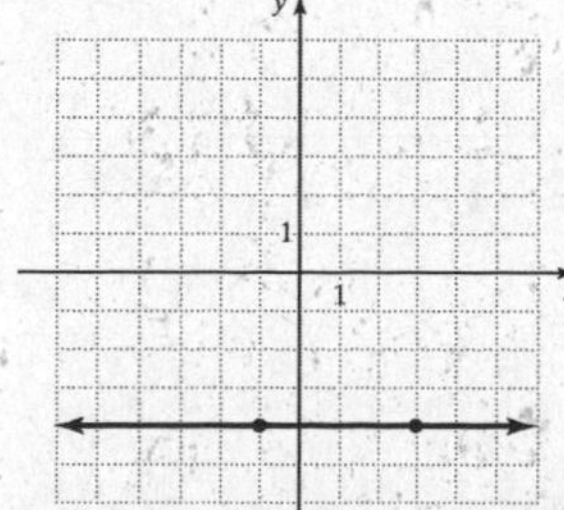

$m=\frac{y_2-y_1}{x_2-x_1}=\frac{-4-(-4)}{3-(-1)}=\frac{0}{4}=0$

9.

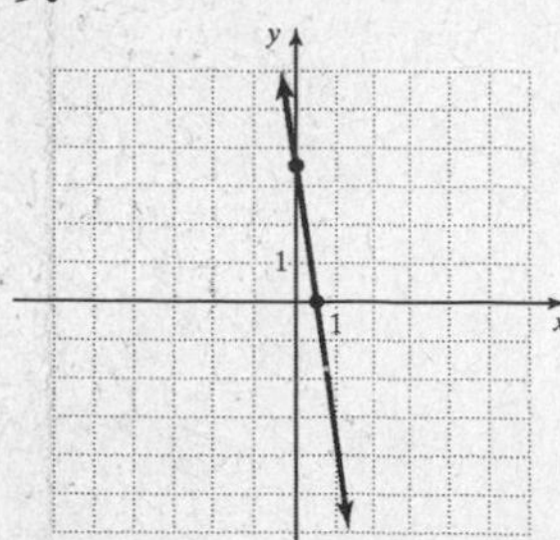

$$m=\frac{y_2-y_1}{x_2-x_1}=\frac{3.5-0}{0-0.5}=\frac{3.5}{-0.5}=-7$$

11. The slope of $\overleftrightarrow{AB}$ is -1.

$$m=\frac{y_2-y_1}{x_2-x_1}=\frac{-3-2}{6-1}=\frac{-5}{5}=-1$$

The slope of $\overleftrightarrow{CD}$ is 2.

$$m=\frac{y_2-y_1}{x_2-x_1}=\frac{8-4}{3-1}=\frac{4}{2}=2$$

13.

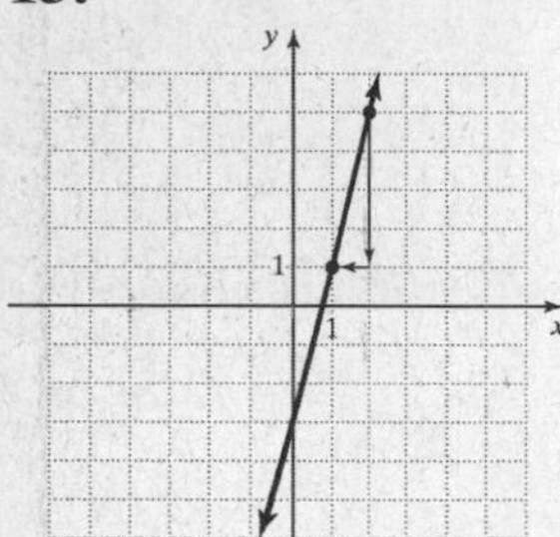

15.

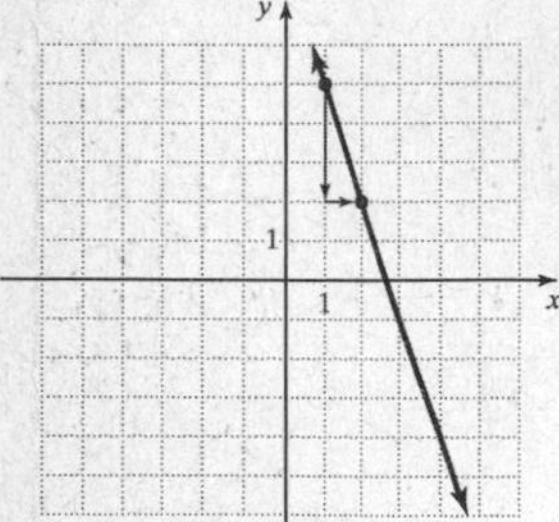

17.

19. Positive slope; neither
21. Negative slope; neither
23. Undefined slope; vertical
25. Zero slope; horizontal

27.a. $\overleftrightarrow{PQ}$

$$m=\frac{y_2-y_1}{x_2-x_1}=\frac{3-(-1)}{1-0}=\frac{4}{1}=4$$

$\overleftrightarrow{RS}$

$$m=\frac{y_2-y_1}{x_2-x_1}=\frac{8-0}{7-5}=\frac{8}{2}=4$$

The lines are parallel.

b. $\overleftrightarrow{PQ}$

$$m=\frac{y_2-y_1}{x_2-x_1}=\frac{4-1}{7-9}=\frac{3}{-2}=-\frac{3}{2}$$

$\overleftrightarrow{RS}$

$$m=\frac{y_2-y_1}{x_2-x_1}=\frac{4-0}{6-0}=\frac{4}{6}=\frac{2}{3}$$

The lines are perpendicular.

Applications

29.
a. The slope is positive.
b. A positive slope indicates that as the temperature of the gas increases, the pressure in the tube increases.

31.
a. Motorcycle A.
b. Motorcycle B.
c. The slope is the change in distance over time, or the average speed of the motorcycles.

33. Since the slopes of the lines are equal, the landfills are growing at the same rate.

35.a.

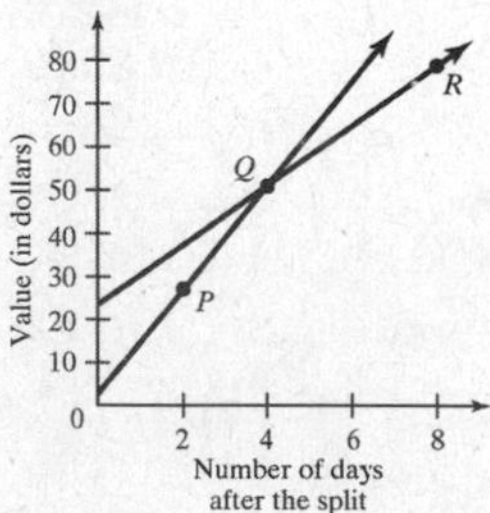

b. $\overleftrightarrow{PQ}$ $m=\frac{y_2-y_1}{x_2-x_1}=\frac{51-27}{4-2}=\frac{24}{2}=12$

$\overleftrightarrow{QR}$ $m=\frac{y_2-y_1}{x_2-x_1}=\frac{79-51}{8-4}=\frac{28}{4}=7$

c. The slopes of the two lines are not equal, therefore the rate of increase did change over time.

37. $\overleftrightarrow{AD}$

$$m=\frac{y_2-y_1}{x_2-x_1}=\frac{2-7}{6-5}=\frac{-5}{1}=-5$$

$\overleftrightarrow{BC}$

$$m=\frac{y_2-y_1}{x_2-x_1}=\frac{3-1}{9-3}=\frac{2}{6}=\frac{1}{3}$$

$$(-5)\left(\frac{1}{3}\right)=-\frac{5}{3}\neq -1$$

The product of the slopes of the two lines is not equal to -1, so $\overleftrightarrow{AD}$ is not the shortest route.

39.a. Graph II: as the car travels, its distance increases with time. This implies a positive slope.

b. Graph I: The car is set for a constant speed. The speed of the car does not change over time. This implies a 0 slope.

3.3 Linear Equations and Their Graphs

Practice 3.3

1.

x	y
0	1
5	11
-3	-5
$-\frac{1}{2}$	0
-2	-3

2.

x	y
-5	3
0	0
5	-3

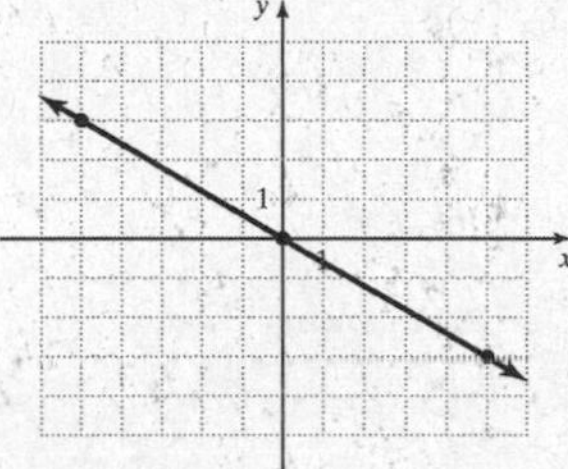

3.

a.

x	y
-1	-7
0	-5
5	5

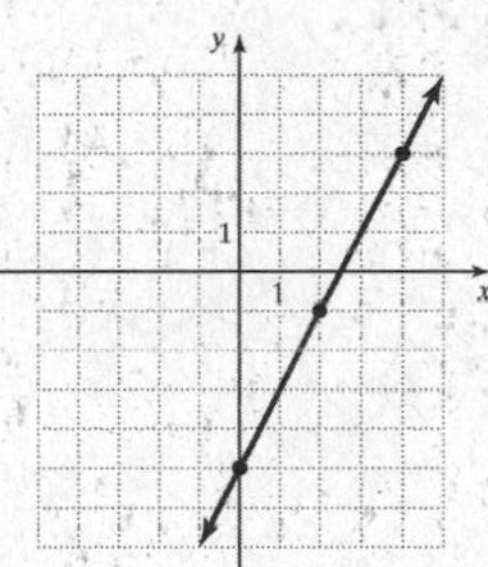

b.

$$m=\frac{5-(-5)}{5-0}=\frac{10}{5}=2$$

4.
x-intercept: (4,0)
$2x-4(0)=8$
$2x=8$
$x=4$
$(4,0)$
y-intercept: (0,-2)
$2(0)-4y=8$
$-4y=8$
$y=-2$

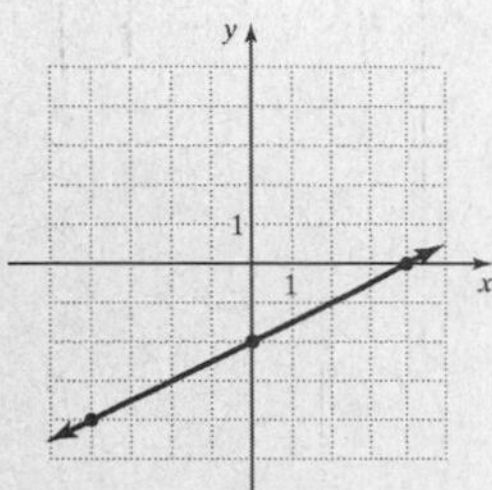

5.

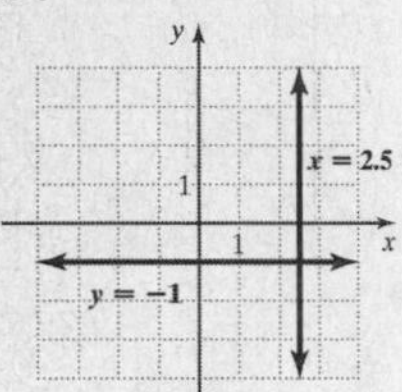

6.
a. $C=0.03x+40$
b.

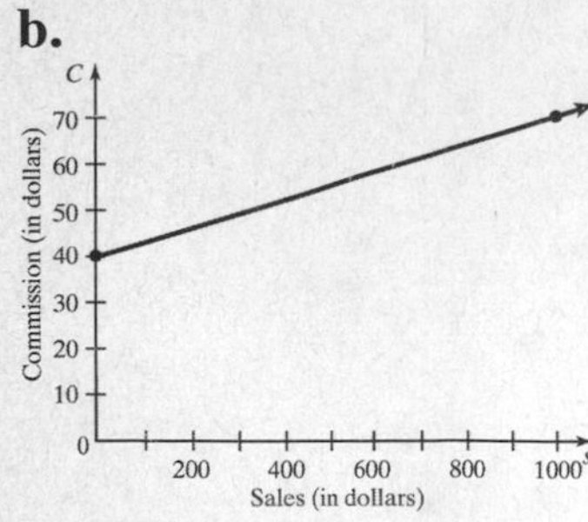

c.
$m=0.03$; for every sale, the commission increases by 0.03 times the value of the sale.
d. If the value of a sale is $500, then the broker's commission would be $55.

7.
a. $w+2t=10$
b.

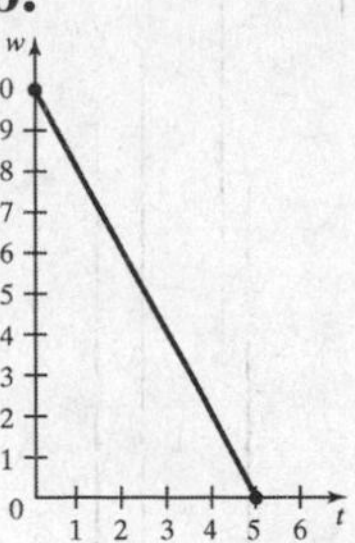

c. For each year the athlete is paid $2 million, the number of years that she could be paid $1 million decreases by 2 years.
d. The t-intercept is the number of years of the contract if she was paid $1 million in each year of the contract. The w-intercept is the number of years of the contract is she was paid $2 million in each year of the contract.

Exercises 3.3

1. $y=3x-8$

x	y
4	4
7	13
$\frac{8}{3}$	0

3. $y=5x$

x	y
3.5	17.5
6	30
$\frac{1}{10}$	$\frac{1}{2}$
$-\frac{8}{5}$	-8

5. $3x+4y=12$

x	y
0	3
-4	6
8	-3
4	0

7. $y=\frac{1}{3}x-1$

x	y
3	0
6	1
-3	-2
0	-1

9. $y=x$

X	y
-5	-5
0	0
5	5

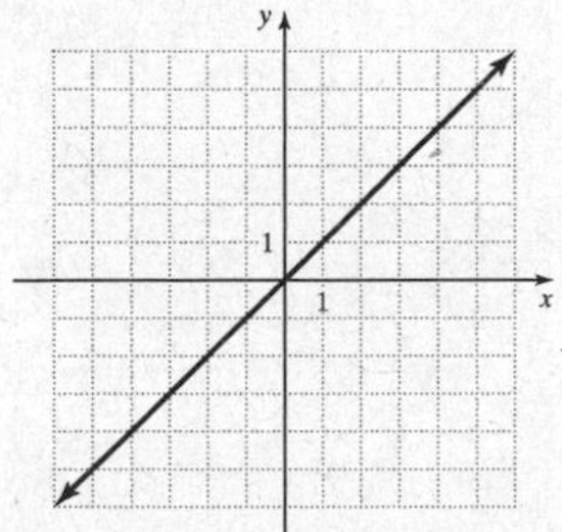

11. $y=\frac{1}{2}x$

X	y
-4	-2
0	0
4	2

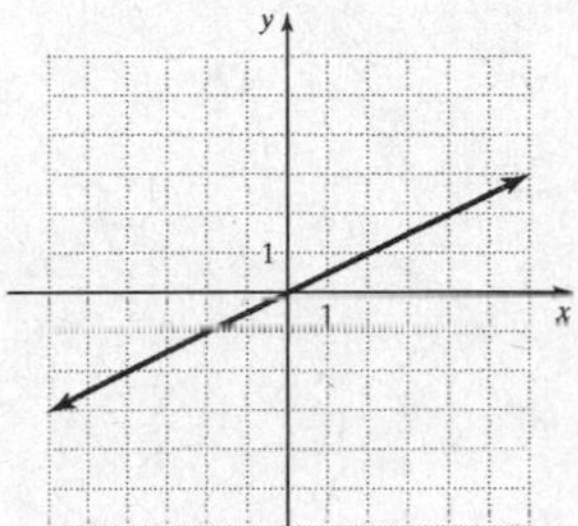

13. $y=-\frac{2}{3}x$

x	y
-3	2
0	0
3	-2

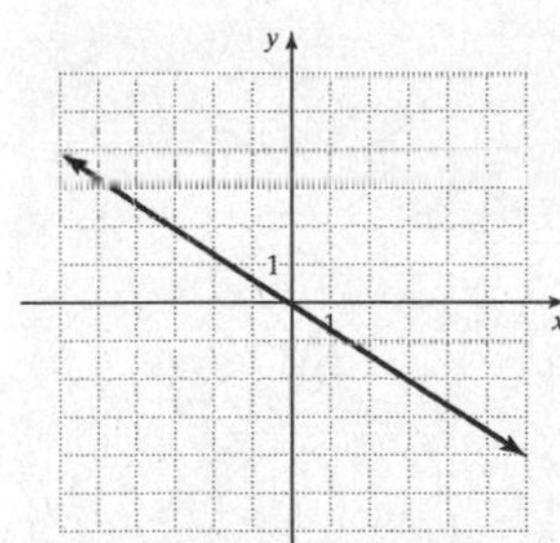

15. $y = 2x + 1$

x	y
-3	-5
0	1
$\frac{1}{2}$	2

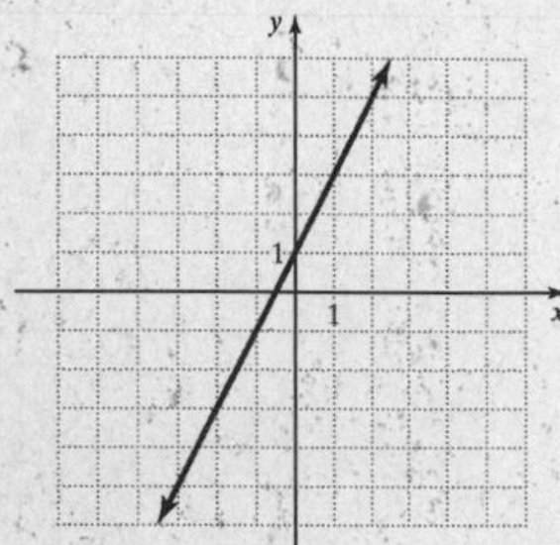

17.

$y = -\frac{1}{3}x + 1$

x	y
-3	2
0	1
6	-1

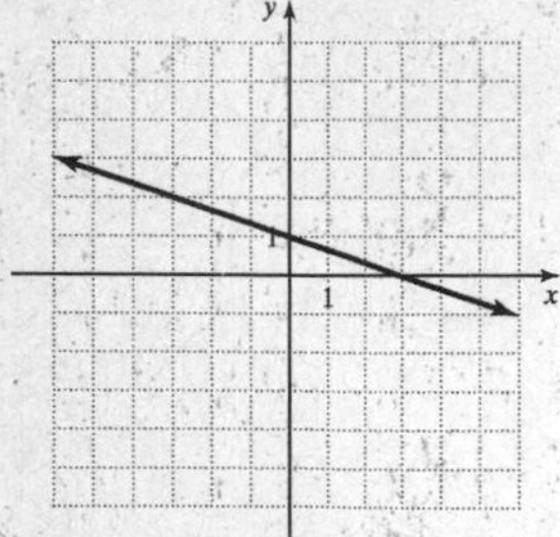

19.

$y - 2x = -3$

$y = 2x - 3$

x	y
-1	-5
0	-3
6	9

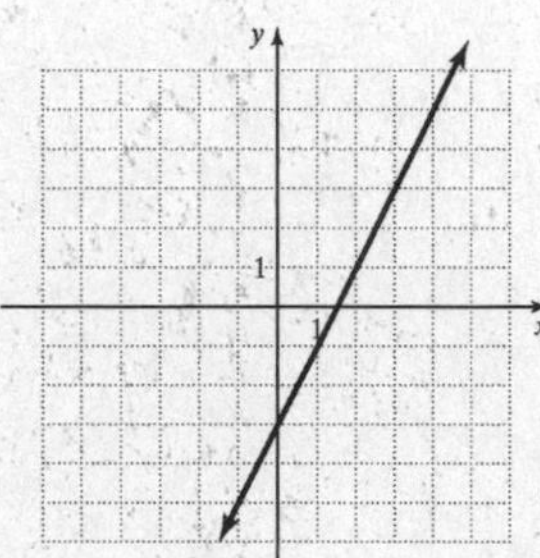

21.

$x + y = 6$

$y = -x + 6$

x	y
0	6
2	4
5	1

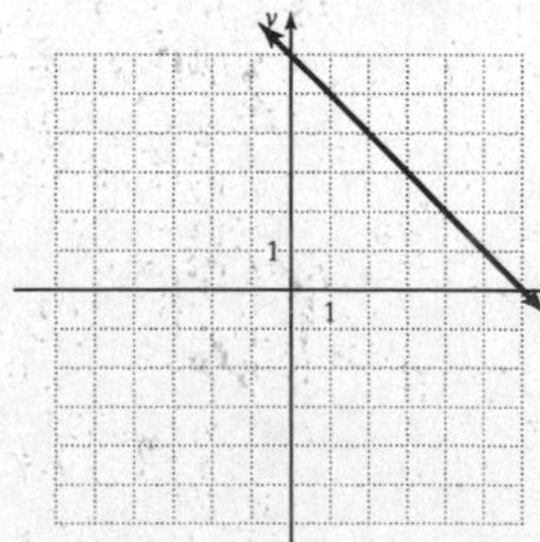

23.

$x-2y=4$

$y=\frac{1}{2}x-2$

x	y
-4	-4
0	-2
6	1

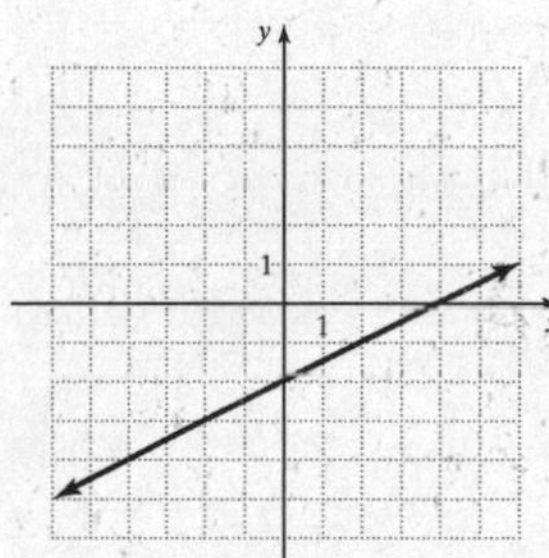

25.

$5x+3y=15$

$y=-\frac{5}{3}x+5$

x-intercept : $(3,0)$

$0=-\frac{5}{3}x+5$

$-5=-\frac{5}{3}x$

$3=x$

y-intercept: $(0,5)$

$y=-\frac{5}{3}(0)+5$

$y=5$

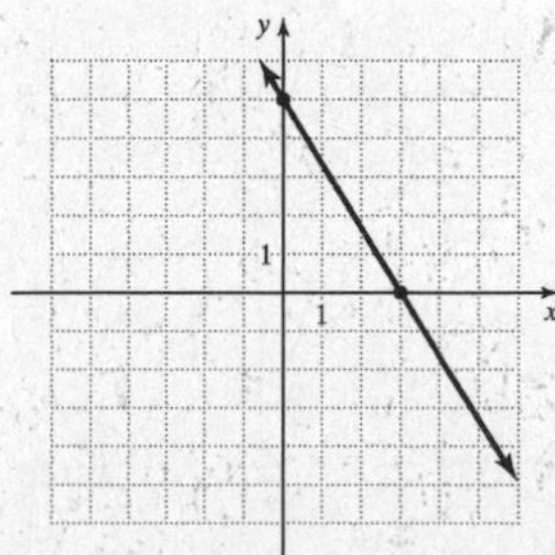

27.

$3x-6y=18$

$y=\frac{1}{2}x-3$

x-intercept : $(6,0)$

$0=\frac{1}{2}x-3$

$3=\frac{1}{2}x$

$6=x$

y-intercept: $(0,-3)$

$y=\frac{1}{2}(0)-3$

$y=-3$

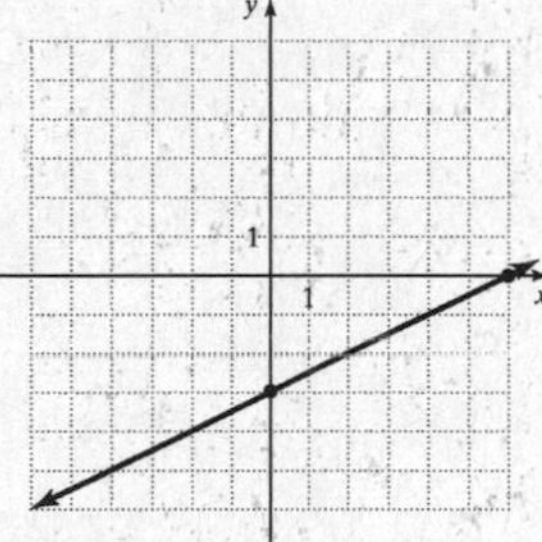

29.

$4y-5x=10$

$y=\frac{5}{4}x+\frac{5}{2}$

y-intercept: $\left(0,\frac{5}{2}\right)$

x-intercept: $(-2,0)$

$4(0)-5x=10$

$-5x=10$

$x=-2$

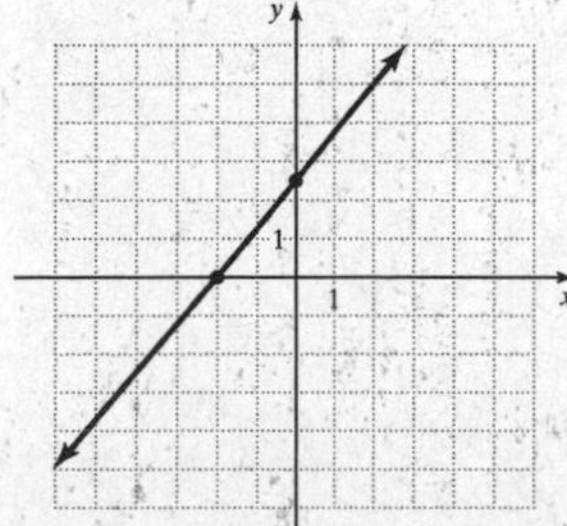

31.

$9y+6x=-9$

$y=-\frac{2}{3}x-1$

x-intercept : $\left(-\frac{3}{2},0\right)$

$0=-\frac{2}{3}x-1$

$1=-\frac{2}{3}x$

$-\frac{3}{2}=x$

y-intercept: $(0,-1)$

$y=-\frac{2}{3}(0)-1$

$y=-1$

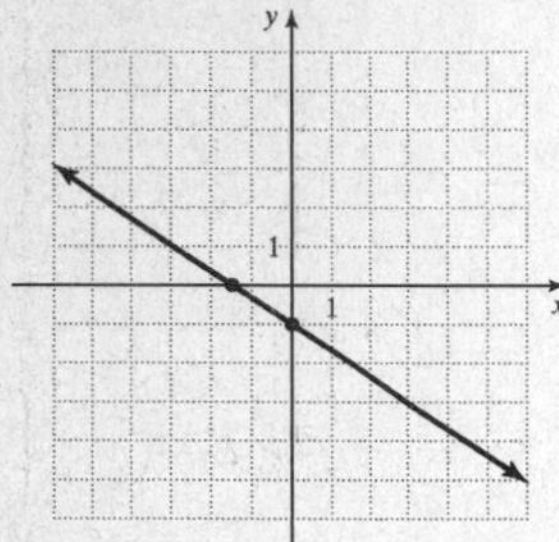

33.

$y=\frac{1}{2}x+2$

y-intercept: $(0,2)$

x-intercept : $(-4,0)$

$0=\frac{1}{2}x+2$

$-2=\frac{1}{2}x$

$-4=x$

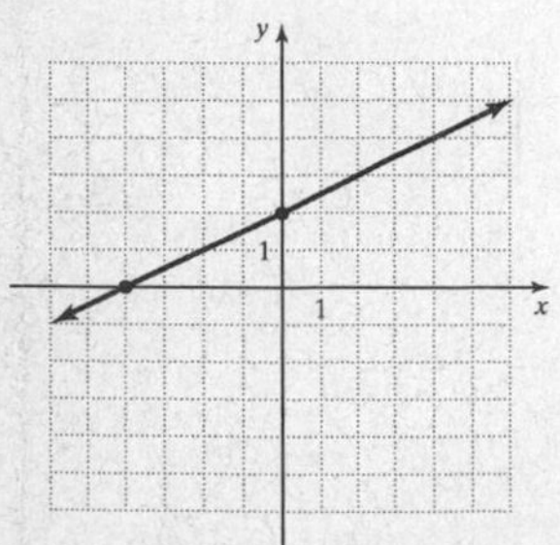

35.

$y=-2$

x-intercept : does not exist

y-intercept: $(0,-2)$

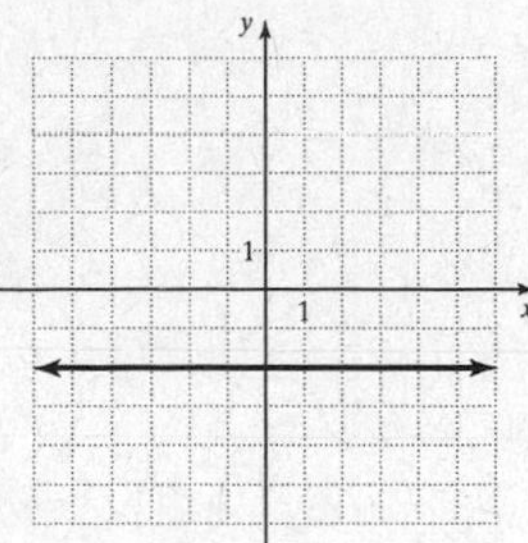

37.

$x=0$

x-intercept : (0,0)

y-intercept: y-axis

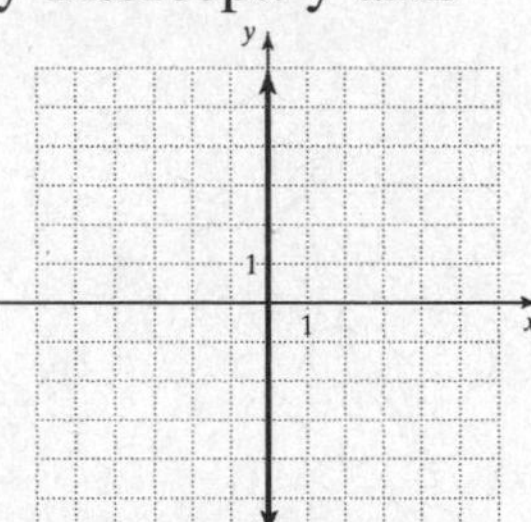

39.

$x=-5.5$

x-intercept : (-5.5,0)

y-intercept: does not exist

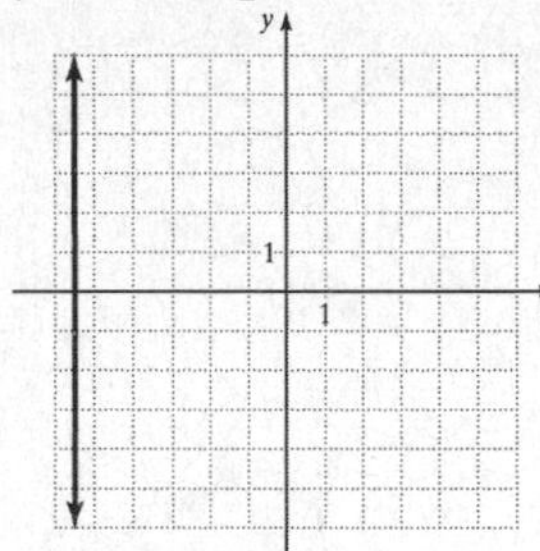

41.

$y=\frac{1}{2}x+3$

x-intercept : (-6, 0)

y-intercept: $(0,3)$

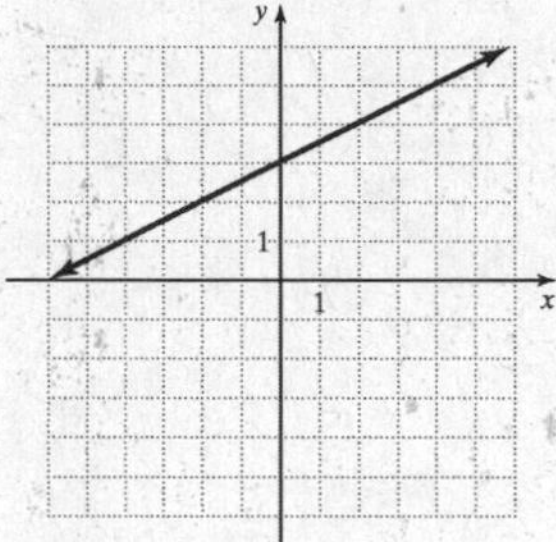

43.

$3x+5y=-15$

$y=-\frac{3}{5}x-3$

y-intercept: $(0,-3)$

x-intercept : $(-5,0)$

$0=-\frac{3}{5}x-3$

$3=-\frac{3}{5}x$

$-5=x$

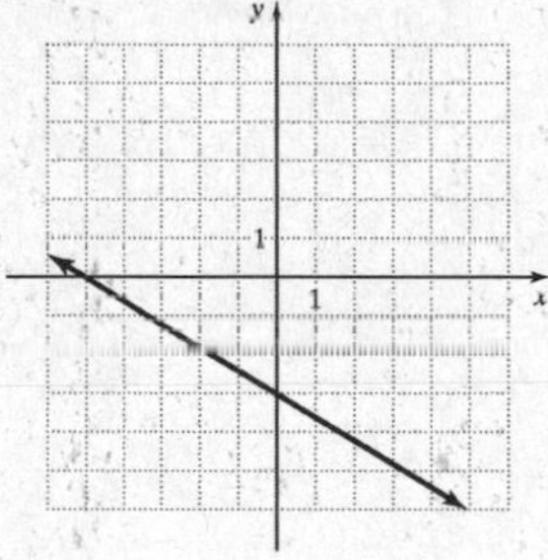

45.

$y=\frac{3}{5}x+\frac{2}{5}$

y-intercept: $\left(0,\frac{2}{5}\right)$

$0=\frac{3}{5}x+\frac{2}{5}$

$-\frac{2}{5}=\frac{3}{5}x$

$-\frac{2}{3}=x$

x-intercept : $\left(-\frac{2}{3},0\right)$

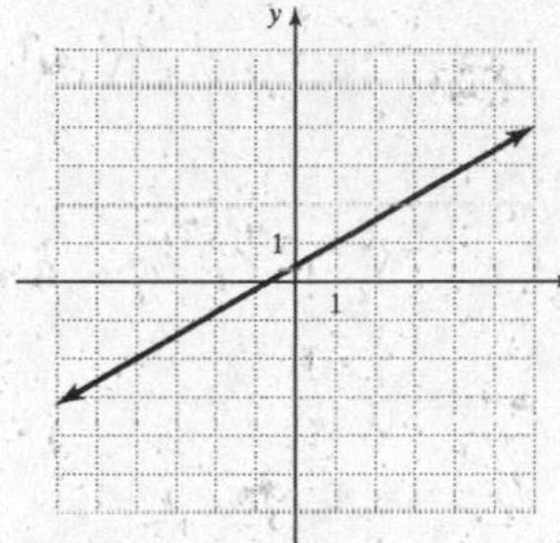

Applications

47. $v=10-32t$

a.

t	v
0	10
0.5	6
1	-22
1.5	-38
2	-54

A positive value of v means that the object is moving upward. A negative value of v means that the object is moving downward.

b.

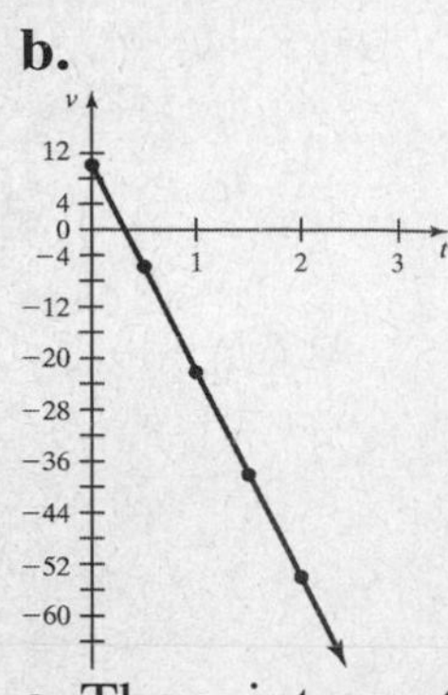

c. The v-intercept is the initial velocity of the object.

d. The t-intercept represents the time when the object changes from an upward motion to a downward motion.

49.

a. $P = 100m + 500$

b.

m	P
1	600
2	700
3	800

c.

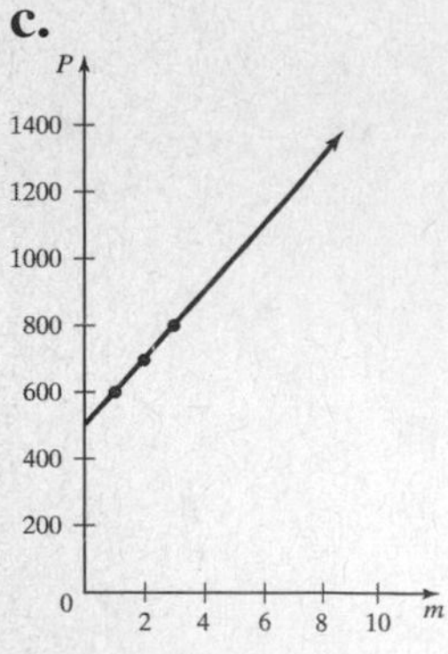

51.

a. $0.05n + 0.1d = 2$

b.

d	n
20	0
10	20
0	40

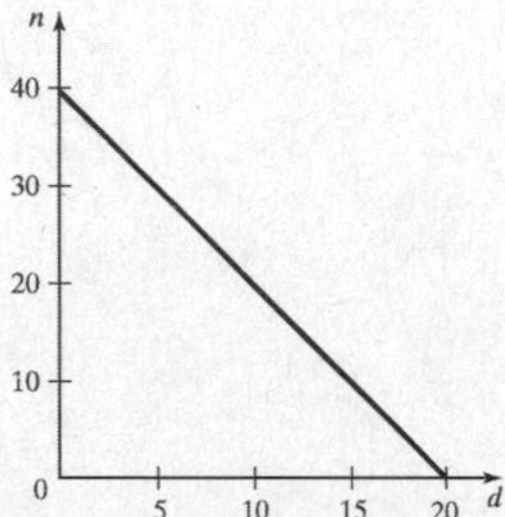

c. Only positive integer values make sense, since you cannot have fractions of nickels or dimes.

53.

a. $F = 5d + 40$

b.

d	F
0	40
8	80
14	110

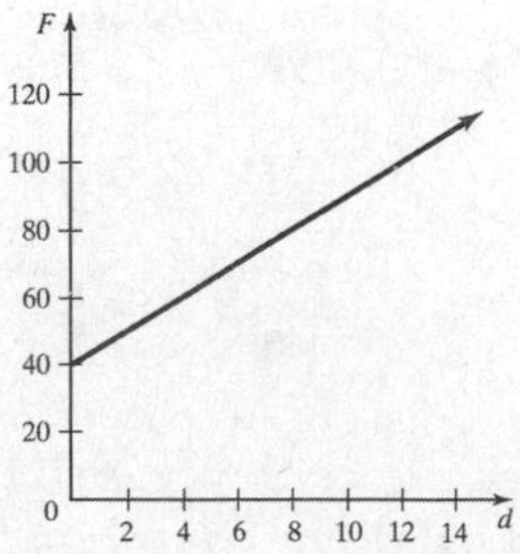

c. The F-intercept represents the cost of renting the computer for 0 days.

3.4 More Linear Equations and Their Graphs

Practice 3.4

1. Slope is -2; y-intercept is (0,3)

2.

$3x-2y=4$

$-2y=-3x+4$

$y=\frac{3}{2}x-2$

3.

$y-2=4(x+1)$

$y-2=4x+4$

$y=4x+6$

4. Slope is -1; y-intercept is (0,0)

5.

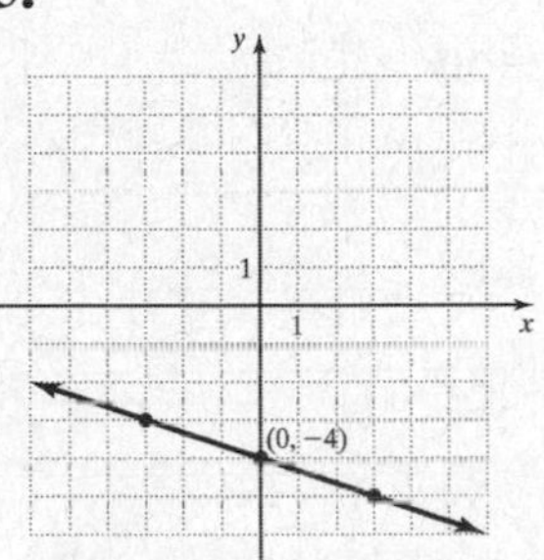

6. $y=1x+2$ or $y=x+2$

7. $m=-2;\quad y=-2x-1$

8. $m=-\frac{1}{2};\quad y=-\frac{1}{2}x-2$

9. $w=45-3t \Rightarrow w=-3t+45$

10. $y-0=2(x-7)$

11.

$m=\frac{7-0}{7-0}=\frac{7}{7}=1$

$y-7=1(x-7)$ or $y-0=1(x-0)$

12. $m=5\quad (b,w)=(4,27)$

Point-slope form: $w-27=5(b-4)$

$w-27=5(b-4)$

$w-27=5b-20$

Slope-intercept form: $w=5b+7$

13.

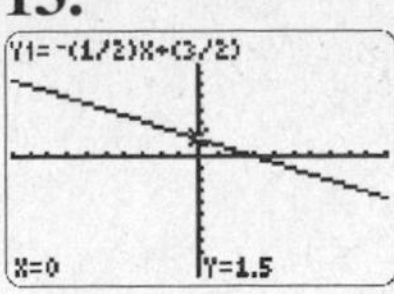

The displayed coordinates of the y-intercepts are x = 0 and y = 1.5.

Exercises 3.4

1.

Equation	Slope, m	y-intercept	Which graph type best describes the line? | /\ —	x-intercept
$y=3x-5$	3	(0, -5)	/	$(\frac{5}{3}, 0)$
$y=-2x$	- 2	(0,0)	\	(0,0)
$y=0.7x+3.5$	0.7	(0,3.5)	/	(-5,0)
$y=\frac{3}{4}x-\frac{1}{2}$	$\frac{3}{4}$	$(0, -\frac{1}{2})$	/	($\left(\frac{2}{3}, 0\right)$
$6x+3y=12$	- 2	(0,4)	\	(2, 0)
$y=-5$	0	(0, 5)	—	No x-intercept
$x=-2$	undefined	No y-intercept	|	(-2,0)

3. $y=-x+2$

Slope: -1; y-intercept: (0, 2)

5. $y=-\frac{1}{2}x$

Slope: $-\frac{1}{2}$; y-intercept: (0, 0)

7.

$x-y=10$

$x-x-y=-x+10$

$-y=-x+10$

$y=x-10$

9.

$x+10y=10$

$x-x+10y=-x+10$

$10y=-x+10$

$y=-\frac{1}{10}x+1$

11.
$$6x+4y=1$$
$$6x-6x+4y=-6x+1$$
$$4y=-6x+1$$
$$y=-\frac{3}{2}x+\frac{1}{4}$$

13.
$$2x-5y=10$$
$$2x-2x-5y=-2x+10$$
$$-5y=-2x+10$$
$$y=\frac{2}{5}x-2$$

15.
$$y+1=3(x+5)$$
$$y+1=3x+15$$
$$y+1-1=3x+15-1$$
$$y=3x+14$$

17. b

19. a

21. slope = 2; y-intercept=(0,1)

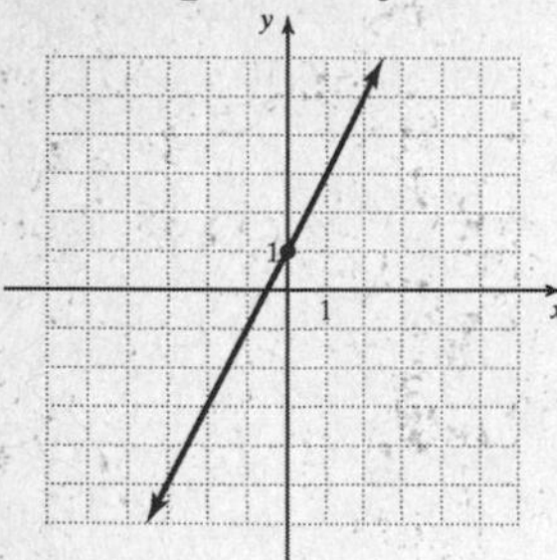

23. slope = $-\frac{2}{3}$; y-intercept=(0,6)

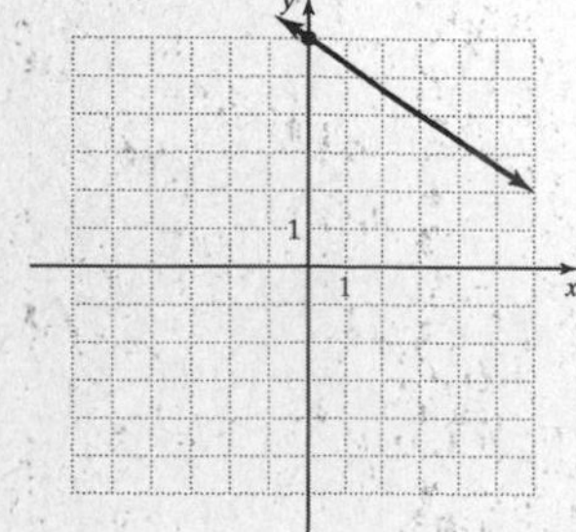

25. slope = -1; y-intercept=(0,1)

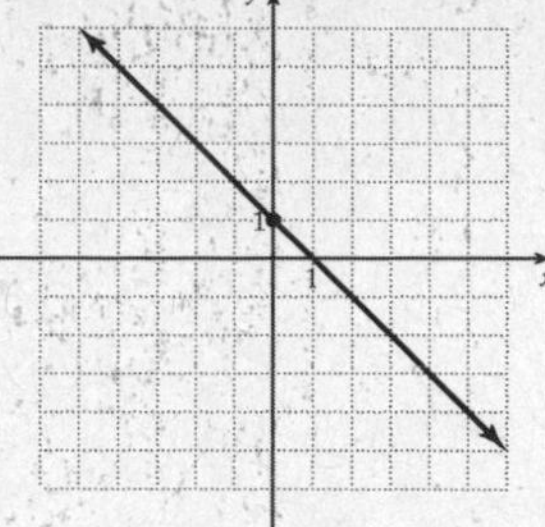

27. slope = $-\frac{3}{4}$; y-intercept=(0,0)

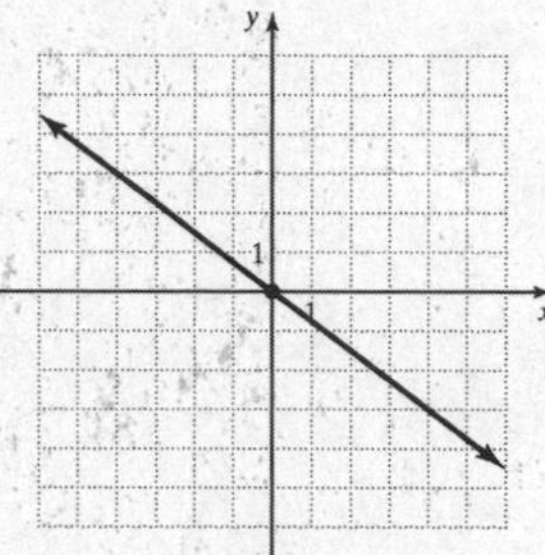

29. slope = $-\frac{1}{2}$; y-intercept=(0,2)

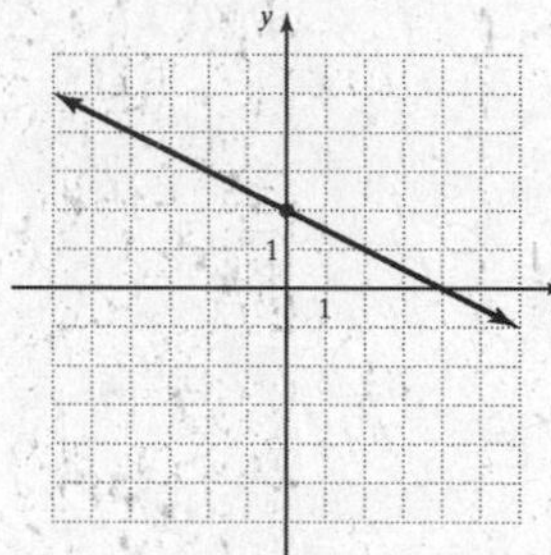

31. slope = 3.735; y-intercept=(0,1.056)

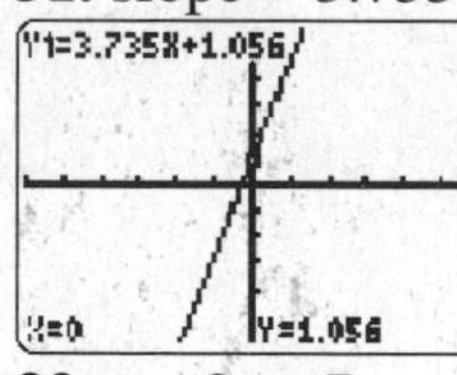

33. $y=3x+7$

35. $y-0=5(x-4)$

37. $y-5=-\frac{1}{2}(x-(-2))$

39. $x=-3$

41. $y=0$

43.
$$m=\frac{2-1}{1-2}=\frac{1}{-1}=-1$$
$$y-1=-1(x-2)$$
$$y-1=-x+2$$
$$y=-x+3$$

45.

$$m=\frac{3-0}{0-(-4)}=\frac{3}{4}$$

$(0,b)=(0,3)$

$$y=\frac{3}{4}x+3$$

47. $y=2$

49.

$$m=\frac{2-(-2)}{-4-0}=\frac{4}{-4}=-1$$

$y-2=-1(x-(-4))$

$y-2=-x-4$

$y=-x-2$

Applications

51.

a. $m=\dfrac{32-14}{0-(-10)}=\dfrac{18}{10}=\dfrac{9}{5}$

b.

$$F=\frac{9}{5}C+32$$

c. Water boils at 100° C.

$$212=\frac{9}{5}C+32$$

$$180=\frac{9}{5}C$$

$$\left(\frac{5}{9}\right)180=\left(\frac{5}{9}\right)\left(\frac{9}{5}C\right)$$

$100=C$

53.

a.

$m=4;\quad (x,y)=(500,3500)$

$y-3500=4(x-500)$

b.

$y-3500=4(x-500)$

$y-3500=4x-2000$

$y=4x+1500$

c. The y-intercept represents the monthly flat fee the utility company charges its residential customers.

55.

a. Point-slope form:

$$m=\frac{2-1}{10-5}=\frac{1}{5};t-1=\frac{1}{5}(L-5)\quad\text{or}$$

$$m=\frac{10-5}{2-1}=5;\ L-5=5(t-1)$$

b. Slope-intercept form:

$$t-1=\frac{1}{5}(L-5)\text{ or }L-5=5(t-1)$$

$$t-1=\frac{1}{5}L-1\text{ or }L-5=5t-5$$

$$t=\frac{1}{5}L\text{ or }L=5t$$

57.

a. $I=0.03S+1500$

b.

S	I
0	1500
5000	1650
10,000	1800

c.

$I=0.03(6200)+1500$

$I=\$1686$

59.

$$(0,1)(-33,2)\quad m=\frac{2-1}{-33-0}=\frac{1}{-33}$$

$$P=-\frac{1}{33}d+1$$

61.

$$(0,10)(6,15)\quad m=\frac{15-10}{6-0}=\frac{5}{6}$$

$$L=\frac{5}{6}F+10$$

3.5 Linear Inequalities and Their Graphs

Practice 3.5

1.

$y < x - 1 \quad (1,3)$

$3 \overset{?}{<} 1 - 1$

$3 \not< 0$

No,(1, 3) is not a solution to the inequality.

2.

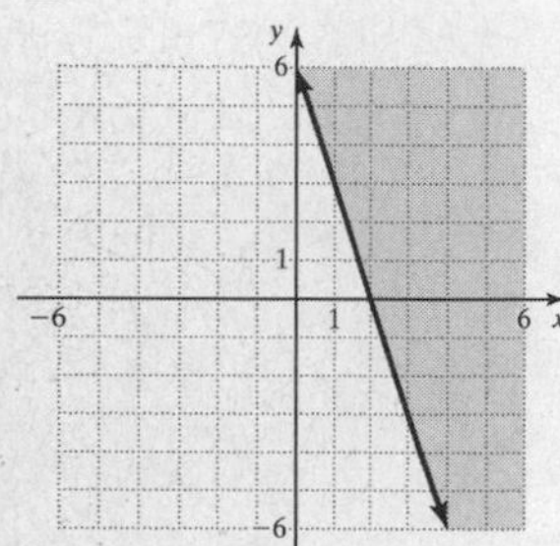

test point:

$y + 3x \geq 6 \quad (0,0)$

$0 \not\geq 6$

The half-plane above and including the line is the graph of the solution.

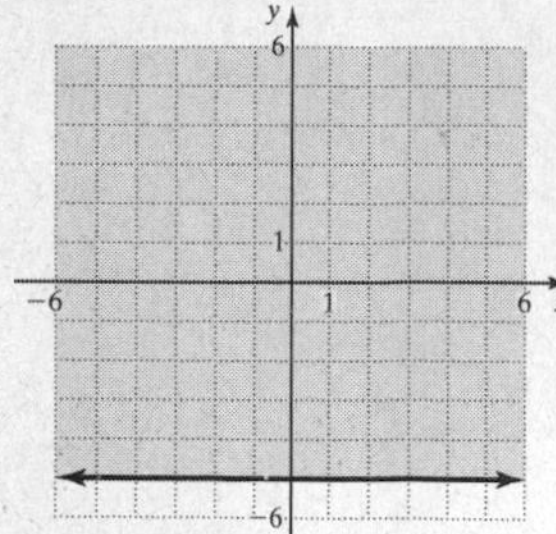

3.

Graph the boundary line y = -5.

test point:

$y \geq -5 \quad (-2,3)$

$3 > -5$

The half-plane above and including the line is the graph of the solution.

4.

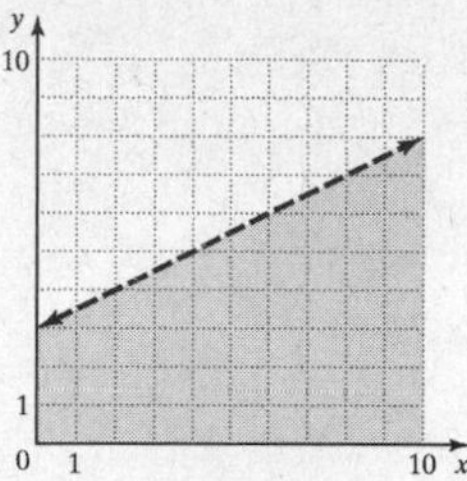

Graph the boundary line $x - 2y = -6$

x	y
0	3
4	5
10	8

test point:

$x - 2y > -6 \quad (5,3)$

$5 - 2(3) \overset{?}{>} -6$

$-1 > -6$

The half-plane below the line is the graph of the solution.

5.

a. $d + g \leq 3000$

b.

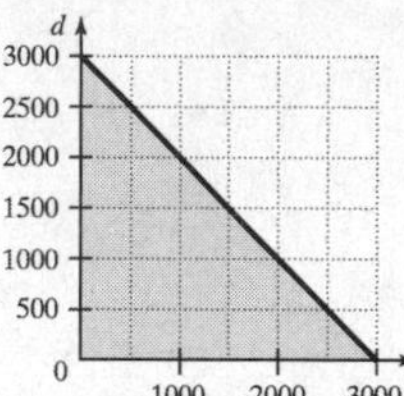

Graph the boundary line $d + g = 3000$

d	g
0	3000
1500	1500
3000	0

test point:

$d + g \leq 3000 \quad (1000, 500)$

$1000 + 500 \overset{?}{\leq} 3000$

$1500 \leq 3000$

The half-plane below and including the line is the graph of the solution.

c. The d-intercept represents the maximum number of gallons of diesel fuel the refinery produces if no gasoline is produced. The g-intercept represents the maximum number of gallons of gasoline the refinery produces if no diesel fuel is produced.

Exercises 3.5

1.

$y < 3x \quad (0,0)$

$0 \overset{?}{<} 3(0)$

$0 \not< 0$

No, not a solution.

3.

$y \ge 2x-1 \quad \left(-\frac{1}{2},-2\right)$

$-2 \overset{?}{\ge} 2\left(-\frac{1}{2}\right)-1$

$-2 \ge -2$

Yes, a solution.

5.

$2x-3y > 10 \quad (10,8)$

$2(10)-3(8) \overset{?}{>} 10$

$-4 \not> 10$

No, not a solution.

7.

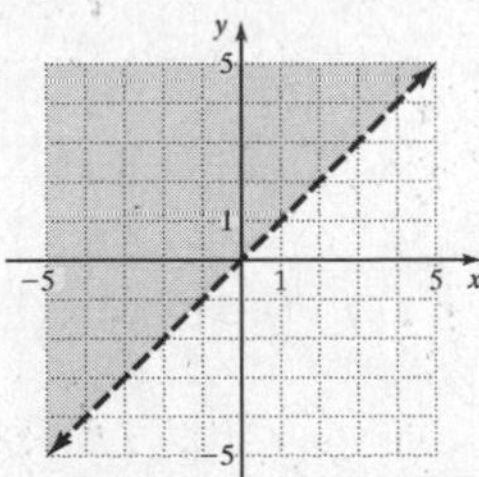

test point:

$y > x \quad (-2,3)$

$3 > -2$

The half-plane above the line is the graph of the solution.

9.

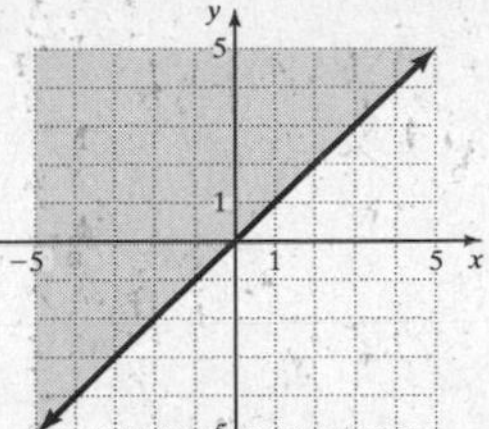

test point:

$x \le y \quad (-2,3)$

$-2 < 3$

The half-plane above and including the line is the graph of the solution.

11.

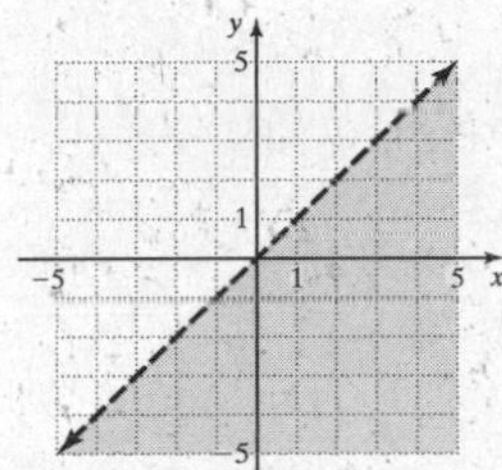

test point:

$y < x \quad (3,1)$

$1 < 3$

The half-plane below the line is the graph of the solution.

13. d

15. b

17.

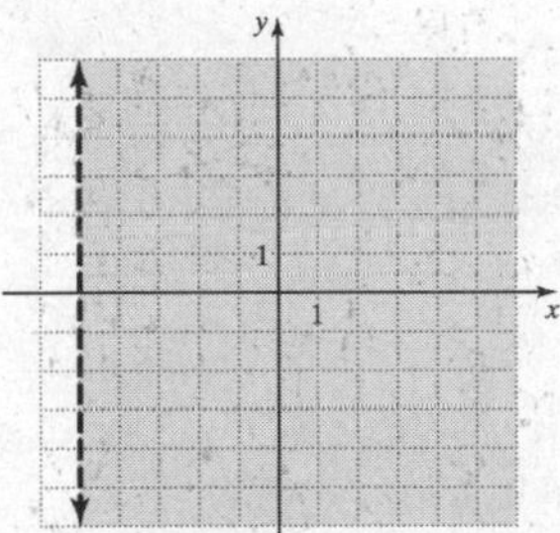

Graph boundary line x = -5.

test point:

$x > -5 \quad (4,-2)$

$4 > -5$

The half-plane to the right of the line is the graph of the solution.

19.

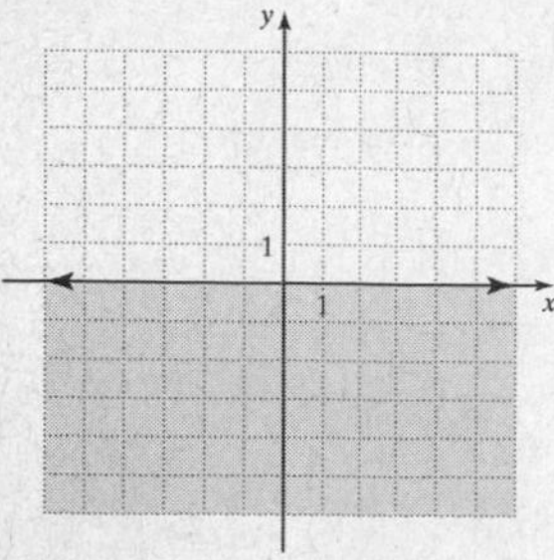

Graph the boundary line y = 0.
test point:
$y < 0 \quad (3, -2)$

$-2 < 0$

The half-plane below the line is the graph of the solution.

21.

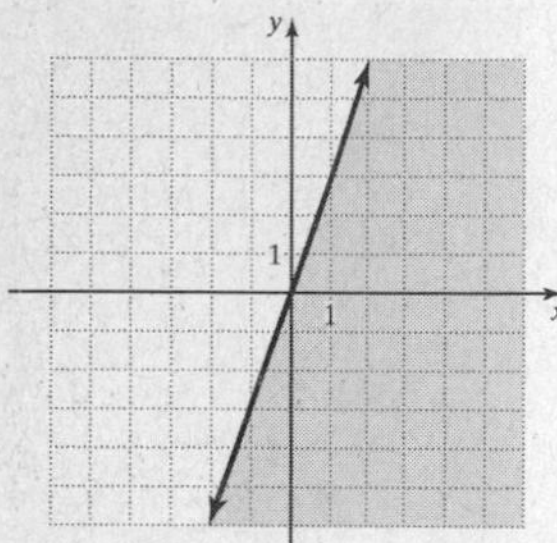

Graph the boundary line y = 3x

x	y
-2	-6
0	0
2	6

test point:
$y \le 3x \quad (4, -3)$

$-3 \overset{?}{\le} 3(4)$

$-3 \le 12$

The half-plane below and including the line is the graph of the solution.

23.

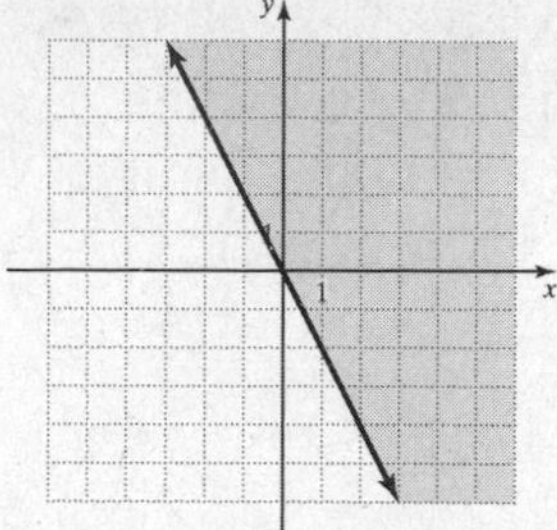

Graph the boundary line $y = -2x$

x	y
-3	6
0	0
2	-4

test point:

$y \ge -2x \quad (2, 3)$

$3 \ge -2(2)$

$3 \ge -4$

The half-plane above and including the line is the graph of the solution.

25.

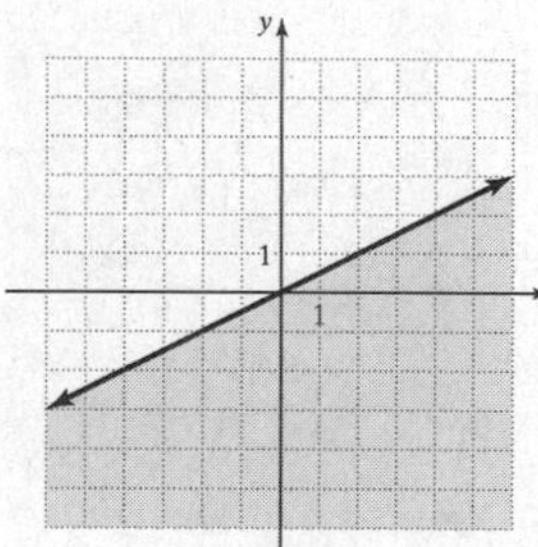

Graph the boundary line $y = \frac{1}{2}x$

x	y
-4	2
0	0
6	3

test point:

$y \le \frac{1}{2}x \quad (4, -3)$

$-3 \overset{?}{\le} \frac{1}{2}(4)$

$-3 \le 2$

The half-plane below and including the line is the graph of the solution.

27.

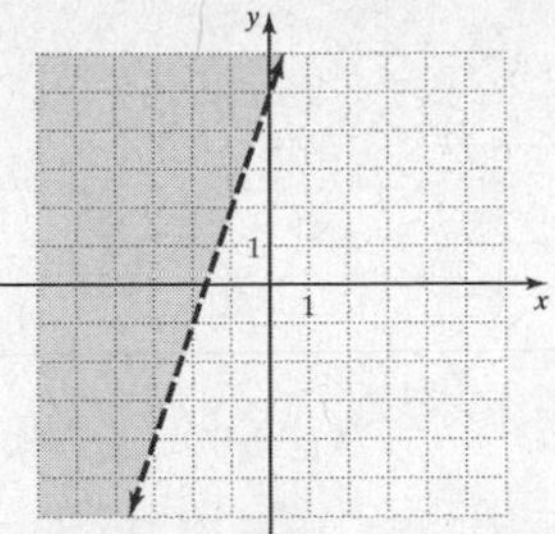

Graph the boundary line $y = 3x + 5$

x	y
-3	-4
-1	2
0	5

test point:

$y > 3x + 5 \quad (-4,5)$

$5 > 3(-4) + 5$

$5 > -7$

The half-plane above the line is the graph of the solution.

29.

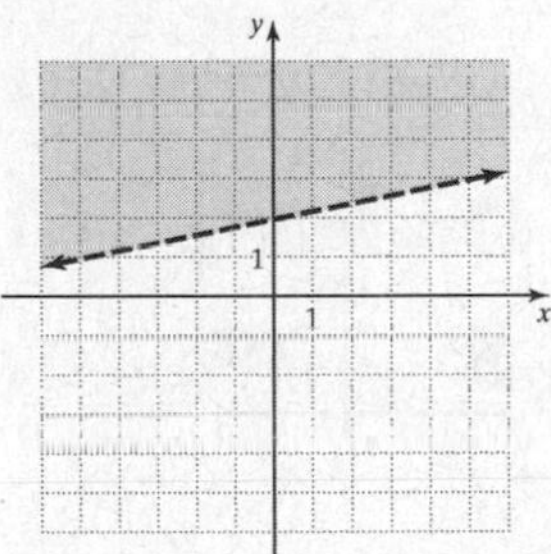

Graph the boundary line $5y - x = 10$

x	y
-5	1
0	2
5	3

test point:

$5y - x > 10 \quad (2,5)$

$5(5) - 2 \overset{?}{>} 10$

$23 > 10$

The half-plane above the line is the graph of the solution.

31.

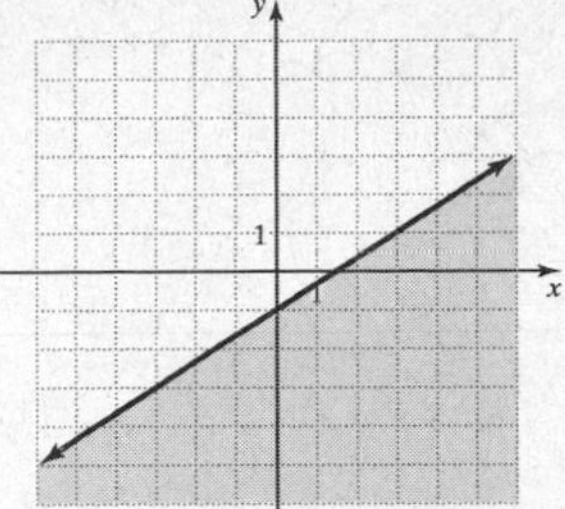

Graph the boundary line $2x - 3y = 3$

x	y
-3	-3
0	-1
6	3

test point:

$2x - 3y \geq 3 \quad (5,0)$

$2(5) - 3(0) \overset{?}{\geq} 3$

$10 \geq 3$

The half-plane below and including the line is the graph of the solution.

Applications

33.

a. $h < \frac{1}{4}i$

b.

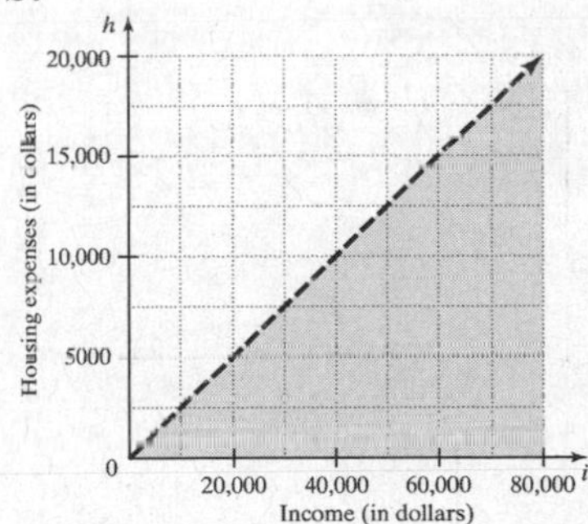

Graph the boundary line $h = \frac{1}{4}i$

i	h
0	0
40,000	10,000
80,000	20,000

test point:

$h < \frac{1}{4}i \quad (20{,}000, 1000)$

$1000 \overset{?}{<} \frac{1}{4}(20{,}000)$

$1000 < 5000$

The half-plane below the line is the graph of the solution.

c. choice of a point may vary. Possible point: (20,000, 2500). The guideline holds since the inequality is true when the values are substituted into the original inequality.

35.

a. $x+y\geq 200$

b.

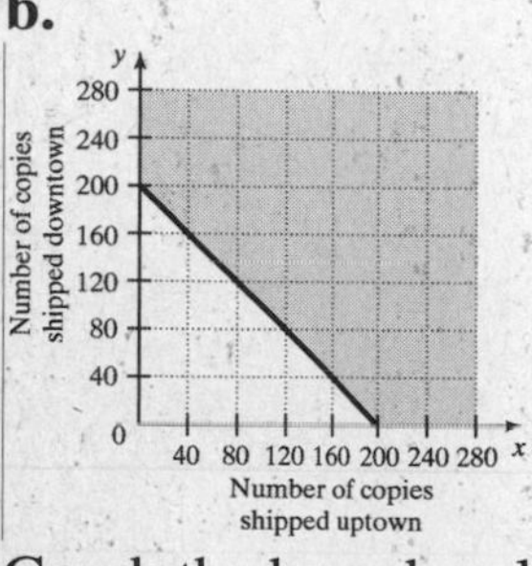

Graph the boundary line $x+y=200$

x	y
0	200
100	100
200	0

test point:

$x+y\geq 200 \quad (250,100)$

$250+100\overset{?}{\geq}200$

$350\overset{?}{\geq}200$

$350\geq 200$

The half-plane above and including the line is the graph of the solution.

c. Choice of point may vary. Possible point: (200, 60) .

$200+60\overset{?}{\geq}200$

$260\geq 200$

37.

a. $30x+75y\geq 1500$

b.

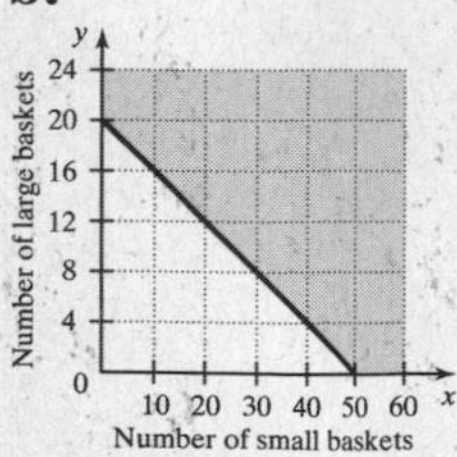

Graph the boundary line $30x+75y=1500$

x	y
0	20
30	8
50	0

test point:

$30x+75y\geq 1500 \quad (50,10)$

$30(50)+75(10)\overset{?}{\geq}1500$

$1500+750\overset{?}{\geq}1500$

$2250\geq 1500$

The half-plane above and including the line is the graph of the solution.

c. Since the point (20, 20) lies in the solution region, selling 20 small and 20 large gift baskets will generate the desired revenue.

39.

$60a+100b<300$

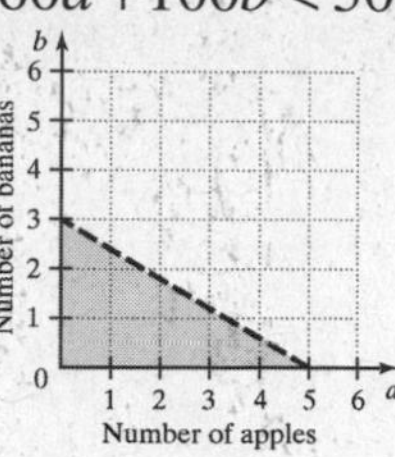

Graph the boundary line $60a+50b=300$

a	b
0	3
3	1.2
5	0

test point:

$60a+50b<300 \quad (1,2)$

$60(1)+50(2)\overset{?}{<}300$

$60+100\overset{?}{<}300$

$160<300$

The half-plane below the line is the graph of the solution.

41.

a. $10w+15m\leq 50,000$

b.

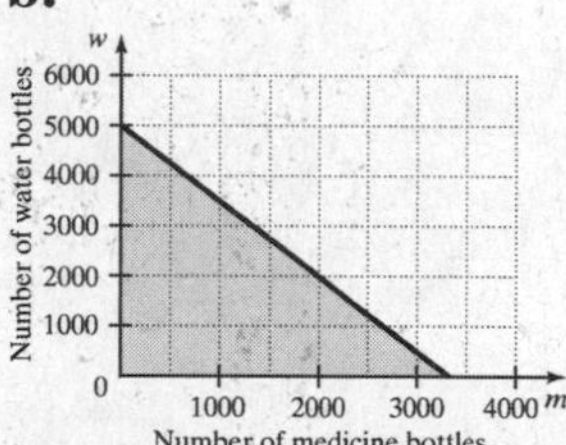

Graph the boundary line $10w+15m=50,000$

m	w
0	5000
2000	2000
3000	500

test point (500,1000)

$10w+15m\leq 50,000$

$10(1000)+15(500)\overset{?}{\leq}50,000$

$10,000+7500\overset{?}{\leq}50,000$

$17,500\leq 50,000$

The half-plane below and including the line is the graph of the solution.

43.

a. $8x+10y \geq 200$

b.

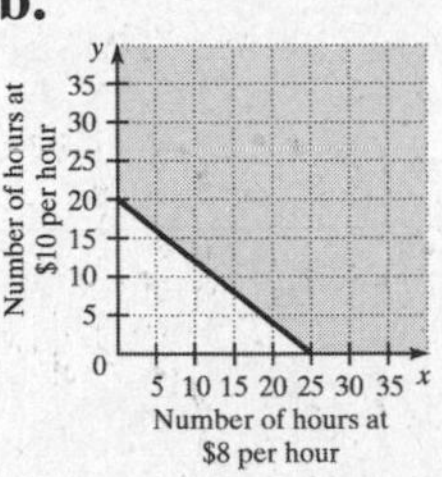

Graph the boundary line $8x+10y=200$

x	y
0	20
15	8
25	0

test point:

$8x+10y \geq 200 \quad (25,30)$

$8(25)+10(30) \overset{?}{\geq} 200$

$200+300 \overset{?}{>} 200$

$500 > 200$

The half-plane above and including the line is the graph of the solution.

c. Answers may vary. Possible answers: 20 hours at the job paying \$8 per hour and 6 hours at the job paying \$10 per hour or 10 hours at the job paying \$8 per hour and 15 hours at the job paying \$10 per hour.

Chapter 3 Review Exercises

1.

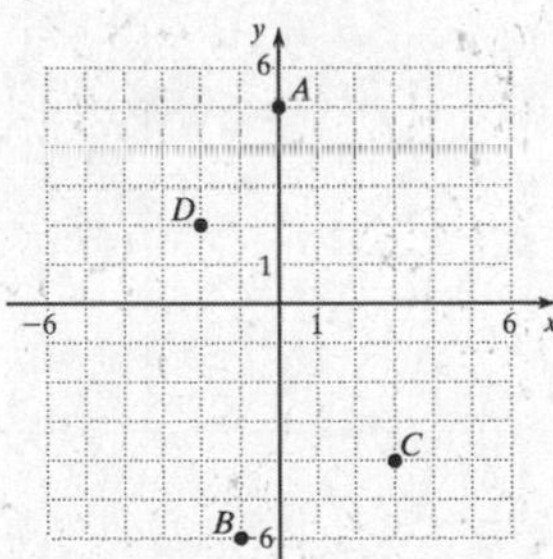

2.

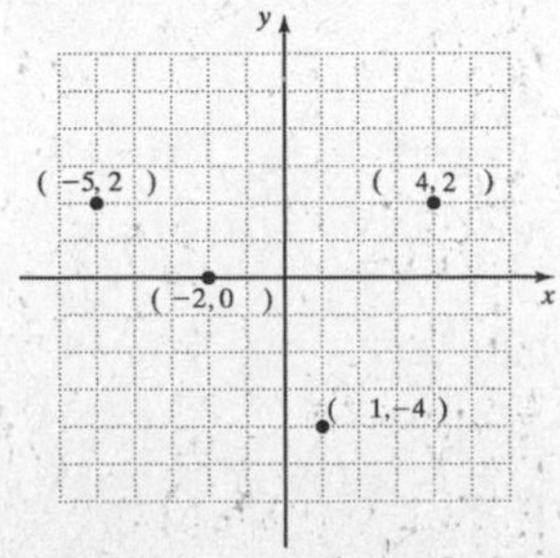

3. Quadrant IV

4. Quadrant III

5. $m=\frac{5-0}{3-2}=\frac{5}{1}=5$

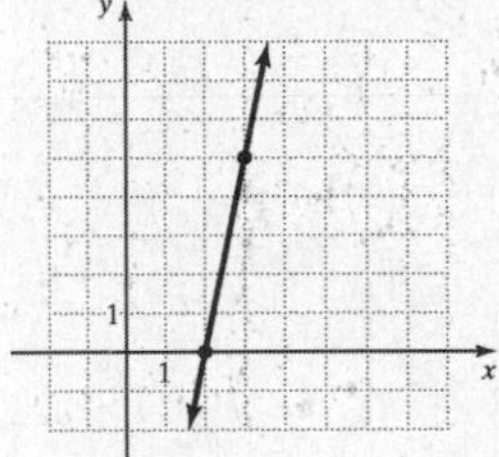

6. $m=\frac{7-7}{2-5}=\frac{0}{-3}=0$

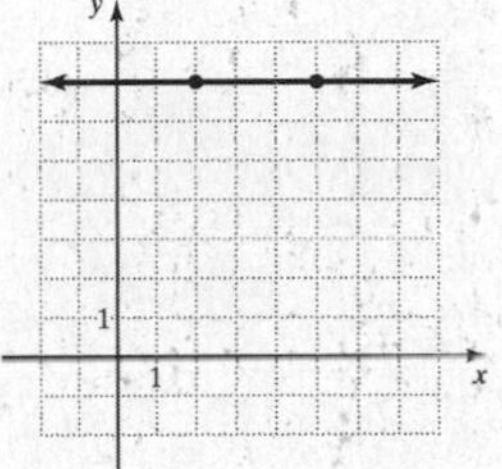

7.

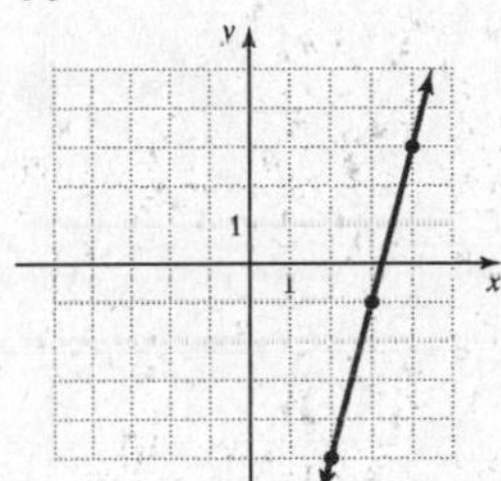

8.

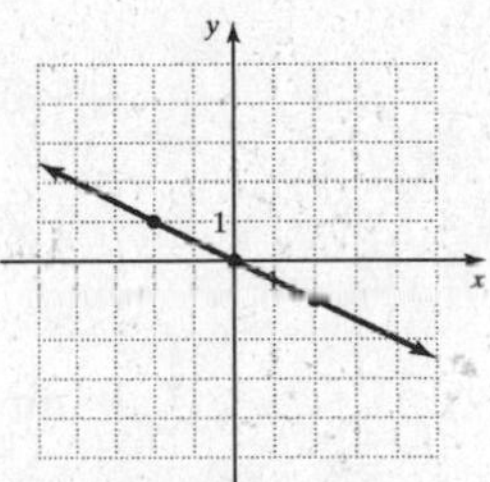

9. Positive slope

10. Undefined slope

11. Negative slope

12. Zero slope

13. Parallel

$\overleftrightarrow{AB} \quad m=\frac{0-0}{3-5}=\frac{0}{-2}=0$

$\overleftrightarrow{CD} \quad m=\frac{-2-(-2)}{1-(-3)}=\frac{0}{4}=0$

14. Perpendicular

$\overleftrightarrow{AB} \quad m=\frac{9-8}{5-4}=\frac{1}{1}=1$

$\overleftrightarrow{CD} \quad m=\frac{-1-(-3)}{0-2}=\frac{2}{-2}=-1$

15. $(30,0)$

16. $(0,50)$

17.

x	y
0	-5
1	-3
$\frac{5}{2}$	0
3	1

18.

x	y
2	1
5	-2
-4	7
8	-5

19.

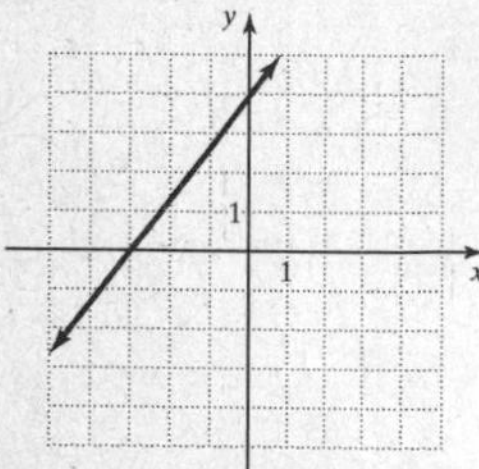

20.

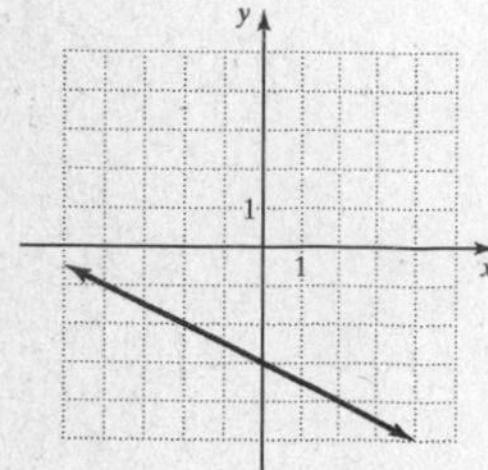

21.

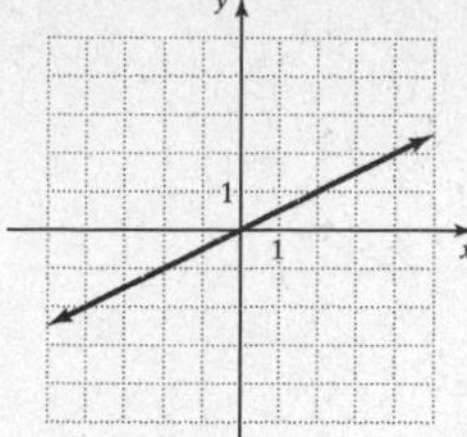

22.

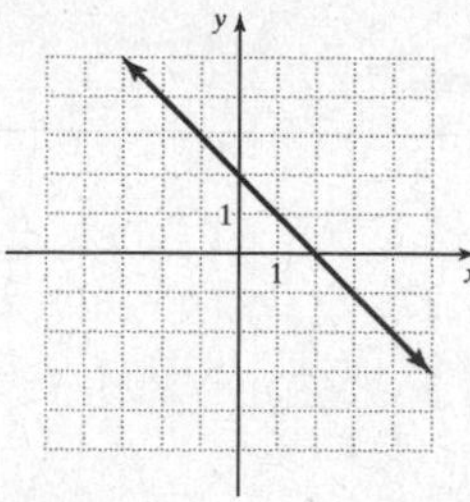

23.

$x-y=10$

$-y=-x+10$

$y=x-10$

24.

$x+2y=-1$

$2y=-x-1$

$y=-\frac{1}{2}x-\frac{1}{2}$

Equation	Slope, m	y-intercept	Which graph type best describes the line? | / \ —	x-intercept
25. $y = 4x - 16$	4	(0, -16)	/	(4, 0)
26. $y = -\frac{1}{3}x$	$-\frac{1}{3}$	(0,0)	\	(0,0)

27.

$5x-10y=20$

$-10y=-5x+20$

$y=\frac{1}{2}x-2$

The slope of a line perpendicular to this line is -2.

28.

$6x-2y=2$

$-2y=-6x+2$

$y=3x-1$

The slope of a line parallel to this line is 3.

29. Point-slope form: $y-5=-(x-3)$;

slope-intercept form: $y=-x+8$

30. $y=0$

31. $m=\frac{5-0}{1-2}=\frac{5}{-1}=-5$

Point-slope form: $y-5=-5(x-1)$;

slope-intercept form: $y=-5x+10$

32. $m=\frac{0-1}{-2-3}=\frac{-1}{-5}=\frac{1}{5}$

Point-slope form: $y-1=\frac{1}{5}(x-3)$;

Slope-intercept form: $y=\frac{1}{5}x+\frac{2}{5}$

33. $m=\frac{3-0}{0-2}=-\frac{3}{2}$

$y=-\frac{3}{2}x+3$

34.

$m=\frac{0-(-3)}{2-0}=\frac{3}{2}$

$y=\frac{3}{2}x-3$

35.

$x+y<1 \quad (-2,7)$

$-2+7\overset{?}{<}1$

$5\not<1$

No,(-2, 7) is not a solution to the inequality.

36.

$2x-3y\geq 14 \quad (1,-4)$

$2(1)-3(-4)\overset{?}{\geq}14$

$14\geq 14$

Yes, (1, -4) is a solution to the inequality.

37.

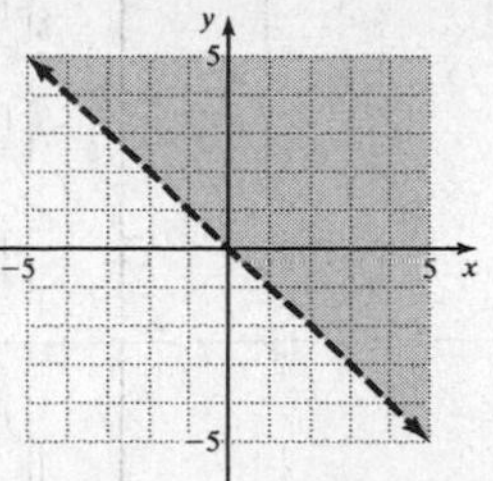

Graph the boundary line $y=-x$

x	y
-3	3
0	0
6	-6

test point:

$y>-x \quad (5,0)$

$0\overset{?}{>}-5$

$0>-5$

The half-plane above the line is the graph of the solution.

38.

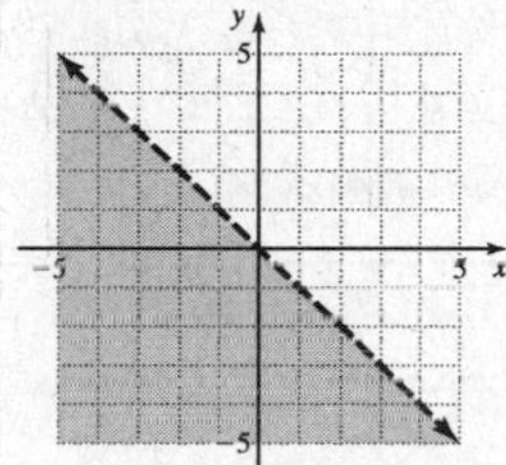

Graph the boundary line $y=-x$

x	y
-3	3
0	0
6	-6

test point:

$y<-x \quad (-5,0)$

$0\overset{?}{<}5$

$0<5$

The half-plane below the line is the graph of the solution.

39.

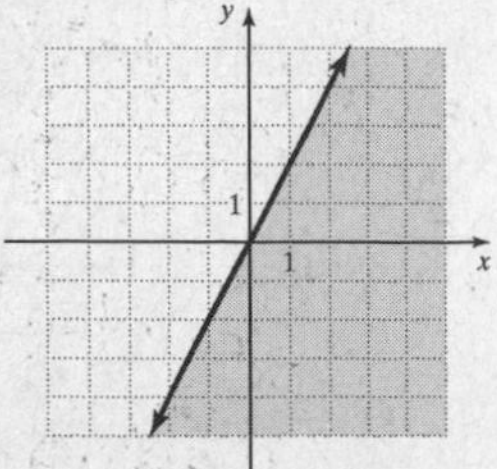

Graph the boundary line $y = 2x$

x	y
-3	-6
0	0
3	6

test point:
$y \le 2x$ (2,3)

$3 \overset{?}{\le} 2(2)$

$3 \le 4$

The half-plane below and including the line is the graph of the solution.

40.

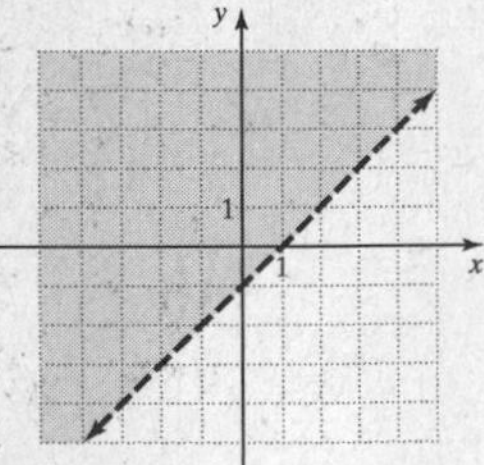

Graph the boundary line $y - x = -1$

x	y
-5	-6
0	-1
6	5

test point:
$y - x > -1$ $(-5,0)$

$0-(-5) \overset{?}{>} -1$

$5 > -1$

The half-plane above the line is the graph of the solution.

41.

a.

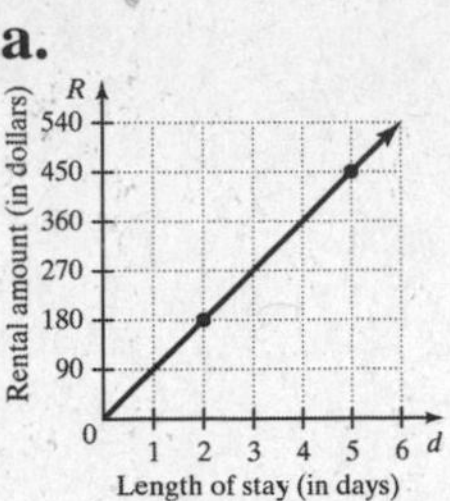

b. The R-intercept is (0,0). The R-intercept means that the cost for renting a room for 0 days is $0.

42.

a.

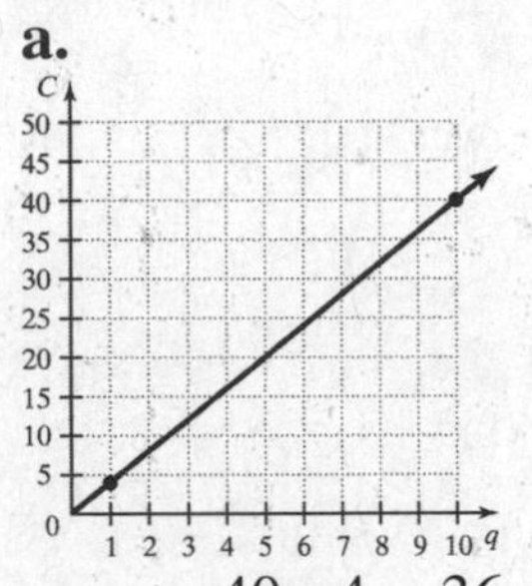

b. $m = \dfrac{40-4}{10-1} = \dfrac{36}{9} = 4$

The slope of the line is 4. The slope represents the rate the print shop charges for each flyer, which is 4 cents.

43. The graph in part (a) could describe this motion. As the child walks toward the wall, the distance between the child and the wall decreases, implying a negative slope. When the child stands still, the distance between the child and the wall remains the same, as indicated by the horizontal line segment. When the child moves toward the wall again, the distance again decreases, implying a negative slope.

44. In the first part of the flight, the airplane takes off and ascends to a particular altitude (line segment slanting up to the right), then it flies at the same altitude during the second and longest part of the flight (horizontal line segment), and finally in the last part of the flight, it descends and lands (line segment slanting down to the right).

45.

a. $i = 20,000 + 0.09s$

b.

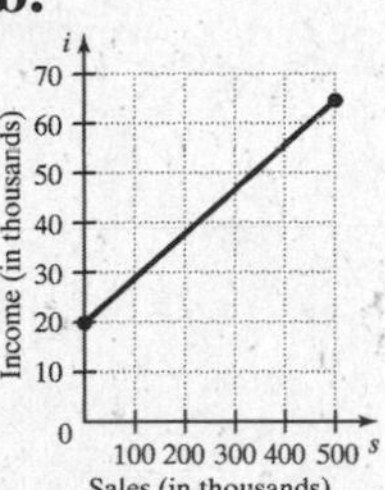

46.

a.

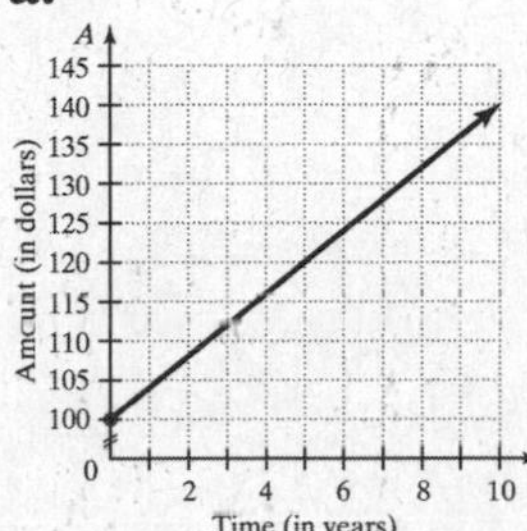

b. The A-intercept of the graph is (0, 100). The A-intercept represents the initial balance in the bank account.

47.

a.

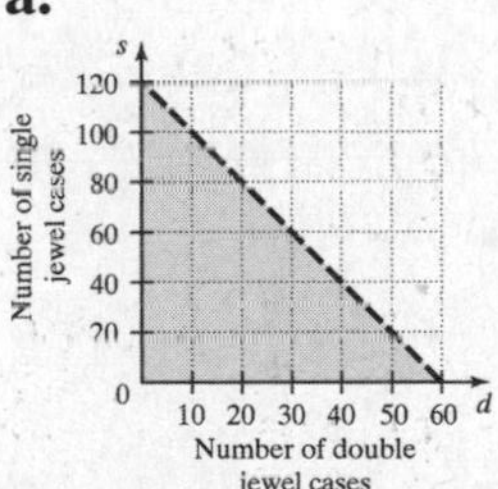

b. Answers may vary. Possible answer: 20 double jewel cases and 30 single jewel cases

48.

a.

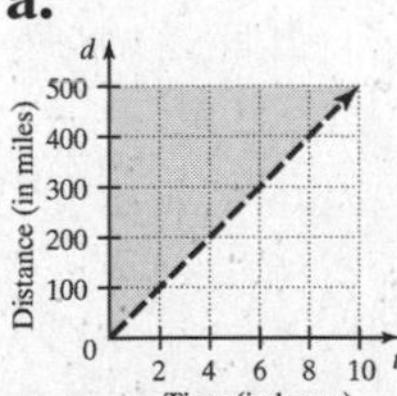

b. Choice of point may vary. Possible answer: (2, 110); the coordinates mean that you would catch up to your friend if you covered a distance of 110 miles in 2 hours.

Chapter 3 Postest

1.

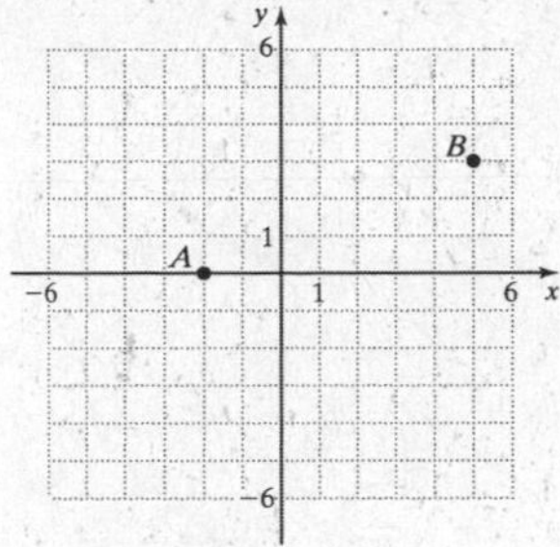

2. Quadrant II

3. $m = \dfrac{-4-1}{3-8} = \dfrac{-5}{-5} = 1$

4. The lines are parallel. The slope of $y = 3x + 1$ is 3 and the slope of $y = 3x - 2$ is 3. Since the slopes of the two lines are equal, their graphs are parallel.

5.

$\overleftrightarrow{AB}$ $m = \dfrac{8-1}{2-0} = \dfrac{7}{2}$

$\overleftrightarrow{CD}$ $m = \dfrac{4-6}{7-0} = \dfrac{-2}{7} = -\dfrac{2}{7}$

The slope of $\overleftrightarrow{AB}$ is $\dfrac{7}{2}$. The slope of $\overleftrightarrow{CD}$ is $-\dfrac{2}{7}$. $\overleftrightarrow{AB}$ is perpendicular to $\overleftrightarrow{CD}$, since the product of their slopes is -1.

6. x-intercept: (-5, 0); y-intercept:(0,2)

7. The slope is positive. As the number of miles driven increases, the rental cost increases.

8. Yes, the points do lie on the same line. The line containing (0,0) and (-2, -4) is y = 2x. The line containing (0,0) and (1,2) is y = 2x.

line $(0,0)(-2,-4)$

$m = \dfrac{-4-0}{-2-0} = \dfrac{-4}{-2} = 2$

$y = 2x$

line $(0,0)(1,2)$

$m = \dfrac{2-0}{1-0} = \dfrac{2}{1} = 2$

$y = 2x$

9.

x	y
-3	10
5	-14
$\frac{1}{3}$	0
1	-2

10.

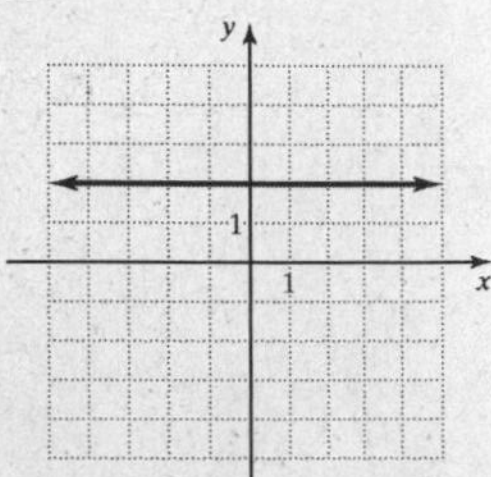

11.

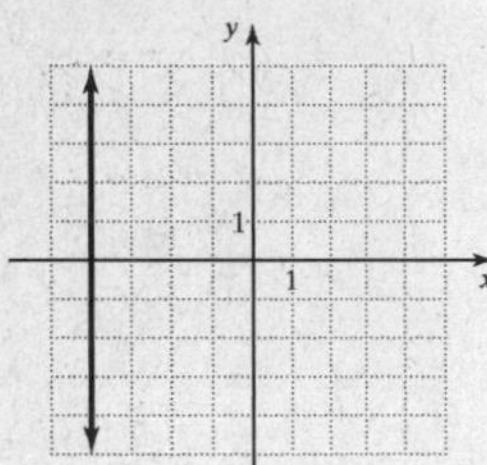

12.

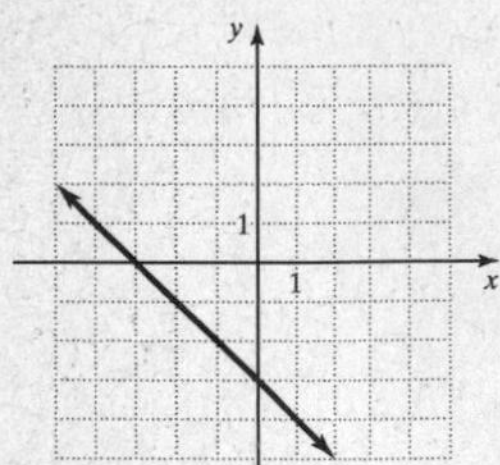

13.

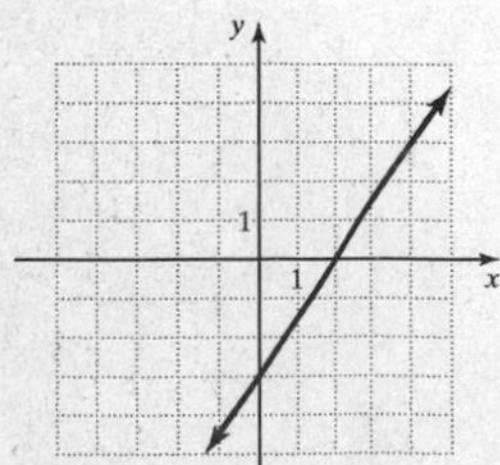

14. Slope: 3; y-intercept: (0,1)

15.

$2x - y = 5$

$-y = -2x + 5$

$y = 2x - 5$

16.

$y-(-3) = -1(x-0)$

$y+3 = -x$

$y = -x-3$

17.

$m = \frac{2-5}{-4-3} = \frac{-3}{-7} = \frac{3}{7}$

Point-slope form: $y-5 = \frac{3}{7}(x-3)$

Slope-intercept form: $y = \frac{3}{7}x + \frac{26}{7}$

$y-5 = \frac{3}{7}(x-3)$

$y-5 = \frac{3}{7}x - \frac{9}{7}$

$y = \frac{3}{7}x + \frac{26}{7}$

18.

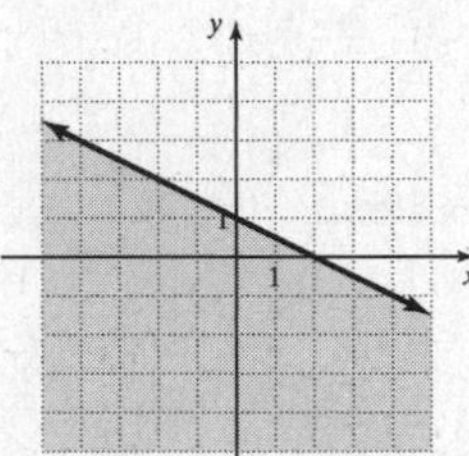

19. $C = 1000 + 30b$

b	C
100	4000
200	7000
300	10,000

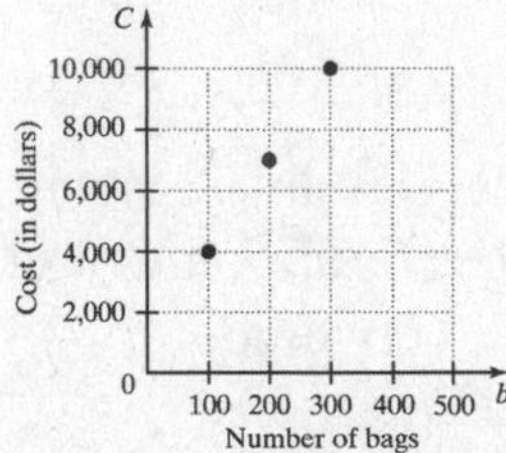

20. $0.04x + 0.08y \geq 500$

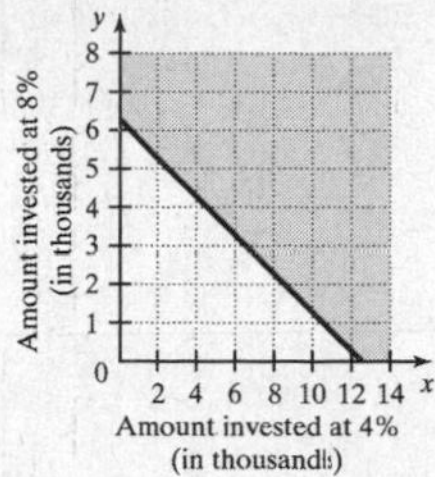

Graph the boundary line $0.04x + 0.08y = 500$

x	y
0	6250
6000	3250
12,000	250

test point:

$0.04x + 0.08y \geq 500 \quad (10,000, 3,000)$

$0.04(10,000) + 0.08(3,000) \overset{?}{\geq} 500$

$400 + 240 \overset{?}{\geq} 500$

$640 \geq 500$

The half-plane above and including the line is the graph of the solution.

Cumulative Review Exercises

1. $2 \cdot 6 - 6^2 = 2 \cdot 6 - 36 = 12 - 36 = -24$

2.

$x^2 - 4x + 1; \;\; x = -3$

$(-3)^2 - 4(-3) + 1 =$

$9 + 12 + 1 = 22$

3.

$2x - 1 = 5x + 11$

$2x = 5x + 12$

$-3x = 12$

$x = -4$

check :

$2(-4) - 1 \overset{?}{=} 5(-4) + 11$

$-9 = -9$

4.

$2x - 3(5 - x) = 4x + 2$

$2x - 15 + 3x = 4x + 2$

$5x - 15 = 4x + 2$

$5x = 4x + 17$

$x = 17$

check :

$2(17) - 3(5 - 17) \overset{?}{=} 4(17) + 2$

$34 - 3(-12) \overset{?}{=} 68 + 2$

$34 + 36 \overset{?}{=} 68 + 2$

$70 = 70$

5.

$3x + 1 > 7$

$3x > 6$

$x > 2$

6.

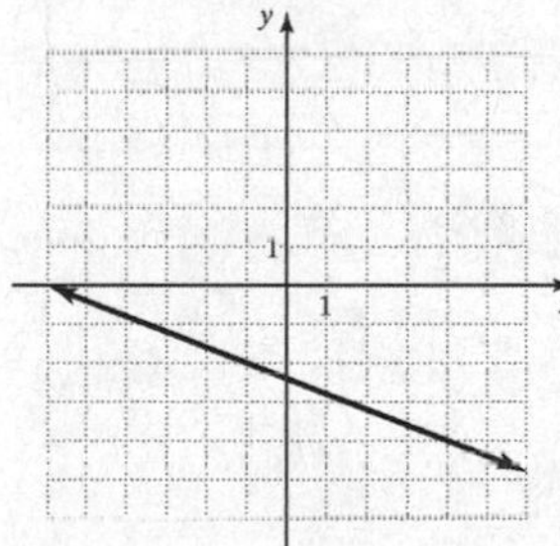

7.

$\frac{NY}{VA} = \frac{160}{400} = \frac{40}{100} = 40\%$

The assets of NY Bank are 40% of those of VA Bank.

8. y = number of years in a score

$1863 - 1776 = 87$

$4y + 7 = 87$

$4y = 80$

$y = 20$ years in a score

9.

$P = 2l + 2w$

$P - 2w = 2l$

$\frac{P - 2w}{2} = l$

$l = \frac{P - 2w}{2}$

10.

a. $y = 6x$

b.

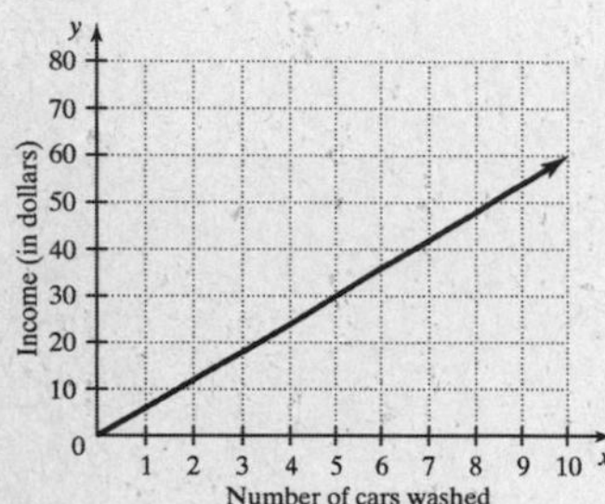

c.

x-intercept: $(0,0)$

y-intercept: $(0,0)$

Chapter 4 Solving Systems of Linear Equations

Chapter 4 Pretest

1.

a. (5, 0) Not a solution

$x+2y=5 \qquad 5x-y=-8$

$5+2(0)\stackrel{?}{=}5 \qquad 5(5)-0\stackrel{?}{=}-8$

$5=5 \qquad 25\neq-8$

b. (-1, 3) A solution

$x+2y=5 \qquad 5x-y=-8$

$-1+2(3)\stackrel{?}{=}5 \qquad 5(-1)-3\stackrel{?}{=}-8$

$5=5 \qquad -8=-8$

c. (1, -3) Not a solution

$x+2y=5 \qquad 5x-y=-8$

$1+2(-3)\stackrel{?}{=}5 \qquad 5(1)-(-3)\stackrel{?}{=}-8$

$-5\neq5 \qquad 8\neq-8$

2. One solution

3. The solution is (-3, 1).

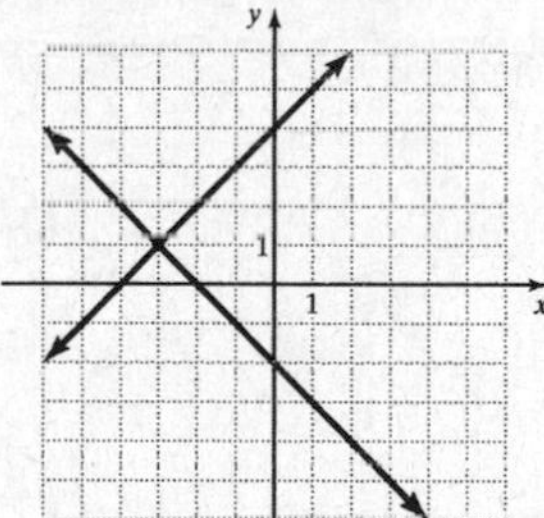

4. The solution is (5, 2).

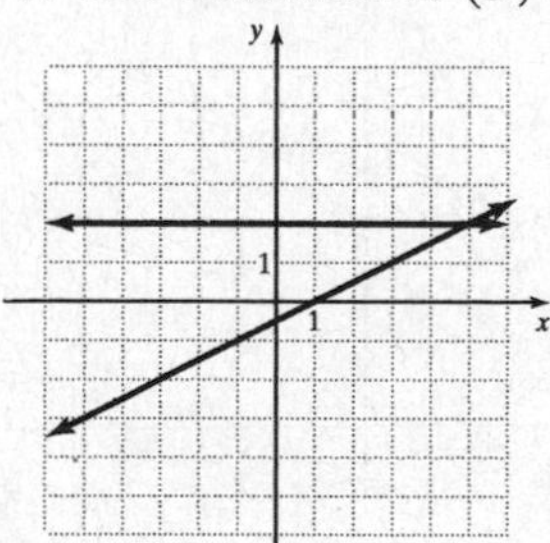

5. The solution is (0, -2).

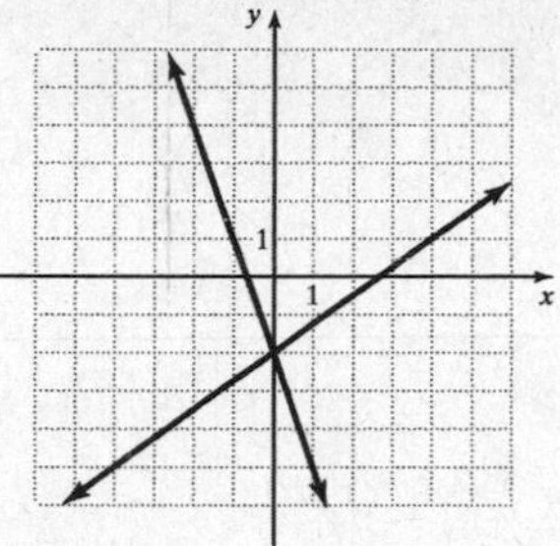

6.

$x-2y=7$

$\underline{y=-11-x}$

$x-2(-11-x)=7$

$x+22+2x=7$

$3x+22=7$

$3x=-15$

$x=-5$

$y=-11-(-5)$

$y=-6 \qquad (-5,-6)$

check :

$-5-2(-6)\stackrel{?}{=}7 \qquad -6\stackrel{?}{=}-11-(-5)$

$-5+12\stackrel{?}{=}7 \qquad -6\stackrel{?}{=}-11+5$

$7=7 \qquad -6=-6$

7.

$7x-4y=10$

$\underline{x-2y=0}$

$x=2y$

$7(2y)-4y=10$

$14y-4y=10$

$10y=10$

$y=1$

$x=2(1)=2 \qquad (2,1)$

check :

$7(2)-4(1)\stackrel{?}{=}10 \qquad 2-2(1)\stackrel{?}{=}0$

$14-4\stackrel{?}{=}10 \qquad 2-2\stackrel{?}{=}0$

$10=10 \qquad 0=0$

8.

$a+3b=-2$

$\underline{a=2b-7}$

$2b-7+3b=-2$

$5b-7=-2$

$5b=5$

$b=1$

$a=2(1)-7$

$a=-5 \qquad (-5,1)$

check :

$-5+3(1)\overset{?}{=}-2 \quad -5\overset{?}{=}2(1)-7$

$-5+3\overset{?}{=}-2 \quad -5\overset{?}{=}2-7$

$-2=-2 \quad -5=-5$

9.

$$\begin{array}{r} 2x+5y=-13 \\ \underline{-2x+6y=-20} \\ 11y=-33 \\ y=-3 \end{array}$$

$2x+5(-3)=-13$

$2x-15=-13$

$2x=2$

$x=1 \qquad (1,-3)$

check :

$2x+5y=-13 \quad -2x+6y=-20$

$2(1)+5(-3)\overset{?}{=}-13 \quad -2(1)+6(-3)\overset{?}{=}-20$

$2+(-15)\overset{?}{=}-13 \quad -2+(-18)\overset{?}{=}-20$

$-13=-13 \quad -20=-20$

10.

$6x-8y=36$

$\underline{1.5x-2y=-9}$

$$\begin{array}{r} 6x-8y=36 \\ \underline{-6x+8y=36} \\ 0\neq 72 \end{array}$$

No solution

11.

$3x-7y=-19$

$\underline{2x+3y=-5}$

$$\begin{array}{r} -6x+14y=38 \\ \underline{6x+9y=-15} \\ 23y=23 \\ y=1 \end{array}$$

$3x-7(1)=-19$

$3x-7=-19$

$3x=-12$

$x=-4 \qquad (-4,1)$

check :

$3x-7y=-19 \quad 2x+3y=-5$

$3(-4)-7(1)\overset{?}{=}-19 \quad 2(-4)+3(1)\overset{?}{=}-5$

$-12-7\overset{?}{=}-19 \quad -8+3\overset{?}{=}-5$

$-19=-19 \quad -5=-5$

12.

$4x+y=0$

$\underline{5y=12-8x}$

$y=-4x$

$5(-4x)=12-8x$

$-20x=12-8x$

$-12x=12$

$x=-1$

$y=-4(-1)$

$y=4 \qquad (-1,4)$

check :

$4(-1)+4\overset{?}{=}0 \quad 5(4)\overset{?}{=}12-8(-1)$

$-4+4\overset{?}{=}0 \quad 20\overset{?}{=}12+8$

$0=0 \quad 20=20$

13.

$-3n+5m=10$

$\underline{-4m=-2(n+1)}$

$-3n+5m=10$

$\underline{2n-4m=-2}$

$-6n+10m=20$

$\underline{6n-12m=-6}$

$-2m=14$

$m=-7$

$-3n+5(-7)=10$

$-3n-35=10$

$-3n=45$

$n=-15 \qquad (-7,-15)$

check :

$-3n+5m=10 \qquad -4m=-2(n+1)$

$-3(-15)+5(-7)\stackrel{?}{=}10 \quad -4(-7)\stackrel{?}{=}-2(-15+1)$

$45-35\stackrel{?}{=}10 \qquad 28\stackrel{?}{=}30-2$

$10=10 \qquad 28=28$

14.

$x+9=2y$

$\underline{8y-13=4x}$

$x=2y-9$

$8y-13=4(2y-9)$

$8y-13=8y-36$

$-13\neq-36$

No solution

15.

$6x+10y-12=0$

$\underline{3x+2.5y-6=0}$

$6x+10y=12$

$\underline{3x+2.5y=6}$

$6x+10y=12$

$\underline{-12x-10y=-24}$

$-6x \quad =-12$

$x=2$

$6(2)+10y=12$

$12+10y=12$

$10y=0$

$y=0 \qquad (2,0)$

check :

$6x+10y-12=0 \qquad 3x+2.5y-6=0$

$6(2)+10(0)-12\stackrel{?}{=}0 \quad 3(2)+2.5(0)-6\stackrel{?}{=}0$

$12+0-12\stackrel{?}{=}0 \qquad 6+0-6\stackrel{?}{=}0$

$0=0 \qquad 0=0$

16.

$a+b=3095$

$\underline{4a=b}$

$a+4a=3095$

$5a=3095$

$a=619$

$b=4(619)$

$b=2476 \qquad (619,2476)$

The college awarded 2475 bachelor's degrees and 619 associate's degrees.

17.

$$a+b=1975$$
$$\underline{10a+6.50b=17,650}$$
$$-10a-10b=-19,750$$
$$\underline{10a+6.5b=\ \ 17,650}$$
$$-3.5b=-2100$$
$$b=600$$
$$a+600=1975$$
$$a=1375 \qquad (1375,600)$$

600 tickets were sold before 5:00 PM and 1375 tickets were sold after 5:00 PM.

18.

$$x=3y$$
$$\underline{2x=5y+50}$$
$$2(3y)=5y+50$$
$$6y=5y+50$$
$$y=50$$

Fifty \$5 tickets were printed.

19.

$$x+y=200,000$$
$$\underline{0.05x+0.06y=11,200}$$
$$x=200,000-y$$
$$0.05(200,000-y)+0.06y=11,200$$
$$10,000-0.05y+0.06y=11,200$$
$$10,000+0.01y=11,200$$
$$0.01y=1200$$
$$y=120,000$$
$$x+120,000=200,000$$
$$x=80,000$$

\$80,000 was invested in the fund at 5% interest, and \$120,000 was invested at 6%.

20.

rate	time	Distance
$r+c$	2	13
$r-c$	2	11

$$r+c=\frac{13}{2}$$
$$\underline{r-c=\frac{11}{2}}$$
$$2r=\frac{24}{2}$$
$$2r=12$$
$$r=6$$
$$6+c=\frac{13}{2}$$
$$c=\frac{1}{2}=0.5$$

The speed of the boat was 6 mph. The speed of the current was 0.5 mph.

4.1 Introduction to Systems of Linear Equations: Solving by Graphing

Practice 4.1

1.

a. (0,2.5) is a solution

$$3x+2y=5 \qquad 4x-2y=-5$$
$$3(0)+2(2.5)\stackrel{?}{=}5 \qquad 4(0)-2(2.5)\stackrel{?}{=}-5$$
$$5=5 \qquad -5=-5$$

b. (1, -1) is not a solution

$$3x+2y=5 \qquad 4x-2y=-5$$
$$3(1)+2(-1)\stackrel{?}{=}5 \qquad 4(1)-2(-1)\stackrel{?}{=}-5$$
$$1\neq 5 \qquad 6\neq -5$$

2.

a. One solution
b. Infinitely many solutions
c. No solution

3.

$x+y=2$

x	y
-4	6
0	2
4	-2

$x-y=4$

x	y
-2	-6
0	-4
6	2

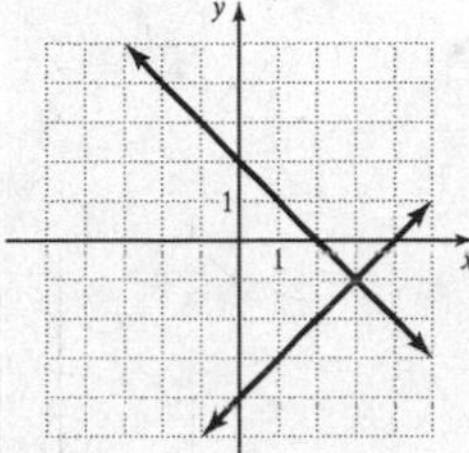

The solution is (3, -1).

4.

$y=x-6$

x	y
-2	-8
0	-6
6	0

$x-y=4$

x	y
-2	-6
0	-4
6	2

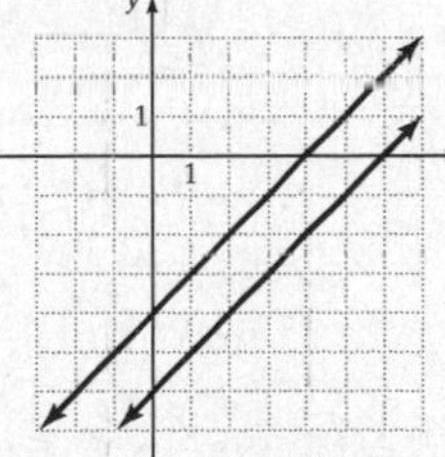

The system has no solution.

5.

$6x=15-3y$

x	y
0	5
3	-1
5	-5

$y=5-2x$

x	y
0	5
2	1
4	-3

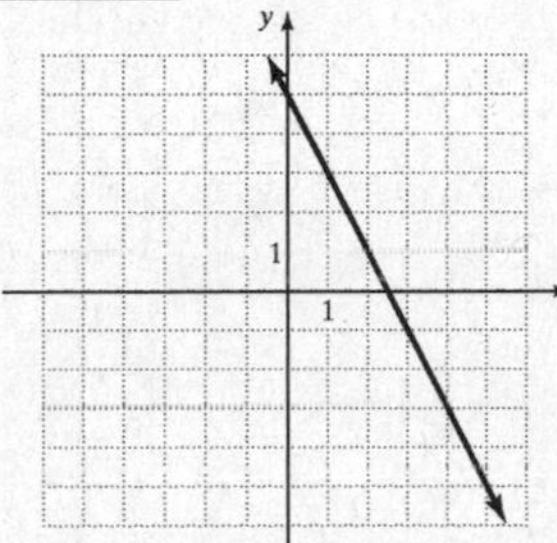

The system has infinitely many solutions.

6.

a.

$m+v=1150$

$v=m-100$

b.

$m+v=1150$

m	v
500	400
600	550
700	450

$v=m-100$

m	v
0	1150
600	500
700	600

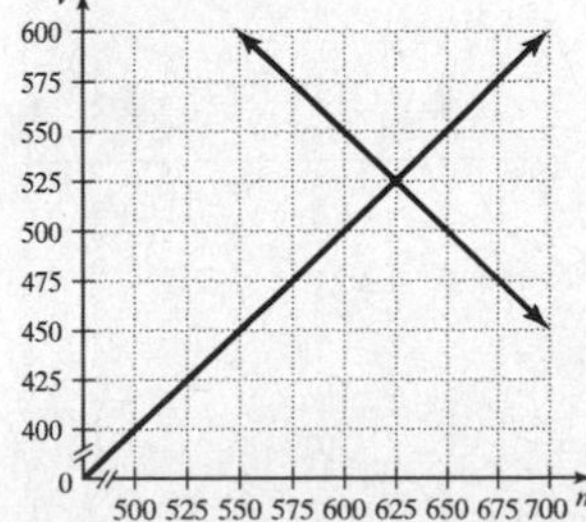

c. The solution is (625, 525).

d. The point of intersection indicates that she got a score of 625 on her test of math skills and a score of 525 on her test of verbal skills.

7.

a. $y = 1.50x + 450$

b. $y = 3x$

c. $y = 1.50x + 450$

x	Y
0	450
250	825
350	975

$y = 3x$

x	Y
0	0
200	600
350	1050

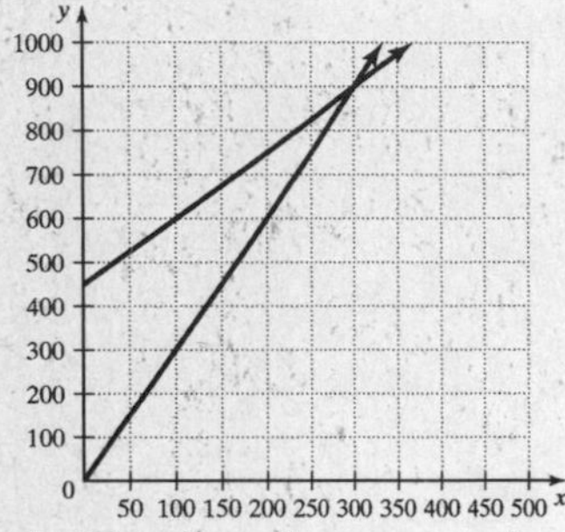

d. The break-even point is (300, 900). So when 300 newsletters are printed, the cost of printing the newsletter and the income from sales will be the same, \$900.

8. The approximate solution is (0.857, 5.857).

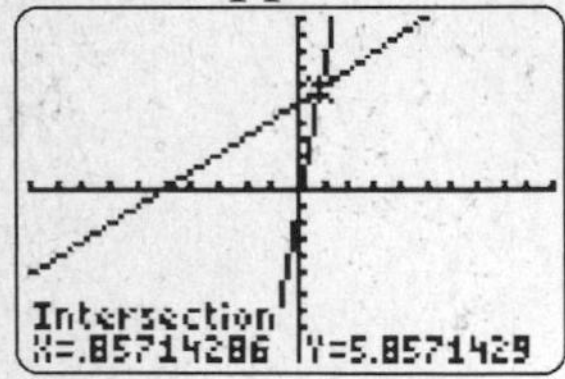

Exercises 4.1

1.

$x + y = 3$

$2x - y = 6$

a.

$(0,3)$

$0 + 3 \stackrel{?}{=} 3 \rightarrow 3 = 3$

$2(0) - 3 \stackrel{?}{=} 6 \rightarrow 0 - 3 \stackrel{?}{=} 6 \rightarrow -3 \neq 6$

Not a solution

b.

$(3,3)$

$3 + 3 \stackrel{?}{=} 3 \rightarrow 6 \neq 3$

$2(3) - 3 \stackrel{?}{=} 6 \rightarrow 6 - 3 \stackrel{?}{=} 6 \rightarrow 3 \neq 6$

Not a solution

c.

$(3,0)$

$3 + 0 \stackrel{?}{=} 3 \rightarrow 3 = 3$

$2(3) - 0 \stackrel{?}{=} 6 \rightarrow 6 - 0 \stackrel{?}{=} 6 \rightarrow 6 = 6$

A solution

3.

$4x + 5y = 0$

$7x - y = 0$

a.

$(1,7)$

$4(1) + 5(7) \stackrel{?}{=} 0 \rightarrow 4 + 35 \stackrel{?}{=} 0 \rightarrow 35 \neq 0$

$7(1) - 7 \stackrel{?}{=} 0 \rightarrow 7 - 7 \stackrel{?}{=} 0 \rightarrow 0 = 0$

Not a solution

b.

$(-5,4)$

$4(-5) + 5(4) \stackrel{?}{=} 0 \rightarrow -20 + 20 \stackrel{?}{=} 0 \rightarrow 0 = 0$

$7(-5) - 4 \stackrel{?}{=} 0 \rightarrow -35 - 4 \stackrel{?}{=} 0 \rightarrow -39 \neq 0$

Not a solution

c.

$(0,0)$

$4(0) + 5(0) \stackrel{?}{=} 0 \rightarrow 0 + 0 \stackrel{?}{=} 0 \rightarrow 0 = 0$

$7(0) - 0 \stackrel{?}{=} 0 \rightarrow 0 - 0 \stackrel{?}{=} 0 \rightarrow 0 = 0$

A solution

5.

a. III

b. IV

c. II

d. I

7.

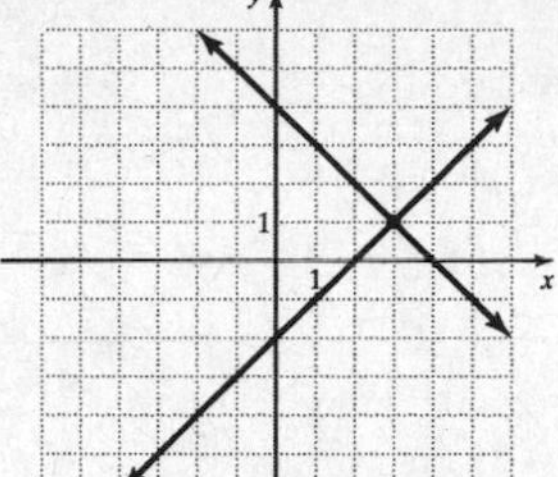

The solution is (3,1).

9.

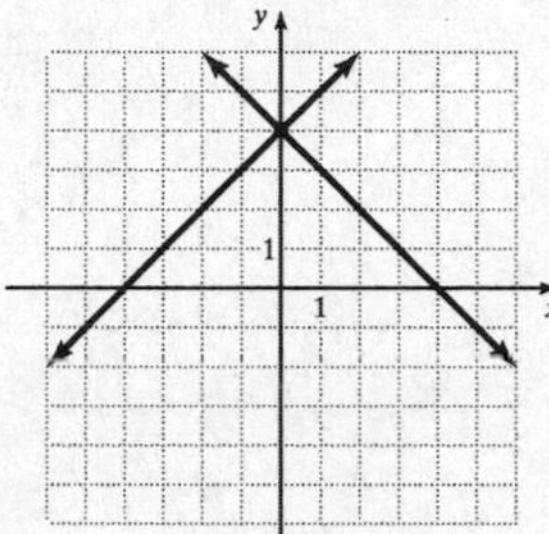

The solution is (0,4).

11.

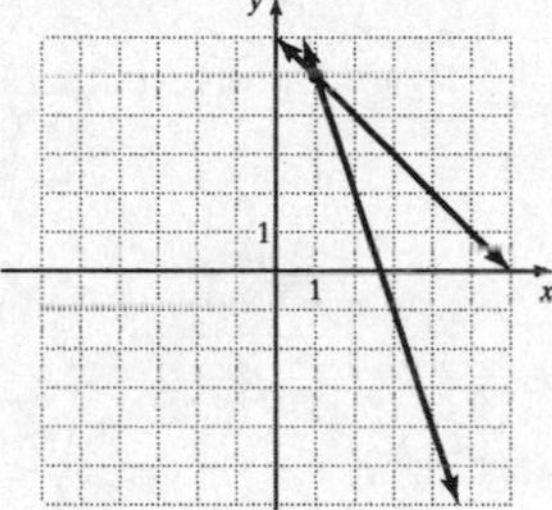

The solution is (1,5).

13.

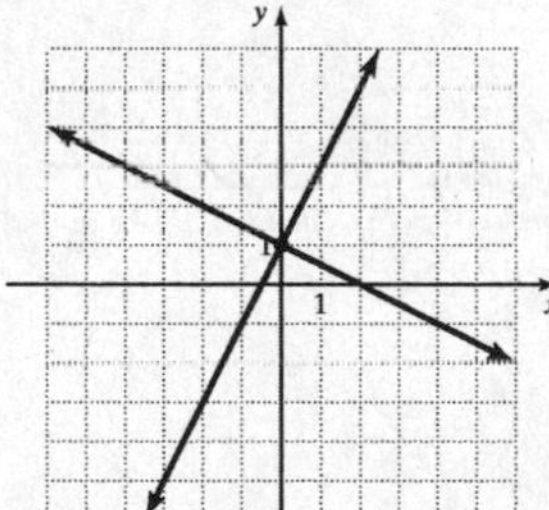

The solution is (0,1).

15.

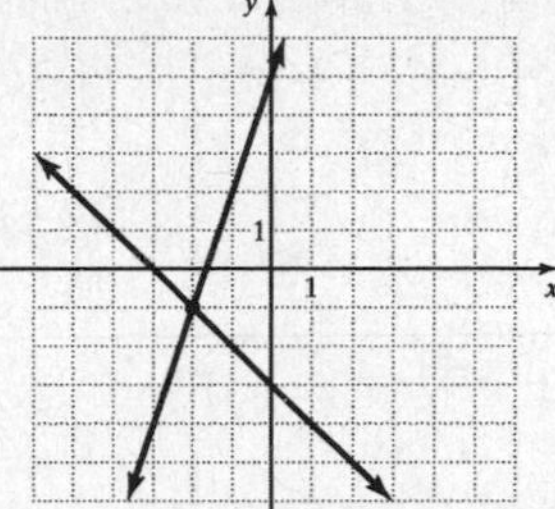

The solution is (-2,-1).

17.

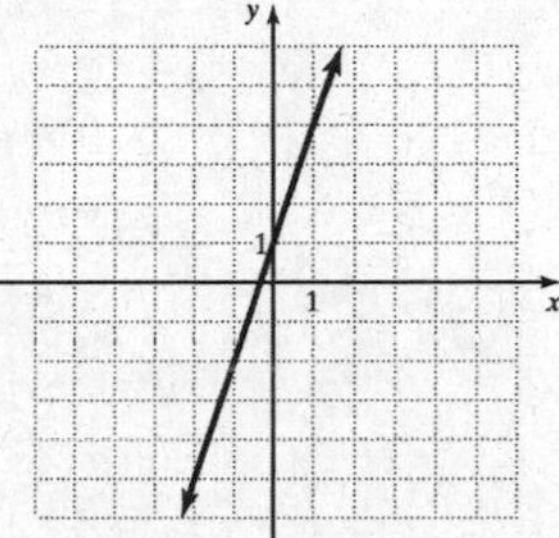

Infinitely many solutions

19.

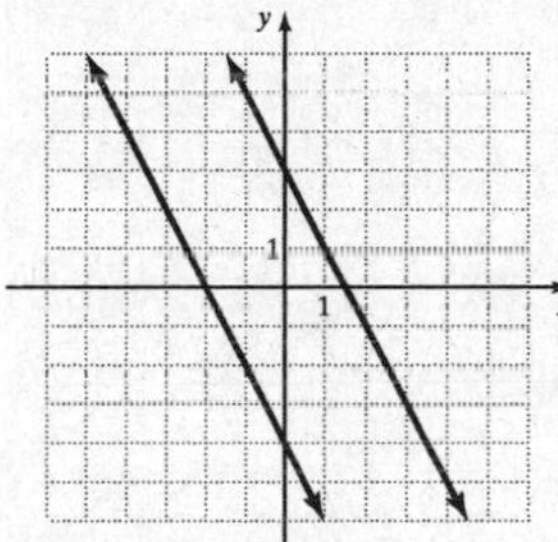

No solution

21.

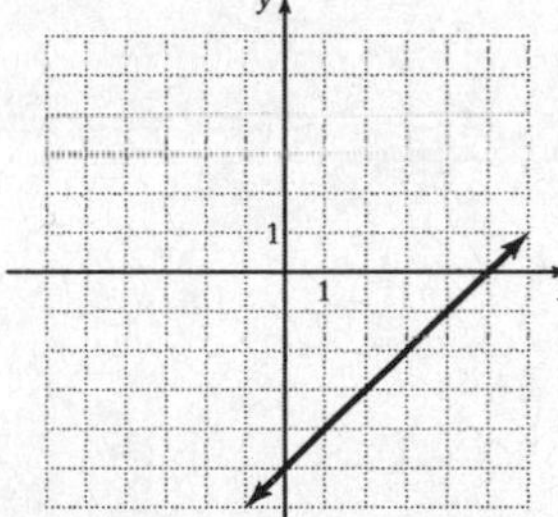

Infinitely many solutions

23.

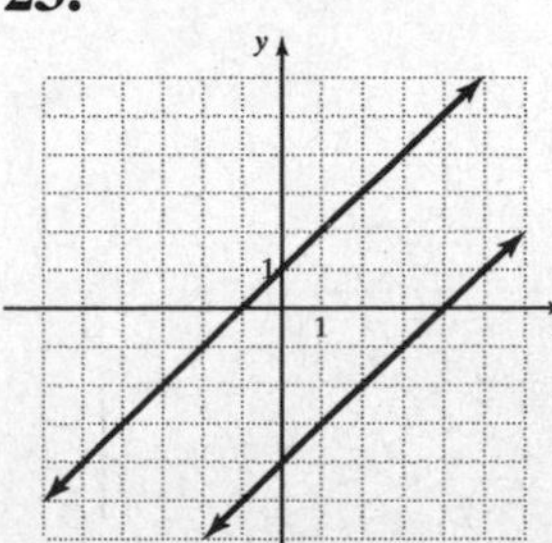

No solution

25.

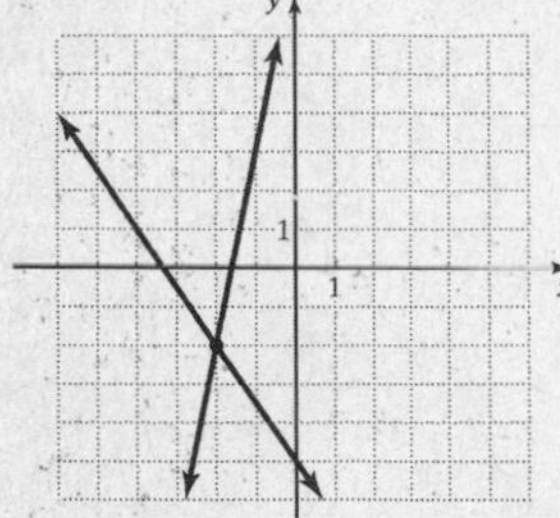

The solution is (-2,-2).

27.

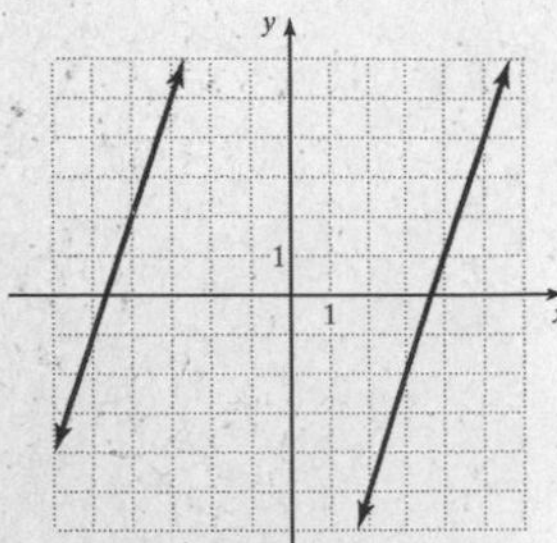

No solution

29.

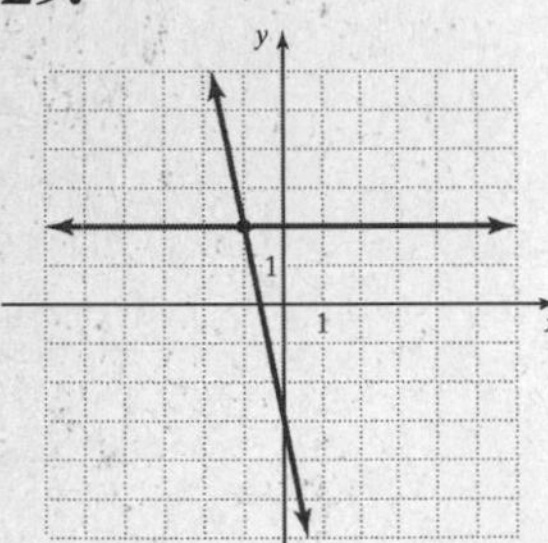

The solution is (-1,2).

31.

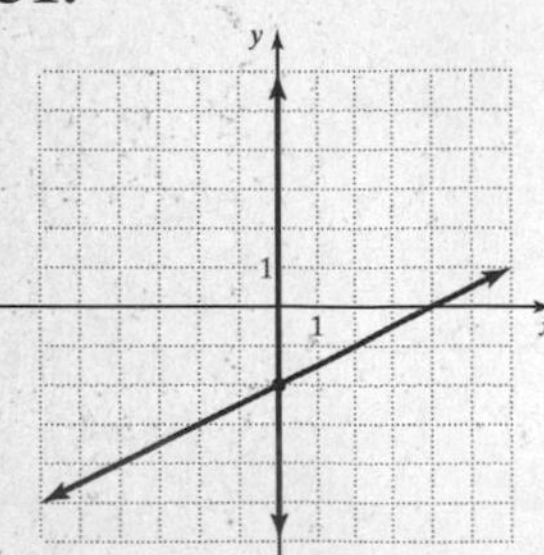

The solution is (0,-2).

33.

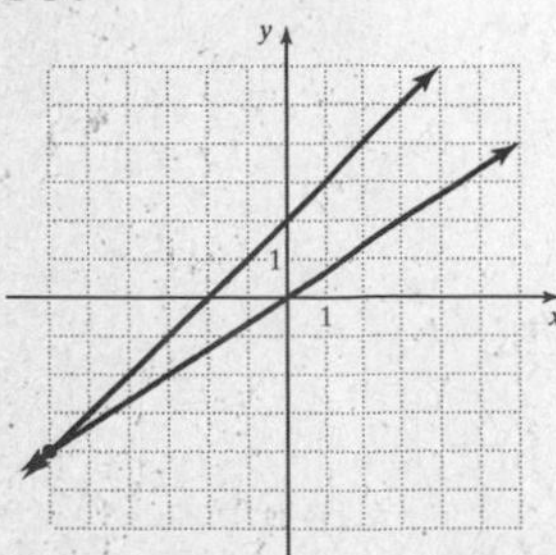

The solution is (-6,-4).

Applications

35.

a.

$$x + y = 57{,}000$$

$$y = x + 3000$$

b. $x + y = 57{,}000$

x	y
0	57,000
30,000	27,000
57,000	0

$y = x + 3000$

x	y
0	3000
30,000	33,000
50,000	53,000

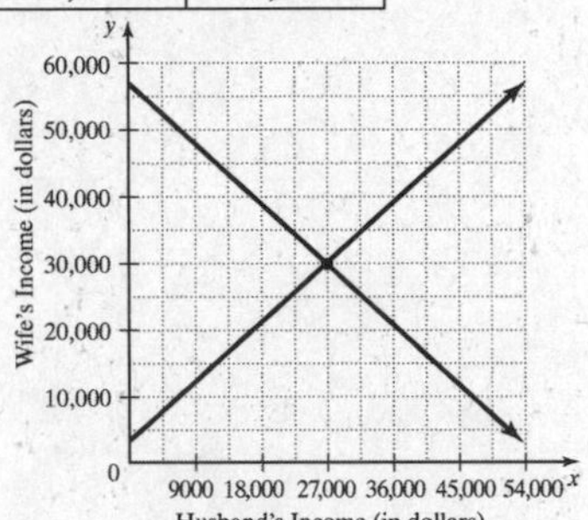

c. The husband made $27,000 and the wife made $30,000.

37.

a.

$$y = 40x + 75 \quad \text{(Mike)}$$

$$y = 30x + 100 \quad \text{(Sally)}$$

b. $y = 40x + 75$

x	y
0	75
3	195
5	275

$y = 30x + 100$

x	y
0	100
3	190
5	250

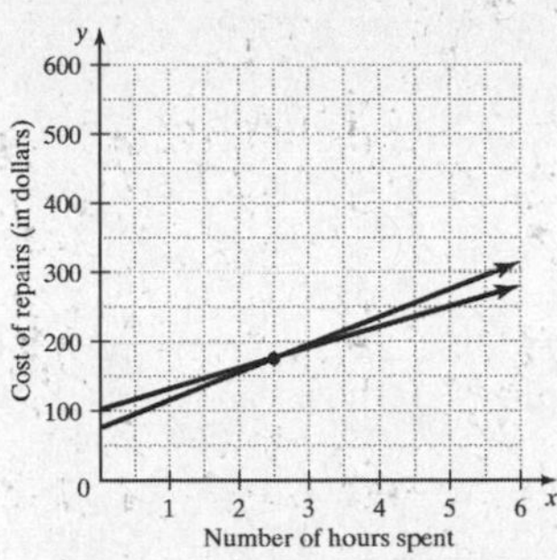

c. The plumbers would charge the same amount for 2.5 hours of work.

d. Sally charges less.

39.

$0.25x + 30 = y$

$1.50x = y$

$0.25x + 30 = y$

x	y
0	30
20	35
40	40

$1.50x = y$

x	y
0	0
20	30
50	75

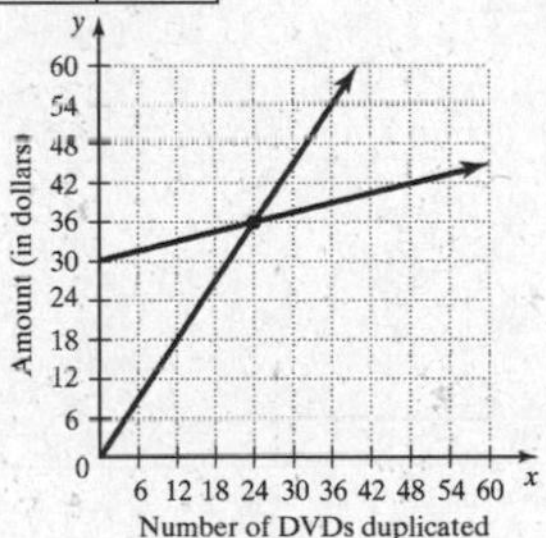

The break-even point for duplicating DVDs is (24, 36).

41.

$x + y = 6$

$2x + 4y = 22$

$x + y = 6$

x	y
0	6
2	4
6	0

$2x + 4y = 22$

x	y
5	3
3	4
1	5

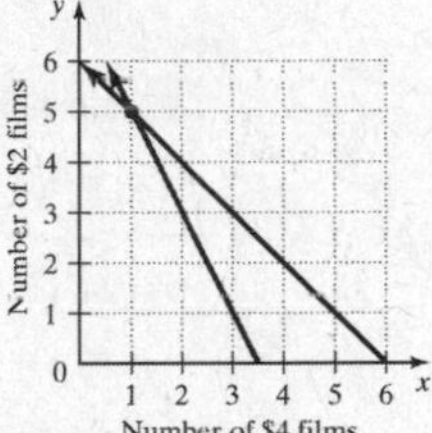

Five $4 films were rented.

43.

$30x + 300 = y$

$25x + 400 = y$

$30x + 300 = y$

x	y
0	300
10	600
20	900

$25x + 400 = y$

x	y
0	400
10	650
20	900

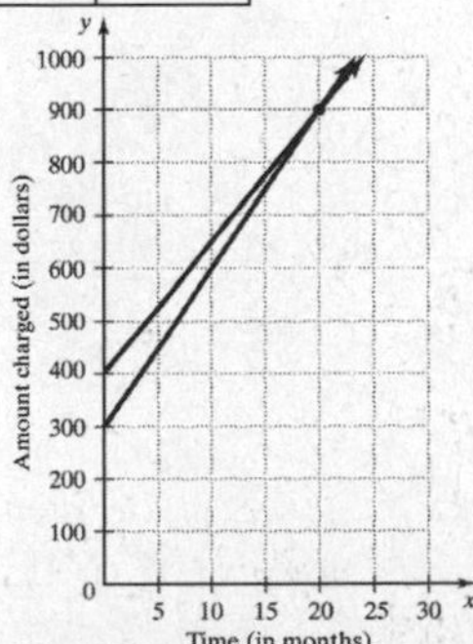

The health clubs charge the same amount ($900) for 20 months.

4.2 Solving Systems of Linear Equations by Substitution

Practice 4.2

1.

$$\begin{aligned}
&(1)\quad x-y=7\\
&(2)\quad \underline{x=-y+1}\\
&(1)\quad -y+1-y=7\\
&\qquad -2y+1=7\\
&\qquad -2y=6\\
&\qquad y=-3\\
&(2)\quad x=-(-3)+1\\
&\qquad x=4 \qquad\qquad (4,-3)
\end{aligned}$$

check :

$$4-(-3)\overset{?}{=}7 \qquad 4\overset{?}{=}-(-3)=1$$

$$7=7 \qquad\qquad 4=4$$

2.

$$\begin{aligned}
&(1)\quad m=-5n+1\\
&(2)\quad \underline{2m+3n=7}\\
&(2)\quad 2(-5n+1)+3n=7\\
&\qquad -10n+2+3n=7\\
&\qquad -7n+2=7\\
&\qquad -7n=5\\
&\qquad n=-\frac{5}{7}\\
&(1)\quad m=-5\left(-\frac{5}{7}\right)+1\\
&\qquad m=\frac{32}{7} \qquad\qquad \left(\frac{32}{7},-\frac{5}{7}\right)
\end{aligned}$$

check :

$$m=-5n+1 \qquad 2m+3n=7$$

$$\frac{32}{7}\overset{?}{=}-5\left(-\frac{5}{7}\right)+1 \qquad 2\left(\frac{32}{7}\right)+3\left(-\frac{5}{7}\right)\overset{?}{=}7$$

$$\frac{32}{7}=\frac{32}{7} \qquad\qquad 7=7$$

3.

$$\begin{aligned}
&(1)\quad 2x-7y=7\\
&(2)\quad \underline{6x-y=1}\\
&(2)\quad y=6x-1\\
&(1)\quad 2x-7(6x-1)=7\\
&\qquad 2x-42x+7=7\\
&\qquad -40x=0\\
&\qquad x=0\\
&(2)\quad y=6(0)-1\\
&\qquad y=-1 \qquad\qquad (0,-1)
\end{aligned}$$

check :

$$2(0)-7(-1)\overset{?}{=}7 \qquad 6(0)-(-1)\overset{?}{=}1$$

$$7=7 \qquad\qquad 1=1$$

4.

$$\begin{aligned}
&(1)\quad 3x+y=10\\
&(2)\quad \underline{-6x-2y=1}\\
&(1)\quad y=-3x+10\\
&(2)\quad -6x-2(-3x+10)=1\\
&\qquad -6x+6x-20=1\\
&\qquad -20\neq 1
\end{aligned}$$

No solution

5.

$$\begin{aligned}
&(1)\quad y=-2x+4\\
&(2)\quad \underline{10x+5y=20}\\
&(2)\quad 10x+5(-2x+4)=20\\
&\qquad 10x-10x+20=20\\
&\qquad 20=20
\end{aligned}$$

Infinitely many solutions

6.

a.

$c = 35n + 20$ (TV Deal)

$c = 25n + 30$ (Movie Deal)

b.

$c = 35n + 20$

$c = 25n + 30$

$35n + 20 = 25n + 30$

$35n = 25n + 30$

$10n = 10$

$n = 1$

$c = 35(1) + 20$

$c = 55$

c. The cost is the same for both cable deals if you sign up for one month.

7.

	% copper	Amt. alloy	Amt. copper
Alloy 1	20	x	.2x
Alloy 2	50	y	.5y
mix	25	15	.25(15)

(1) $x + y = 15$

(2) $0.2x + 0.5y = 3.75$

(1) $x = 15 - y$

(2) $0.2(15 - y) + 0.5y = 3.75$

$3 - 0.2y + 0.5y = 3.75$

$3 + 0.3y = 3.75$

$0.3y = 0.75$

$y = 2.5$

(1) $x + 2.5 = 15$

$x = 12.5$

12.5 oz. of the 20% copper alloy and 2.5 oz. of the 50% copper alloy are required.

8.

(1) $x + y = 198{,}000$

(2) $0.04x = 0.05y$

(1) $x = 198{,}000 - y$

(2) $0.04(198{,}000 - y) = 0.05y$

$7920 - 0.04y = 0.05y$

$7920 = 0.09y$

$0.09y = 7920$

$y = 88{,}000$

(1) $x + 88{,}000 = 198{,}000$

$x = 110{,}000$

The manager put $110,000 in the investment that pays 4% and $88,000 in the investment that pays 5%.

Exercises 4.2

1.

(1) $x + y = 10$

(2) $y = 2x + 1$

(1) $x + (2x + 1) = 10$

$3x + 1 = 10$

$3x = 9$

$x = 3$

(2) $y = 2(3) + 1$

$y = 7$

$(3, 7)$

check :

$x + y = 10 \quad y = 2x + 1$

$3 + 7 \stackrel{?}{=} 10 \quad 7 \stackrel{?}{=} 2(3) + 1$

$10 = 10 \quad 7 = 7$

3.

(1) $y=-3x-15$

(2) $y=-x-7$

(1) $-x-7=-3x-15$

$2x=-8$

$x=-4$

(2) $y=-(-3)-7$

$y=-3$

$(-4,-3)$

check:

$y=-3x-15 \qquad y=-x-7$

$-3\overset{?}{=}-3(-4)-15 \quad -3\overset{?}{=}-(-3)-7$

$-3=-3 \qquad -3=-3$

5.

(1) $-x-y=8$

(2) $x=-3y$

(1) $-(-3y)-y=8$

$2y=8$

$y=4$

(2) $x=-3(4)$

$x=-12$

$(-12,4)$

check:

$-x-y=8 \qquad x=-3y$

$-(-12)-4\overset{?}{=}8 \quad -12\overset{?}{=}-3(4)$

$8=8 \qquad -12=-12$

7.

(1) $4x+2y=10$

(2) $x=2$

(1) $4(2)+2y=10$

$8+2y=10$

$2y=2$

$y=1$

$(2,1)$

check:

$4x+2y=10 \qquad x=2$

$4(2)+2(1)\overset{?}{=}10 \quad (2)\ 2=2$

$10=10$

9.

(1) $-x+20y=0$

(2) $x-y=0$

$x=y$

(1) $-y+20y=0$

$19y=0$

$y=0$

(2) $x-0=0$

$x=0$

$(0,0)$

check:

$-x+20y=0 \qquad x-y=0$

$-(0)+20(0)\overset{?}{=}0 \quad 0-0\overset{?}{=}0$

$0=0 \qquad 0=0$

11.

(1) $6x+4y=2$

(2) $2x+y=0$

$y=-2x$

(1) $6x+4(-2x)=2$

$6x-8x=2$

$-2x=2$

$x=-1$

(2) $2(-1)+y=0$

$-2+y=0$

$y=2$

$(-1,2)$

check :

$6x+4y=2 \qquad 2x+y=0$

$6(-1)+4(2)\stackrel{?}{=}2 \quad 2(-1)+2\stackrel{?}{=}0$

$2=2 \qquad 0=0$

13.

(1) $3x+5y=-12$

(2) $x+2y=-6$

$x=-2y-6$

(1) $3(-2y-6)+5y=-12$

$-6y-18+5y=-12$

$-y=6$

$y=-6$

(2) $x+2(-6)=-6$

$x-12=-6$

$x=6$

$(6,-6)$

check :

$3x+5y=-12 \qquad x+2y=-6$

$3(6)+5(-6)\stackrel{?}{=}-12 \quad 6+2(-6)\stackrel{?}{=}-6$

$-12=-12 \qquad -6=-6$

15.

(1) $7x-3y=26$

(2) $3x-y=11$

$y=3x-11$

(1) $7x-3(3x-11)=26$

$7x-9x+33=26$

$-2x=-7$

$x=\frac{7}{2}$

(2) $3\left(\frac{7}{2}\right)-y=11$

$\frac{21}{2}-y=11$

$y=-\frac{1}{2}$

$\left(\frac{7}{2},-\frac{1}{2}\right)$

check :

$7x-3y=26 \qquad 3x-y=11$

$7\left(\frac{7}{2}\right)-3\left(-\frac{1}{2}\right)\stackrel{?}{=}26 \quad 3\left(\frac{7}{2}\right)-\left(-\frac{1}{2}\right)\stackrel{?}{=}11$

$26=26 \qquad 11=11$

17.

(1) $8x+2y=-1$

(2) $y=-4x+1$

(1) $8x+2(-4x+1)=-1$

$8x-8x+2=-1$

$2\neq-1$

No solution

19.

(1) $6x-2y=2$

(2) $y=3x-1$

(1) $6x-2(3x-1)=2$

$6x-6x+2=2$

$2=2$

Infinitely many solutions

21.

(1) $p+2q=13$

(2) $q+7=4p$

$q=4p-7$

(1) $p+2(4p-7)=13$

$p+8p-14=13$

$9p-14=13$

$9p=27$

$p=3$

(2) $q+7=4(3)$

$q+7=12$

$q=5$

$(3,5)$

check :

$p+2q=13 \quad q+7=4p$

$3+2(5)\stackrel{?}{=}13 \quad 5+7\stackrel{?}{=}4(3)$

$13=13 \qquad 12=12$

23.

(1) $s-3t+5=0$

(2) $-4s+t-9=0$

$t=4s+9$

(1) $s-3(4s+9)+5=0$

$s-12s-27+5=0$

$-11s=22$

$s=-2$

(2) $-4(-2)+t-9=0$

$-1+t=0$

$t=1$

$s=-2; t=1$

check :

$(1)-2-3(1)+5\stackrel{?}{=}0 \quad (2)-4(-2)+1-9\stackrel{?}{=}0$

$0=0 \qquad 0=0$

Applications

25.

a.

$c=1.25m+3$

$c=1.50m+2$

b.

(1) $c=1.25m+3$

(2) $c=1.50m+2$

$1.25m+3=1.50m+2$

$-0.25m=-1$

$m=4$

(1) $c=1.25(4)+3$

$c=8$

$m=4 \quad c=8$

The solution indicates that both companies charge the same amount ($8) for a 4 mile taxi ride.

27.

F = number of full-price tickets purchased
D = number of discount-price tickets purchased

(1) $F+D=172$

(2) $\underline{310F+210D=44,120}$

(1) $D=172-F$

(2) $310F+210(172-F)=44,120$

(2) $310F+36,120-210F=44,120$

$100F+36,120=44,120$

$100F=8000$

$F=80$

80 full-price tickets were sold.

29.

	Amt. of liquid	% water	Amt water
Antiseptic	x	30	0.30x
Antiseptic	y	70	0.70y
mix	10 L	60	.6(10)=6L

(1) $x+y=10$

(2) $0.3x+.07y=6$

(1) $y=10-x$

(2) $0.3x+0.7(10-x)=6$

$0.3x+7-0.7x=6$

$-0.4x=-1$

$x=2.5$

$(1) 2.5+y=10$

$y=7.5L$

She can combine 2.5L of the antiseptic that is 30% alcohol with 7.5L of the antiseptic that is 70% alcohol to get the desired concentration.

31.

	# employees	% women	number. Women
Dept 1	x	5	0.05x
Dept 2	y	80	0.8y
Merge	150	50	75

(1) $x+y=150$

(2) $0.05x+0.8y=75$

(1) $x=150-y$

$(2) 0.05(150-y)+0.8y=75$

$7.5-0.05y+0.8y=75$

$7.5+0.75y=75$

$0.75y=67.5$

$y=90$

$x=150-90=60$

80% of 90=72

5% of 60=3

There were 3 woman in one department and 72 women in the other department.

33.

	P	R	I
Loan 1	x	6%	0.06x
Loan 2	y	7%	0.07y

(1) $x+y=5000$

(2) $0.06x+0.07y=310$

(1) $x=5000-y$

(2) $0.06(5000-y)+0.07y=310$

$300-0.06y+0.07y=310$

$300+0.01y=310$

$0.01y=10$

$y=1000$

$x=5000-1000=4000$

She borrowed \$1000 at 7% and \$4000 at 6%.

35.

	P	r	I
Invest 1	x	7%	0.07x
Invest 2	y	9%	0.09y
Total	40,000		3140

(1) $x+y=40,000$

(2) $0.07x+0.09y=3140$

(1) $x=40,000-y$

(2) $0.07(40,000-y)+0.09y=3140$

$2800-0.07y+0.09y=3140$

$0.02y=340$

$y=17,000$

$x=40,000-17,000=23,000$

\$23,000 was invested at 7% and \$17,000 was invested at 9%.

4.3 Solving Systems of Linear Equations by Elimination

Practice 4.3

1.

$$\begin{array}{ll}(1) & x+y=6 \\ (2) & \underline{x-y=-10} \\ & 2x \quad =-4 \\ & x=-2 \\ (2) & -2+y=6 \\ & y=8 \qquad (-2,8)\end{array}$$

check :

$$x+y=6 \quad x-y=-10$$

$$-2+8\stackrel{?}{=}6 \quad -2-8\stackrel{?}{=}-10$$

$$6=6 \qquad -10=-10$$

2.

$$\begin{array}{ll}(1) & 4x+3y=-7 \\ (2) & \underline{5x+3y=-5} \\ (1) & -4x-3y=7 \\ (2) & \underline{\ 5x+3y=-5} \\ & x \quad =2 \\ (2) & 5(2)+3y=-5 \\ & 3y=-15 \\ & y=-5 \qquad (2,-5)\end{array}$$

check :

$$4x+3y=-7 \qquad 5x+3y=-5$$

$$4(2)+3(-5)\stackrel{?}{=}-7 \quad 5(2)+3(-5)\stackrel{?}{=}-5$$

$$-7=-7 \qquad -5=-5$$

3.

$$\begin{array}{ll}(1) & x-3y=-18 \\ (2) & \underline{5x+2y=12} \\ (1) & -5x+15y=90 \\ (2) & \underline{\ 5x+2y=12} \\ & 17y \quad =102 \\ & y=6 \\ (1) & x-3(6)=-18 \\ & x-18=-18 \\ & x=0 \qquad (0,6)\end{array}$$

check :

$$x-3y=-18 \quad 5x+2y=12$$

$$0-3(6)\stackrel{?}{=}-18 \quad 5(0)+2(6)\stackrel{?}{=}12$$

$$-18=-18 \qquad 12=12$$

4.

$$\begin{array}{ll}(1) & 5x-7y=24 \\ (2) & \underline{3x-5y=16} \\ (1) & -15x+21y=-72 \\ (2) & \underline{\ 15x-25y=80} \\ & -4y=8 \\ & y=-2 \\ (1) & 5x-7(-2)=24 \\ & 5x+14=24 \\ & 5x=10 \\ & x=2 \qquad (2,-2)\end{array}$$

check :

$$5x-7y=24 \qquad 3x-5y=16$$

$$5(2)-7(-2)\stackrel{?}{=}24 \quad 3(2)-5(-2)\stackrel{?}{=}16$$

$$24=24 \qquad 16=16$$

5.

$(1)\quad -2x+5y=20$

$(2)\quad \underline{3x=7y-26}$

$(1)\quad -2x+5y=20$

$(2)\quad \underline{3x-7y=-26}$

$(1)\quad -6x+15y=60$

$(2)\quad \underline{6x-14y=-52}$

$y\ =8$

$(1)\quad -2x+5(8)=20$

$-2x=-20$

$x=10 \qquad (10,8)$

check :

$-2x+5y=20 \qquad 3x=7x-26$

$-2(10)+5(8)\overset{?}{=}20 \quad 3(10)\overset{?}{=}7(8)-26$

$20=20 \qquad 30=30$

6.

$(1)\quad 3x=4+y$

$(2)\quad \underline{9x-3y=12}$

$(1)\quad 3x-y=4$

$(2)\quad \underline{9x-3y=12}$

$(1)\quad -9x+3y=-12$

$(2)\quad \underline{9x-3y=12}$

$0=0$

Infinitely many solutions

7.

rate	time	distance
$r+c$	2	80
$r-c$	2	40

$r+c=40$

$\underline{r-c=20}$

$2r=60$

$r=30$

$30+c=40$

$c=10$

The whale's speed in calm water is 30mph and the speed of the current is 10 mph.

8.

$(1)\quad a+c=175$

$(2)\quad \underline{10a+6c=1450}$

$(1)\quad -10a-10c=-1750$

$(2)\quad \underline{10a+6c=1450}$

$-4c=-300$

$c=75$

$(1)\quad a+75=175$

$a=100 \qquad (100,75)$

100 adults and 75 children attended the game.

Exercises 4.3

1.

$(1)\quad x+y=3$

$(2)\quad \underline{x-y=7}$

$2x\ =10$

$x=5$

$(2)\quad 5-y=7$

$-y=2$

$y=-2 \qquad (5,-2)$

check :

$x+y=3 \quad -x+3y=7$

$5+(-2)\overset{?}{=}3 \quad 5-(-2)\overset{?}{=}7$

$3=3 \qquad 7=7$

3.

$$(1)\quad x+y=-4$$
$$(2)\quad \underline{-x+3y=-6}$$
$$4y=-10$$
$$y=-\frac{5}{2}$$
$$(1)\quad x+\left(-\frac{5}{2}\right)=-4$$
$$x=-\frac{3}{2} \qquad \left(-\frac{3}{2},-\frac{5}{2}\right)$$

check :

$$x+y=-4 \qquad -x+3y=-6$$
$$-\frac{3}{2}+\left(-\frac{5}{2}\right)\overset{?}{=}-4 \quad -\left(-\frac{3}{2}\right)+3\left(-\frac{5}{2}\right)\overset{?}{=}-6$$
$$-4=-4 \qquad -6=-6$$

5.

$$(1)\quad 10p-q=-14$$
$$(2)\quad \underline{-4p+q=-4}$$
$$6p \quad =-18$$
$$p=-3$$
$$(2)\quad -4(-3)+q=-4$$
$$q=-16 \qquad p=-3, q=-16$$

check :

$$10p-q=-14 \qquad -4p+q=-4$$
$$10(-3)-(-16)\overset{?}{=}-14 \quad -4(-3)+(-16)\overset{?}{=}-4$$
$$-14=-14 \qquad -4=-4$$

7.

$$(1)\quad 3x+y=-3$$
$$(2)\quad \underline{4x+y=-4}$$
$$(1)\quad -3x-y=3$$
$$(2)\quad \underline{4x+y=-4}$$
$$x=-1$$
$$(1)\quad 3(-1)+y=-3$$
$$y=0 \qquad x=-1, y=0$$

check :

$$3x+y=-3 \quad 4x+0=-4$$
$$3(-1)+0\overset{?}{=}-3 \quad 4(-1)+0\overset{?}{=}-4$$
$$-3=-3 \qquad -4=-4$$

9.

$$(1)\quad 3x+5y=10 \qquad\qquad 3x+5y=10$$
$$(2)\quad \underline{3x+5y=-5} \quad \xrightarrow{\text{multiply by } -1} \quad \underline{-3x-5y=5}$$
$$0\neq 15$$

No solution

11.

$$(1)\quad 9x+6y=-15 \qquad\qquad 9x+6y=-15$$
$$(2)\quad \underline{-3x-2y=5} \quad \xrightarrow{\text{multiply by } 3} \quad \underline{-9x-6y=15}$$
$$0=0$$

Infinitely many solutions

13.

$$5x+2y=-9$$
$$\underline{-5x+2y=11}$$
$$4y=2$$
$$y=\frac{1}{2}$$
$$5x+2\left(\frac{1}{2}\right)=-9$$
$$5x+1=-9$$
$$5x=-10$$
$$x=-2$$

check:

$$5x+2y=-9 \qquad -5x+2y=11$$
$$5(-2)+2\left(\frac{1}{2}\right)\stackrel{?}{=}-9 \quad -5(-2)+2\left(\frac{1}{2}\right)\stackrel{?}{=}11$$
$$-10+1\stackrel{?}{=}-9 \qquad 10+1\stackrel{?}{=}11$$
$$-9=-9 \qquad 11=11$$

15.

$$(1)\quad 2s+d=-2 \xrightarrow{\text{multiply by }-3} -6s-3d=6$$
$$(2)\quad \underline{5s+3d=-6} \qquad \underline{5s+3d=-6}$$
$$-s=0$$
$$s=0$$

$$(1)\quad 2(0)+d=-2$$
$$0+d=-2$$
$$d=-2 \qquad s=0\ d=-2$$

check:

$$2s+d=-2 \qquad 5s+3d=-6$$
$$2(0)+(-2)\stackrel{?}{=}-2 \quad 5(0)+3(-2)\stackrel{?}{=}-6$$
$$-2=-2 \qquad -6=-6$$

17.

$$(1)\quad 3x-5y=1 \xrightarrow{\text{multiply by }8} 24x-40y=8$$
$$(2)\quad \underline{7x-8y=17} \xrightarrow{\text{multiply by }-5} \underline{-35x+40y=-85}$$
$$-11x=-77$$
$$x=7$$

$$(1)\quad 3(7)-5y=1$$
$$21-5y=1$$
$$-5y=-20$$
$$y=4 \qquad (7,4)$$

check:

$$3x-5y=1 \qquad 7x-8y=17$$
$$3(7)-5(4)\stackrel{?}{=}1 \quad 7(7)-8(4)\stackrel{?}{=}17$$
$$1=1 \qquad 17=17$$

19.

$$(1)\quad 5x+2y=-1 \xrightarrow{\text{multiply by }-4} -20x-8y=4$$
$$(2)\quad \underline{4x-5y=-14} \xrightarrow{\text{multiply by }5} \underline{20x-25y=-70}$$
$$-33y=-66$$
$$y=2$$

$$(1)\quad 5x+2(2)=-1$$
$$5x=-5$$
$$x=-1 \qquad (-1,2)$$

check:

$$5x+2y=-1 \qquad 4x-5y=-14$$
$$5(-1)+2(2)\stackrel{?}{=}-1 \quad 4(-1)-5(2)\stackrel{?}{=}-14$$
$$-1=-1 \qquad -14=-14$$

21.

$$\begin{aligned}
&(1)\quad 7p+3q=15 \xrightarrow{\text{multiply by } 5} 35p+15q=75\\
&(2)\quad \underline{-5p-7q=16} \xrightarrow{\text{multiply by } 7} \underline{-35p-49q=112}\\
&\qquad\qquad -34q=187\\
&\qquad\qquad q=-\frac{11}{2}
\end{aligned}$$

$$\begin{aligned}
&(1)\quad 7p+3\left(-\frac{11}{2}\right)=15\\
&\qquad 7p=\frac{63}{2}\\
&\qquad p=\frac{9}{2} \qquad \left(\frac{9}{2},-\frac{11}{2}\right)
\end{aligned}$$

check :

$$7x+3y=15 \qquad -5x-7y=16$$

$$7\left(\frac{9}{2}\right)+3\left(-\frac{11}{2}\right)\stackrel{?}{=}15 \qquad -5\left(\frac{9}{2}\right)-7\left(-\frac{11}{2}\right)\stackrel{?}{=}16$$

$$15=15 \qquad 16=16$$

23.

$$\begin{aligned}
&(1)\quad 6x+5y=-8.5 \xrightarrow{\text{multiply by } -2} -12x-10y=17\\
&(2)\quad \underline{8x+10y=-3} \qquad \underline{8x+10y=-3}\\
&\qquad\qquad -4x\quad =14\\
&\qquad\qquad x=-3.5
\end{aligned}$$

$$\begin{aligned}
&(1)\quad 6(-3.5)+5y=-8.5\\
&\qquad -21+5y=-8.5\\
&\qquad 5y=12.5\\
&\qquad y=2.5 \qquad (-3.5,2.5)
\end{aligned}$$

check :

$$6x+5y=-8.5 \qquad 8x+10y=-3$$

$$6(-3.5)+5(2.5)\stackrel{?}{=}-8.5 \quad 8(-3.5)+10(2.5)\stackrel{?}{=}-3$$

$$-8.5=-8.5 \qquad -3=-3$$

25.

$$\begin{aligned}
&(1)\quad 3.5x+5y=-3\\
&(2)\quad \underline{2x=-2y}\\
&(1)\quad 3.5x+5y=-3 \xrightarrow{\text{multiply by } 2} 7x+10y=-6\\
&(2)\quad \underline{2x+2y=0} \xrightarrow{\text{multiply by } -5} \underline{-10x-10y=0}\\
&\qquad\qquad -3x\quad =-6\\
&\qquad\qquad x=2
\end{aligned}$$

$$\begin{aligned}
&(1)\quad 3.5(2)+5y=-3\\
&\qquad 5y=-10\\
&\qquad y=-2 \qquad (2,-2)
\end{aligned}$$

check :

$$3.5x+5y=-3 \qquad 2x=-2y$$

$$3.5(2)+5(-2)\stackrel{?}{=}-3 \quad 2(2)\stackrel{?}{=}-2(-2)$$

$$-3=-3 \qquad 4=4$$

27.

$$\begin{aligned}
&(1)\quad 2x-4=-y\\
&(2)\quad \underline{x+2y=0}\\
&(1)\quad 2x+y=4 \xrightarrow{\text{multiply by } -2} -4x-2y=-8\\
&(2)\quad \underline{x+2y=0} \qquad \underline{x+2y=0}\\
&\qquad\qquad -3x\quad =-8\\
&\qquad\qquad x=\frac{-8}{-3}=\frac{8}{3}
\end{aligned}$$

$$\begin{aligned}
&(1)\quad 2\left(\frac{8}{3}\right)-4=-y\\
&\qquad y=-\frac{4}{3} \qquad \left(\frac{8}{3},-\frac{4}{3}\right)
\end{aligned}$$

check :

$$2x-4=-y \qquad x+2y=0$$

$$2\left(\frac{8}{3}\right)-4\stackrel{?}{=}-\left(-\frac{4}{3}\right) \quad \left(\frac{8}{3}\right)+2\left(-\frac{4}{3}\right)\stackrel{?}{=}0$$

$$\frac{4}{3}=\frac{4}{3} \qquad 0=0$$

29.

$(1)\ 8x+10y=1$

$(2)\ \underline{-4x-5y+6=0}$

$(1)\ 8x+10y=1 \qquad\qquad 8x+10y=1$

$(2)\ \underline{-4x-5y=-6} \xrightarrow{\text{multiply by 2}} \underline{-8x-10y=-12}$

$0 \neq -11$

No solution

31.

	r	t	d
with wind	s + w	2.5	40
against wind	s – w	2	20

$(1)\quad s+w=16$

$(2)\quad \underline{s-w=10}$

$2s=26$

$s=13$

$(1)\quad 13+w=16$

$w=3$

The speed of the pass if there were no wind would be 13 yards per second.

33.

f = number of full-price admissions
h = number of half-price admissions

$(1)\quad f+h=223 \xrightarrow{\text{multiply by }-5} -5f-5h=-1115$

$(2)\quad \underline{5f+2.5h=765} \qquad \underline{5f+2.5h=765}$

$-2.5h=-350$

$h=140$

$(1)\quad f+140=223$

$f=83$

The zoo collected 83 full-price admissions and 140 half-price admissions.

35.

$s+c=227{,}000$

$\underline{s-c=23{,}200}$

$2s=250{,}200$

$s=125{,}100$

$125{,}100+c=227{,}000$

$c=101{,}900$

The salary of a senator is $125,100 and the salary of a congressman is $101,900.

37.

$(1)\quad 5s+7p=43 \xrightarrow{\text{multiply by 2}} 10s+14p=86$

$(2)\quad \underline{2s+9p=42} \xrightarrow{\text{multiply by }-5} \underline{-10s-45p=-210}$

$-31p=-124$

$p=4$

$(1)\quad 5s+7(4)=43$

$5s=15$

$s=3$

It takes the computer 3 nanoseconds to carry out one sum and 4 nanoseconds to carry out one product.

39.

$(1)\ 3f+5h=6075 \xrightarrow{\text{multiply by 4}} 12f+20h=24{,}300$

$(2)\ \underline{4f+4h=6380} \xrightarrow{\text{multiply by }-5} \underline{-20f-20h=-31{,}900}$

$-8f=-7600$

$f=950$

$(1)\quad 3(950)+5h=6075$

$5h=3225$

$h=645$

The rate for full-page ads is $950 and for half-page ads is $645.

Chapter 4 Review Exercises

1. (2, -3) is not a solution

$x+2y=-4 \qquad 3x-y=3$

$2+2(-3)\stackrel{?}{=}-4 \quad 3(2)-(-3)\stackrel{?}{=}3$

$-4=-4 \qquad 9\neq 3$

2.

a. No solution

b. One solution

c. Infinitely many solutions

3.

a. III

b. IV

c. II

d. I

4.

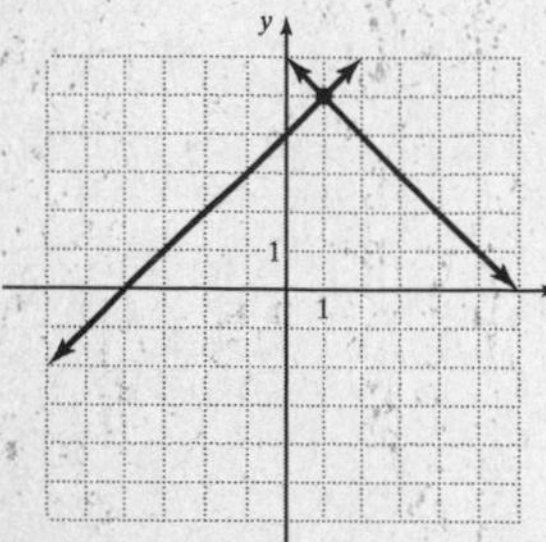

The solution is (1,5).

5.

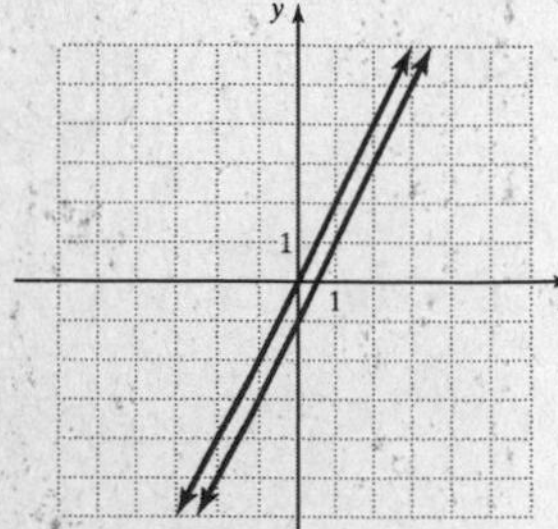

No solution

6.

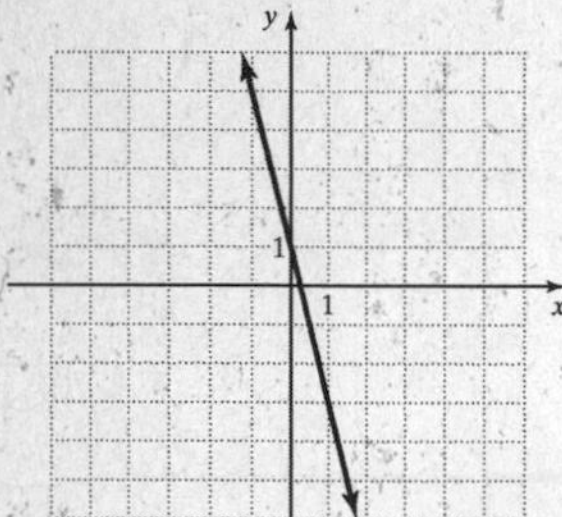

Infinitely many solutions

7.

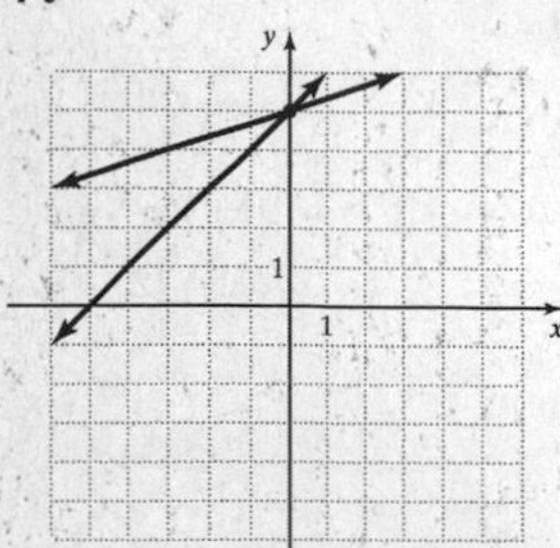

The solution is (0,5).

8.

(1) $x + y = 3$

(2) $\underline{y = 2x + 6}$

(1) $x + 2x + 6 = 3$

$3x + 6 = 3$

$3x = -3$

$x = -1$

(2) $y = 2(-1) + 6$

$y = 4 \qquad (-1, 4)$

check :

$x + y = 3 \qquad \underline{y = 2x + 6}$

$-1 + 4 \stackrel{?}{=} 3 \qquad 4 \stackrel{?}{=} 2(-1) + 6$

$3 = 3 \qquad 4 = 4$

9.

(1) $a = 3b - 4$

(2) $\underline{a + 4b = 10}$

(2) $3b - 4 + 4b = 10$

$7b - 4 = 10$

$7b = 14$

$b = 2$

(1) $a = 3(2) - 4$

$a = 2 \qquad (2, 2)$

check :

$a = 3b - 4 \qquad \underline{a + 4b = 10}$

$2 \stackrel{?}{=} 3(2) - 4 \qquad 2 + 4(2) \stackrel{?}{=} 10$

$2 = 2 \qquad 10 = 10$

10.

(1) $x - 3y = 1$

(2) $\underline{-2x + 6y = 7}$

(1) $x = 3y + 1$

(2) $\underline{-2x + 6y = 7}$

(2) $-2(3y + 1) + 6y = 7$

$-6y - 2 + 6y = 7$

$-2 \neq 7$

No solution

11.

$$(1)\quad 10x+2y=14$$
$$(2)\quad \underline{-y=5x-7}$$
$$(1)\quad 10x+2y=14$$
$$(2)\quad \underline{y=-5x+7}$$
$$(1)\quad 10x+2(-5x+7)=14$$
$$10x-10x+14=14$$
$$14=14$$

Infinitely many solutions

12.

$$(1)\quad x+y=1$$
$$(2)\quad \underline{x-y=7}$$
$$2x \quad =8$$
$$x=4$$
$$(2)\quad 4-y=7$$
$$-y=3$$
$$y=-3 \qquad (4,-3)$$

check :

$$x+y=1 \qquad x-y=7$$
$$4+(-3)\overset{?}{=}1 \qquad 4-(-3)\overset{?}{=}7$$
$$1=1 \qquad 7=7$$

13.

$$(1)\quad 2x+3y=8 \xrightarrow{\text{multiply by } -2} -4x-6y=-16$$
$$(2)\quad \underline{4x+6y=16} \qquad \underline{4x+6y=16}$$
$$0=0$$

Infinitely many solutions

14.

$$(1)\ 4x=9-5y$$
$$(2)\ \underline{2x+3y=3}$$
$$(1)\ 4x+5y=9 \qquad 4x+5y=9$$
$$(2)\ \underline{2x+3y=3} \xrightarrow{\text{multiply by } -2} \underline{-4x-6y=-6}$$
$$-y=3$$
$$y=-3$$
$$(1)\quad 4x=9-5(-3)$$
$$4x=24$$
$$x=6 \qquad (6,-3)$$

check :

$$4x=9-5y \qquad 2x+3y=3$$
$$4(6)\overset{?}{=}9-5(-3) \qquad 2(6)+3(-3)\overset{?}{=}3$$
$$24=24 \qquad 3=3$$

15.

$$(1)\quad 3x+2y=-4 \xrightarrow{\text{multiply by } -4} -12x-8y=16$$
$$(2)\ \underline{4x-3y=23} \xrightarrow{\text{multiply by } 3} \underline{12x-9y=69}$$
$$-17y=85$$
$$y=-5$$
$$(1)\quad 3x+2(-5)=-4$$
$$3x=6$$
$$x=2 \qquad (2,-5)$$

check :

$$3x=2y=-4 \qquad 4x-3y=23$$
$$3(2)+2(-5)\overset{?}{=}-4 \qquad 4(2)-3(-5)\overset{?}{=}23$$
$$-4=-4 \qquad 23=23$$

Mixed Applications

16.

a.

$$y=0.50x+1750$$
$$y=5.50x$$

b.

$y = 0.50x + 1750$

$y = 5.50x$

$0.50x + 1750 = 5.50x$

$1750 = 5x$

$350 = x$

The student must type 350 pages in order to break even.

17.

a.

$s = 10h$

$s = 8h + 50$

b.

$s = 10h$

$s = 8h + 50$

$10h = 8h + 50$

$2h = 50$

$h = 25$

The applicant would have to work 25 hours per week in order to earn the same amount of money at each position.

18.

$l - w = 26$

$\underline{2l + 2w = 332}$

$2l - 2w = 52$

$\underline{2l + 2w = 332}$

$4l = 384$

$l = 96$

$96 - w = 26$

$-w = -70$

$w = 70$

$A = lw = (96)(70) = 6720$

The area of the screen is 6720 square feet.

19.

$l - w = 51$

$\underline{2l + 2w = 228}$

$2l - 2w = 102$

$\underline{2l + 2w = 228}$

$4l = 330$

$l = 82.5$

$82.5 - w = 51$

$-w = -31.5$

$w = 31.5$

The tennis court is 31.5 feet wide and 82.5 feet long.

20.

(1) $n + d = 350$

(2) $0.05n + 0.10d = 25$

$d = 350 - n$

(2) $0.05n + 0.10(350 - n) = 25$

$0.05n + 35 - 0.1n = 25$

$35 - 0.05x = 25$

$-0.05n = -10$

$n = 200$

(1) $200 + d = 350$

$d = 150$

The box contained 200 nickels and 150 dimes.

21.

$x - y = 5$

$\underline{4x + 4y = 500}$

$x = 5 + y$

$4(5 + y) + 4y = 500$

$20 + 4y + 4y = 500$

$20 + 8y = 500$

$8y = 480$

$y = 60$

$x - 60 = 5$

$x = 65$

One train travels at a rate of 60 miles per hour and the other travels at a rate of 65 miles per hour.

22.

$x+y=1085$

$\underline{2x+3y=2437}$

$x=1085-y$

$2(1085-y)+3y=2437$

$2170-2y+3y=2437$

$2170+y=2437$

$y=267$

$x+267=1085$

$x=818$

The team made 818 two-point baskets and 267 three-point baskets.

23.

$x+y=200$

$\underline{0.10x+0.30y=.25(200)}$

$x=200-y$

$\underline{0.10x+0.30y=50}$

$0.10(200-y)+0.30y=50$

$20-0.10y+0.30y=50$

$20+0.20y=50$

$0.20y=30$

$y=150$

$x=200-150$

$x=50$

The pharmacist should mix 150ml of the 30% solution and 50ml of the 10% solution.

24.

$I+w=2000$

$\underline{0.50I+1w=.65(2000)}$

$w=2000-I$

$\underline{0.50I+1w=1300}$

$0.50I+2000-I=1300$

$-0.50I+2000=1300$

$-0.50I=-700$

$I=1400$

$w=2000-1400=600$

1400 L of the 50% solution and 600 L of water are needed to fill the tank.

25.

$x+y=50,000$

$\underline{0.06x+0.08y=3200}$

$x=50,000-y$

$\underline{0.06x+0.08y=3200}$

$0.06(50,000-y)+0.08y=3200$

$3000-0.06y+0.08y=3200$

$3000+0.02y=3200$

$0.02y=200$

$y=10,000$

$x=50,000-10,000$

$x=40,000$

The client puts \$40,000 in municipal bonds and \$10,000 in corporate stocks.

26.

$x+y=10,000$

$\underline{0.12x+0.02y=900}$

$x=10,000-y$

$\underline{0.12x+0.02y=900}$

$0.12(10,000-y)+0.02y=900$

$1200-0.12y+0.02y=900$

$1200-0.1y=900$

$-0.1y=-300$

$y=3,000$

$x=10,000-3,000$

$x=7,000$

\$7000 was invested in the high-risk fund and \$3000 was invested in the low-risk fund.

27.

$x - y = 100$

$\underline{2x + 2y = 1800}$

$x = 100 + y$

$\underline{2x + 2y = 1800}$

$2(100 + y) + 2y = 1800$

$200 + 2y + 2y = 1800$

$200 + 4y = 1800$

$4y = 1600$

$y = 400$

The speed of the slower plane was 400mph.

28.

$r + w = \frac{13}{0.5}$

$\underline{r - w = \frac{8}{0.5}}$

$r + w = 26$

$\underline{r - w = 16}$

$2r = 42$

$r = 21$

$21 + w = 26$

$w = 5$

The bird flies at a rate of 21mph in calm air and the speed of the wind is 5 mph.

29.

a.

$f + a = 100$

$\underline{f - a = 14}$

$2f = 114$

$f = 57$

57 senators voted for the treaty

b.

$57 + a = 100$

$a = 43$

43 senators voted against the treaty

30.

a.

$y = 3x + 2$ (copper)

$y = -4x + 86$ (iron)

$3x + 2 = -4x + 86$

$7x = 84$

$x = 12$

The two metals will be the same temperature after 12 minutes.

b.

$3x + 2 = -4x + 86 + 14$

$3x + 2 = -4x + 100$

$7x = 98$

$x = 14$

The iron will be 14° colder than the copper after 14 minutes.

Chapter 4 Posttest

1.

a. (0, -1) A solution

$x + y = -1 \qquad 3x - y = 1$

$0 + (-1) \stackrel{?}{=} -1 \qquad 3(0) - (-1) \stackrel{?}{=} 1$

$-1 = -1 \qquad 1 = 1$

b. (-1, 0) Not a solution

$x + y = -1 \qquad 3x - y = 1$

$-1 + 0 \stackrel{?}{=} -1 \qquad 3(-1) - 0 \stackrel{?}{=} 1$

$-1 = -1 \qquad -3 \neq 1$

c. (2, -3) Not a solution

$x + y = -1 \qquad 3x - y = 1$

$2 + (-3) \stackrel{?}{=} -1 \qquad 3(2) - (-3) \stackrel{?}{=} 1$

$-1 = -1 \qquad 9 \neq 1$

2. The system has no solution.

3. The solution is (3, 0).

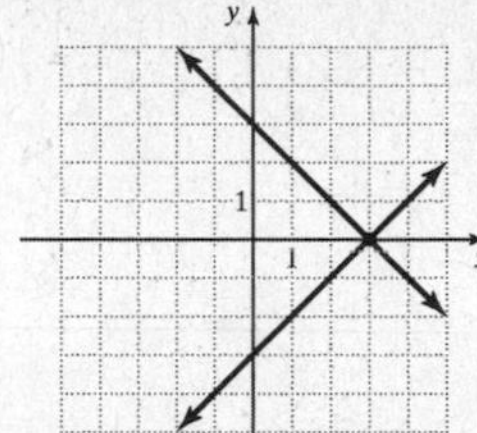

4. The system has no solution.

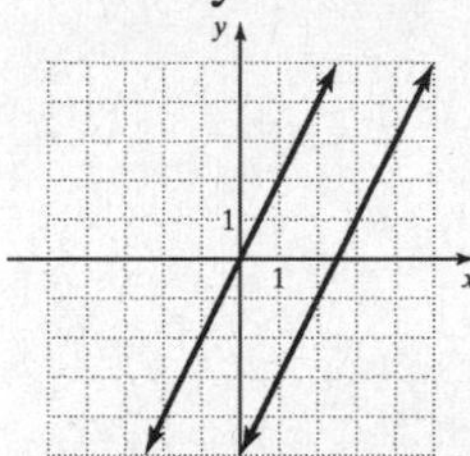

5. The solution is (-1, 2).

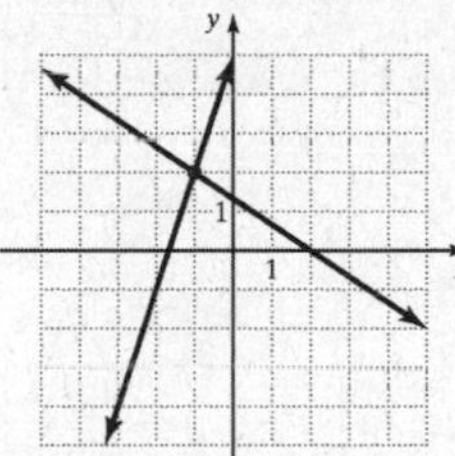

6.

$$x = 3y - 7$$
$$\underline{y = x + 5}$$
$$x = 3(x+5) - 7$$
$$x = 3x + 15 - 7$$
$$x = 3x + 8$$
$$-2x = 8$$
$$x = -4$$
$$y = -4 + 5$$
$$y = 1 \qquad (-4,1)$$

check :

$$-4 \stackrel{?}{=} 3(1) - 7 \qquad 1 \stackrel{?}{=} -4 + 5$$
$$-4 = -4 \qquad 1 = 1$$

7.

$$3x - 5y = -12$$
$$\underline{x + 2y = 7}$$
$$x = -2y + 7$$
$$3(-2y+7) - 5y = -12$$
$$-6y + 21 - 5y = -12$$
$$-11y = -33$$
$$y = 3$$
$$x = -2(3) + 7 = 1 \qquad (1,3)$$

check :

$$3(1) - 5(3) \stackrel{?}{=} -12 \qquad 1 + 2(3) \stackrel{?}{=} 7$$
$$3 - 15 \stackrel{?}{=} -12 \qquad 1 + 6 \stackrel{?}{=} 7$$
$$-12 = -12 \qquad 7 = 7$$

8.

$$u - 3v = -12$$
$$\underline{5u + v = 8}$$
$$u = 3v - 12$$
$$\underline{5u + v = 8}$$
$$5(3v - 12) + v = 8$$
$$15v - 60 + v = 8$$
$$16v = 68$$
$$v = \frac{17}{4}$$
$$u = 3\left(\frac{17}{4}\right) - 12$$
$$u = \frac{3}{4} \qquad \left(\frac{3}{4}, \frac{17}{4}\right)$$

check :

$$\frac{3}{4} - 3\left(\frac{17}{4}\right) \stackrel{?}{=} -12 \qquad 5\left(\frac{3}{4}\right) + \frac{17}{4} \stackrel{?}{=} 8$$
$$-12 = -12 \qquad 8 = 8$$

9.

$$4x + y = 3$$
$$\underline{7x - y = 19}$$
$$11x = 22$$
$$x = 2$$
$$4(2) + y = 3$$
$$8 + y = 3$$
$$y = -5 \qquad (2,-5)$$

check :

$$4x + y = 3 \qquad 7x - y = 19$$
$$4(2)+(-5)\stackrel{?}{=}3 \quad 7(2)-(-5)\stackrel{?}{=}19$$
$$8+(-5)\stackrel{?}{=}3 \qquad 14+5\stackrel{?}{=}19$$
$$3=3 \qquad 19=19$$

10.

$$x - y = 5$$
$$\underline{2x - 2y = 5}$$
$$-2x + 2y = -10$$
$$\underline{2x - 2y = 5}$$
$$0 \neq -5$$

No solution

11.

$$-5p + 2q = 1$$
$$\underline{4p + 3q = 1.5}$$
$$-20p + 8q = 4$$
$$\underline{20p + 15q = 7.5}$$
$$23q = 11.5$$
$$q = 0.5$$
$$-5p + 2(0.5) = 1$$
$$-5p + 1 = 1$$
$$-5p = 0$$
$$p = 0 \qquad (0,0.5)$$

check :

$$-5p + 2q = 1 \qquad 4p + 3q = 1.5$$
$$-5(0)+2(0.5)\stackrel{?}{=}1 \quad 4(0)+3(0.5)\stackrel{?}{=}1.5$$
$$0+1\stackrel{?}{=}1 \qquad 0+1.5\stackrel{?}{=}1.5$$
$$1=1 \qquad 1.5=1.5$$

12.

$$5x = 3y$$
$$\underline{-3x + 2y = 9}$$
$$5x - 3y = 0$$
$$\underline{-3x + 2y = 9}$$
$$15x - 9y = 0$$
$$\underline{-15x + 10y = 45}$$
$$y = 45$$
$$5x = 3(45)$$
$$5x = 135$$
$$x = 27 \qquad (27,45)$$

check :

$$5(27)\stackrel{?}{=}3(45) \quad -3(27)+2(45)\stackrel{?}{=}9$$
$$135 = 135 \qquad 9 = 9$$

13.

$$4l = -(m+3)$$
$$\underline{8l + 2m = -6}$$
$$4l + m = -3$$
$$\underline{8l + 2m = -6}$$
$$-8l - 2m = 6$$
$$\underline{8l + 2m = -6}$$
$$0 = 0$$

Infinitely many solutions

14.

$5x+2y=-1$

$\underline{x-1=y}$

$5x+2(x-1)=-1$

$5x+2x-2=-1$

$7x-2=-1$

$7x=1$

$x=\frac{1}{7}$

$\frac{1}{7}-1=y$

$y=-\frac{6}{7} \qquad \left(\frac{1}{7},-\frac{6}{7}\right)$

check :

$5\left(\frac{1}{7}\right)+2\left(-\frac{6}{7}\right)\stackrel{?}{=}-1 \qquad \frac{1}{7}-1\stackrel{?}{=}-\frac{6}{7}$

$-1=-1 \qquad -\frac{6}{7}=-\frac{6}{7}$

15.

$5x+3y-9=0$

$\underline{2x-7y-20=0}$

$5x+3y=9$

$\underline{2x-7y=20}$

$10x+6y=18$

$\underline{-10x+35y=-100}$

$41y=-82$

$y=-2$

$5x+3(-2)-9=0$

$5x-15=0$

$5x=15$

$x=3 \qquad (3,-2)$

check :

$5x+3y-9=0 \qquad 2x-7y-20=0$

$5(3)+3(-2)-9\stackrel{?}{=}0 \quad 2(3)-7(-2)-20\stackrel{?}{=}0$

$15-6-9\stackrel{?}{=}0 \qquad 6+14-20\stackrel{?}{=}0$

$0=0 \qquad 0=0$

16.

$w+l=6306$

$\underline{2l=w}$

$2l+l=6306$

$3l=6306$

$l=2102$

$w=2(2102)$

$w=4204$

The winning candidate got 4204 votes.

17.

(1) $135t+99s=333$

(2) $\underline{3t+3s=9}$

$3t+3s-9$

$t+s=3$

$s=3-t$

(1) $135t+99(3-t)=333$

$135t+297-99t=333$

$36t=36$

$t-1$

$s=3\text{-}t$

$s=2$

One serving of turkey and two servings of salmon would be needed to get 9g of fat and 333 calories.

18.

$x=2y$

$\underline{0.075x+0.06y=840}$

$0.075(2y)+0.06y=840$

$0.15y+0.06y=840$

$0.21y=840$

$y=4000$

$x=2(4000)=8000$

\$8000 was invested at 7.5% and \$4000 was invested at 6%.

19.

$$x+y=4$$
$$\underline{0.20x+0.60y=0.50(4)}$$
$$x+y=4$$
$$\underline{0.20x+0.60y=2}$$
$$x=4-y$$
$$0.20(4-y)+0.60y=2$$
$$0.8-0.20y+0.60y=2$$
$$0.40y=1.2 \quad y=3$$
$$x+3=4 \qquad x=1$$

One gallon of the 20% iodine solution and 3 gallons of the 60% iodine solution are mixed to produce 4 gallons of the 50% iodine solution.

20.

$$r+w=170$$
$$\underline{r-w=130}$$
$$2r=300$$
$$r=150$$
$$150+w=170$$
$$w=20$$

The speed of the wind was 20 mph. The rate of the plane was 150 mph.

Cumulative Review Exercises

1.

$$-4\div 2+3(-1)(8)=$$
$$-2+3(-1)(8)=$$
$$-2+(-24)=-26$$

2. True

3. No, 5 is not a solution.

$$3p+1=9-p$$
$$3(5+1)\stackrel{?}{=}9-5$$
$$18\neq 4$$

4.

$$5(x+1)-(x-2)=x-2$$
$$5x+5-x+2=x-2$$
$$4x+7=x-2$$
$$3x=-9$$
$$x=-3$$

check :

$$5(-3+1)-(-3-2)\stackrel{?}{=}-3-2$$
$$-10+5\stackrel{?}{=}-5$$
$$-5=-5$$

5.

$$3x-7<4(x-2)$$
$$3x-7<4x-8$$
$$-x<-1$$
$$x>1$$

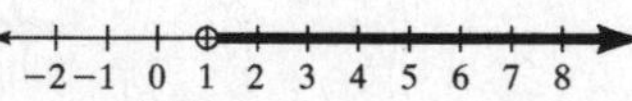

6. $m=\dfrac{2-(-4)}{5-(-3)}=\dfrac{6}{8}=\dfrac{3}{4}$

7.

$$3x+6y=12$$
$$6y=-3x+12$$
$$y=-\frac{1}{2}x+2$$

Slope: $m=-\frac{1}{2}$; y-intercept: $(0,2)$

8.

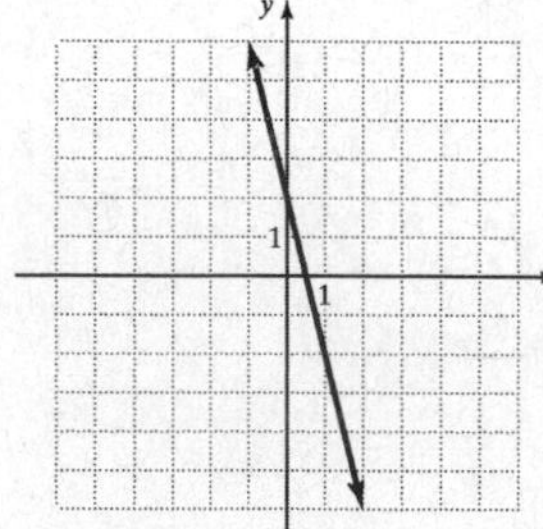

9.

$$M = \frac{S}{740}$$

$$2.1 = \frac{S}{740}$$

$$(740)2.1 = \frac{S}{740}(740)$$

$$1554 = S$$

The plane can fly 1554 mph (S=1554).

10.

$$c = 0.20m + 35$$

$$\underline{c = 0.15m + 50}$$

$$0.20m + 35 = 0.15m + 50$$

$$0.05m + 35 = 50$$

$$0.05m = 15$$

$$m = 300$$

$$c = 0.20(300) + 35$$

$$c = 60 + 35 = 95$$

The two companies charge the same amount $(\$95)$ for a one-day rental if the car is driven 300 miles.

Chapter 5 Exponents and Polynomials

Chapter 5 Pretest

1. $x^5 \bullet x^4 = x^{5+4} = x^9$

2. $y^7 \div y^3 = y^{7-3} = y^4$

3. $-3a^0 = -3(1) = -3$

4. $(4x^4y^3)^2 = 4^2 \bullet x^{4\bullet 2}y^{3\bullet 2} = 16x^8y^6$

5. $(\frac{a}{b^5})^3 = \frac{a^3}{b^{5\bullet 3}} = \frac{a^3}{b^{15}}$

6.

$(5x^{-1}y^4)^{-2} = 5^{-2}x^2y^{-8} =$

$\frac{x^2}{5^2y^8} = \frac{x^2}{25y^8}$

7.

a. $6x^4, 5x^3, x^2, -7x,$ and 8

b. 6, 5, 1, -7, and 8

c. 4

d. 8

8.

$(2n^2 + 7n - 10) + (n^2 - 6n + 12) =$

$3n^2 + n + 2$

9.

$(8x^2 - 9) - (7x^2 - x - 5) = -(7x^2 - x - 5)$

10.

$(6a^2b + ab - a^2) + (2a^2 + 3a^2b - 5b^2) =$

$-6a^2 + 9a^2b + 3ab - 4b^2$

11.

$3x^2(x^2 - 4x + 9) =$

$3x^4 - 12x^3 + 27x^2$

12.

$(n+3)(2n^2 + n - 6) =$

$n(2n^2 + n - 6) + 3(2n^2 + n - 6) =$

$2n^3 + n^2 - 6n + 6n^2 + 3n - 18 =$

$2n^3 + 7n^2 - 3n - 18$

13.

$(4x+9)(x-3) =$

$4x(x-3) + 9(x-3) =$

$4x^2 - 12x + 9x - 27 =$

$4x^2 - 3x - 27$

14.

$(3y-7)(3y+7) =$

$3y(3y+7) - 7(3y+7) =$

$9y^2 + 21y - 21y - 49 =$

$9y^2 - 49$

15.

$(5-2n)^2 = (5-2n)(5-2n) =$

$5(5-2n) - 2n(5-2n) =$

$25 - 10n - 10n + 4n^2 =$

$25 - 20n + 4n^2$

16.

$\frac{9t^4 - 18t^3 - 45t^2}{9t^2} = \frac{9t^2(t^2 - 2t - 5)}{9t^2} =$

$t^2 - 2t - 5$

17.

$(4x^2 - 3x - 10) \div (x - 2) =$

$\frac{4x^2 - 3x - 10}{x-2} = \frac{\overset{1}{\cancel{(x-2)}}(4x+5)}{\underset{1}{\cancel{x-2}}} =$

$4x + 5$

18.

$200 \bullet (6 \times 10^{23}) =$

$(2 \times 10^2) \bullet (6 \times 10^{23}) =$

$(12 \times 10^{25}) = 1.2 \times 10^{26}$

200 moles of hydrogen will contain 1.2×10^{26} molecules.

19.

$t = 2$

$-0.535(2)^2 + 2.64(2) + 45.3 = 48.44$

The average monthly cellular telephone bill was $48.44 in 2002.

20.

$A = (2x-5)(2x-5+10)$

$A = (2x-5)(2x+5)$

$A = 4x^2 - 25$

5.1 Laws of Exponents

Practice 5.1

1.
a. $10^4 = (10)(10)(10)(10) = 10{,}000$
b. $(-\frac{1}{2})^5 = -\frac{1}{32}$
c. $(-y)^6 = (-y)(-y)(-y)(-y)(-y)(-y) = y^6$
d. $(-y)^3 = (-y)(-y)(-y) = -y^3$
2.
a. $8^0 = 1$
b. $(-\frac{2}{3})^0 = 1$
c. $y^0 = 1$
d. $-y^0 = -1(1) = -1$
3.
a. $10^8 \cdot 10^4 = 10^{8+4} = 10^{12}$
b. $(-4)^3 \cdot (-4)^3 = (-4)^{3+3} = (-4)^6$
c. $n^3 \cdot n^7 = n^{3+7} = n^{10}$
d. $y^5 \cdot y^0 = y^{5+0} = y^5$
e. Cannot apply the product rule because the bases are not the same
4.
a. $\frac{7^7}{7^2} = 7^{7-2} = 7^5$
b. $(-9)^6 \div (-9)^5 = (-9)^{6-5} = (-9)^1 \text{ or } -9$
c. $\frac{s^{10}}{s^{10}} = s^{10-10} = s^0 = 1$
d. $\frac{r^8}{r} = r^{8-1} = r^7$
e. Cannot apply the quotient rule because the bases are not the same.
5.
a. $y^2 \cdot y^3 \cdot y^4 = y^{2+3+4} = y^9$
b. $(x^3y^3)(x^2y^3) = x^3 \cdot x^2 \cdot y^3 \cdot y^3 = x^5y^6$
c. $\frac{a^7}{a \cdot a^4} = \frac{a^7}{a^5} = a^2$
6.
a. $9^{-2} = \frac{1}{9^2} = \frac{1}{81}$
b. $n^{-5} = \frac{1}{n^5}$
c. $-(3y)^{-1} = -\frac{1}{3y}$
d. $(5)^{-3} = \frac{1}{5^3} = \frac{1}{125}$
7.
a. $8^{-1}s = \frac{s}{8}$
b. $3x^{-1} = \frac{3}{x}$
c. $\frac{r^3}{r^9} = r^{-6} = \frac{1}{r^6}$
d. $3^2 \cdot g^{-1} \cdot g^{-4} = \frac{3^2}{g^{1+4}} = \frac{9}{g^5}$
e. $\frac{1}{x^{-3}} = x^3$
8.
a. $\frac{1}{a^{-3}} = a^3$
b. $\frac{2}{5x^{-2}} = \frac{2x^2}{5}$
c. $\frac{r^3}{2s^{-1}} = \frac{r^3s}{2}$
9.
a. $(2^3)(2^{20}) = 2^{23}$
b. $(2^{20})(2^{10}) = 2^{30}$

Exercises 5.1

1. $5^3 = (5)(5)(5) = 125$
3. $-(0.5)^2 = -(0.5)(0.5) = -0.25$
5. $(-2)^3 = (-2)(-2)(-2) = -8$
7. $(-\frac{1}{2})^3 = (-\frac{1}{2})(-\frac{1}{2})(-\frac{1}{2}) = -\frac{1}{8}$
9. $(-x)^4 = (-x)(-x)(-x)(-x) = x^4$
11. $(pq)^1 = pq$
13. $(-3)^0 = 1$

15. $-a^0 = (-1)(1) = -1$

17. $10^9 \cdot 10^2 = 10^{9+2} = 10^{11}$

19. $a^4 \cdot a^2 = a^{4+2} = a^6$

21. Cannot be simplified.

23. $n^6 \cdot n = a^{6+1} = a^7$

25. Cannot be simplified

27. $\frac{8^5}{8^3} = 8^{5-3} = 8^2$

29. $\frac{y^6}{y^5} = y^{6-5} = y^1 \text{ or } y$

31. $\frac{a^{10}}{a^4} = a^{10-4} = a^6$

33. Cannot be simplified.

35. $\frac{x^6}{x^6} = x^{6-6} = x^0 = 1$

37. $y^2 \cdot y^3 \cdot y = y^{2+3+1} = y^6$

39. $(p^2q^3)(p^5q^2) = p^{2+5}q^{3+2} = p^7q^5$

41. $(yx^2)(xz^2)(yz) = x^{2+1}y^{1+1}z^{2+1} = x^3y^2z^3$

43. $\frac{a^2 \cdot a^3}{a^4} = \frac{a^{2+3}}{a^4} = \frac{a^5}{a^4} = a^{5-4} = a^1$ or a

45. $\frac{x^2 \cdot x^4}{x^3 \cdot x} = \frac{x^{2+4}}{x^{3+1}} = \frac{x^6}{x^4} = x^{6-4} = x^2$

47. $5^{-1} = \frac{1}{5}$

49. $x^{-1} = \frac{1}{x}$

51. $(-3a)^{-1} = \frac{1}{-3a} \text{ or } -\frac{1}{3a}$

53. $2^{-4} = \frac{1}{2^4} = \frac{1}{16}$

55. $(-4)^{-2} = \frac{1}{(-4)^2} = \frac{1}{16}$

57. $-3^{-4} = -\frac{1}{3^4} = -\frac{1}{81}$

59. $8n^{-3} = \frac{8}{n^3}$

61. $(-x)^{-2} = \frac{1}{(-x)^2} = \frac{1}{x^2}$

63. $-x^{-2} = -\frac{1}{x^2}$

65. $-3^{-2}x = -\frac{x}{3^2} = -\frac{x}{9}$

67. $x^{-2}y^3 = \frac{y^3}{x^2}$

69. $qr^{-1} = \frac{q}{r^1} = \frac{q}{r}$

71. $4x^{-1}y^2 = \frac{4y^2}{x}$

73. $p^{-2} \cdot p^{-3} = p^{-2+(-3)} = p^{-5} = \frac{1}{p^5}$

75. $p^{-1} \cdot p^4 = p^{-1+4} = p^3$

77. $\frac{a^3}{a^4} = a^{3-4} = a^{-1} = \frac{1}{a}$

79. $\frac{2}{n^{-4}} = 2n^4$

81. $\frac{1}{2p^{-5}} = \frac{p^5}{2}$

83. $\frac{p^4}{q^{-1}} = p^4q$

85. $\frac{t^{-2}}{t^3} = t^{-2-3} = t^{-5} = \frac{1}{t^5}$

87. $\frac{x^5}{x^{-2}} = x^{5-(-2)} = x^7$

89. $\frac{a^{-4}}{a^{-5}} = a^{-4-(-5)} = a$

91. $\frac{a^{-3}}{b^{-3}} = \frac{b^3}{a^3}$

93.

a. $35 \cdot 2^5 = 1120$ people were ill on the sixth day of the epidemic; $35 \cdot 2^9 = 17{,}920$ people were ill on the tenth day.

b. $\frac{35 \cdot 2^9}{35 \cdot 2^5} = 2^{9-5} = 2^4 = 16$ The number of people ill on the tenth day was 2^4, or 16, times as great as the number ill on the sixth day .

95. $60 \times (0.95)^{11}$ ppm

97.

a. Volume of the small box = $(2x)^3 = 8x^3$

Volume of the large box = $(5x)^3 = 125x^3$

b. $\frac{125x^3}{8x^3} = 15\frac{5}{8}$ The volume of the large box is $15\frac{5}{8}$ times the volume of the small box.

5.2 More Laws of Exponents and Scientific Notation

Practice 5.2

1.

a. $(2^3)^2 = 2^{3\cdot 2} = 2^6 = 64$

b. $(7^3)^{-1} = 7^{3\cdot(-1)} = 7^{-3} = \frac{1}{7^3} = \frac{1}{343}$

c. $(q^2)^4 = q^{2\cdot 4} = q^8$

d. $-(p^3)^{-5} = -(p^{3\cdot(-5)}) =$
$-(p^{-15}) = \frac{-1}{p^{15}} = -\frac{1}{p^{15}}$

2.

a. $(7a)^2 = 7^2 \cdot a^2 = 49a^2$

b. $(-4x)^3 = (-4)^3 \cdot x^3 = -64x^3$

c. $-(4x)^3 = -(4^3) \cdot x^3 = -64x^3$

3.

a. $(-6a^9)^2 = (-6)^2(a^9)^2 = 36a^{18}$

b. $(q^8r^{10})^2 = (q^8)^2(r^{10})^2 = q^{16}r^{20}$

c. $-2(ab^7)^3 = -2a^3b^{21}$

d. $(7a^{-1}c^{-5})^2 = (7)^2(a^{-1})^2(c^{-5})^2 =$

$49a^{-2}c^{-10} = \frac{49}{a^2c^{10}}$

4.

a. $(\frac{y}{3})^2 = \frac{y^2}{3^2} = \frac{y^2}{9}$

b. $(\frac{-u}{v})^{10} = \frac{u^{10}}{v^{10}}$

c. $(\frac{3}{y})^{-2} = \frac{3^{-2}}{y^{-2}} = \frac{y^2}{3^2} = \frac{y^2}{9}$

d. $(\frac{-10a^5}{3b^2c})^2 = \frac{(-10a^5)^2}{(3b^2c)^2} = \frac{100a^{10}}{9b^4c^2}$

e. $(\frac{5x}{y^{-2}})^3 = \frac{(5x)^3}{(y^{-2})^3} = \frac{125x^3}{y^{-6}} = 125x^3y^6$

5.

a. $(\frac{5}{a})^{-2} = (\frac{a}{5})^2 = \frac{a^2}{5^2} = \frac{a^2}{25}$

b. $(\frac{4u}{v})^{-1} = (\frac{v}{4u})^1 = \frac{v}{4u}$

c. $(\frac{a^5}{b^3})^{-2} = \frac{a^{-10}}{b^{-6}} = \frac{b^6}{a^{10}}$

6. 253.9

7. 0.0000000043

8. 8×10^{12}

9. 7.1×10^{-11}

10.

a. $(7\times10^{-2})(3.52\times10^3) =$

$(7\times3.52)(10^{-2}\times10^3) =$

$24.64\times10^1 = 2.464\times10^2$

b. $(2.4\times10^3)\div(6\times10^{-9}) =$

$\frac{2.4\times10^3}{6\times10^{-9}} = \frac{2.4}{6}\times\frac{10^3}{10^{-9}} =$

$0.4\times10^{12} = 4\times10^{11}$

11. $(1.5\times10^5)(6\times10^9) = 9\times10^{14}$

12. $\frac{2\times10^{-10}}{2\times10^{-7}} = \frac{2}{2}\times\frac{10^{-10}}{10^{-7}} = 1\times10^{-3}$

13. $(5\times10^{-9})^2 = 2.5E^{-}17$

14. $7.3E^{-}10$

15. $4.6\times10^8 = 460{,}000{,}000$

Exercises 5.2

1. $(2^2)^4 = 2^8 = 256$

3. $(10^5)^2 = 10^{10} = 10{,}000{,}000{,}000$

5. $(4^{-2})^2 = 4^{-4} = \frac{1}{4^4} = \frac{1}{256}$

7. $(x^4)^6 = x^{24}$

9. $(y^4)^2 = y^8$

11. $(x^{-2})^3 = x^{-6} = \frac{1}{x^6}$

13. $(n^{-2})^{-2} = n^4$

15. $(4x)^3 = 4^3 \cdot x^3 = 64x^3$

17. $(-8y)^2 = (-8)^2 \cdot y^2 = 64y^2$

19. $-(4n^5)^3 = -(4)^3(n^5)^3 = -64n^{15}$

21.
$4(-2y^2)^4 = 4(-2)^4(y^2)^4 =$
$4(16)y^8 = 64y^8$

23. $(3a)^{-2} = \frac{1}{(3a)^2} = \frac{1}{9a^2}$

25. $(pq)^{-7} = \frac{1}{(pq)^7} = \frac{1}{p^7q^7}$

27. $(r^2t)^6 = (r^2)^6(t)^6 = r^{12}t^6$

29. $(-2p^5q)^2 = (-2)^2(p^5)^2q^2 = 4p^{10}q^2$

31. $-2(m^4n^8)^3 = -2(m^4)^3(n^8)^3 = -2m^{12}n^{24}$

33.
$(-4m^5n^{-10})^3 = (-4)^3(m^5)^3(n^{-10})^3 = \frac{-64m^{15}}{n^{30}}$

35. $(a^3b^2)^{-4} = \frac{1}{(a^3b^2)^4} = \frac{1}{a^{12}b^8}$

37. $(4x^{-2}y^3)^2 = 16x^{-4}y^6 = \frac{16y^6}{x^4}$

39. $\left(\frac{5}{b}\right)^3 = \frac{125}{b^3}$

41. $\left(\frac{c}{b}\right)^2 = \frac{c^2}{b^2}$

43. $-\left(\frac{a}{b}\right)^7 = -\frac{a^7}{b^7}$

45. $(\frac{a^2}{3})^3 = \frac{a^6}{27}$

47. $(-\frac{p^3}{q^2})^5 = -\frac{p^{15}}{q^{10}}$

49. $(\frac{a}{4})^{-1} = \frac{4}{a}$

51. $(\frac{2x^5}{y^2})^3 = -\frac{(2x^5)^3}{(y^2)^3} = \frac{8x^{15}}{y^6}$

53. $(\frac{pq}{p^2q^2})^5 = (\frac{1}{pq})^5 = \frac{1}{(pq)^5} = \frac{1}{p^5q^5}$

55. $(\frac{3x}{y^{-3}})^4 = (3xy^3)^4 = 81x^4y^{12}$

57. $(\frac{-u^2v^3}{4vu^4})^2 = (\frac{-v^2}{4u^2})^2 = \frac{v^4}{16u^4}$

59. $(\frac{x^{-2}y}{2z^{-4}})^4 = (\frac{yz^4}{2x^2})^4 = \frac{y^4z^{16}}{16x^8}$

61. $(\frac{r^5}{t^6})^{-2} = (\frac{t^6}{r^5})^2 = \frac{t^{12}}{r^{10}}$

63. $(\frac{-2a^4}{b^2})^{-3} = (\frac{b^2}{-2a^4})^3 = \frac{b^6}{-8a^{12}}$

65. 317,000,000

67. 0.000001

69. 6,200,000

71. 0.00004025

73. 4.2×10^8

75. 3.5×10^{-6}

77. 2.17×10^{11}

79. 7.31×10^{-9}

81.

Standard Notation	Scientific Notation (written)	Scientific Notation (displayed on calculator)
975,000,000	9.75×10^8	9.75*E*8
487,000,000	4.87×10^8	4.87*E*8
0.0000000001652	1.652×10	1.652*E*⁻10
0.000000067	6.7×10^{-8}	6.7 *E*⁻8
0.0000000000001	1×10^{-13}	1*E*⁻13
3,281,000,000	3.281×10^9	3.281*E*9

83. 9×10^7

85. 2.075×10^{-4}

87. 3.784×10^{-2}

89. 1.25×10^{10}

91. 3×10^2
93. 3×10^8
95. 4,000,000,000 bytes and 17,000,000,000 bytes
97. 7×10^{-7} m
99. 2×10^{11} cells
101. 0.0000000000000000000000017 g
103. 780,000,000
105. $10^2(3.2\times10^4)(5\times10^6)=1.6\times10^{13}$ red blood cells
107.
a. 1.86×10^5 mi per sec
b. $(4.7\times10^6)\div(1.86\times10^5)\approx2.5\times10$; It will take about 25 seconds.

5.3 Basic Concepts of Polynomials

Practice 5.3

1. Terms: $-10x^2$, $4x$, and 20
Coefficients: -10, 4, and 20
2.

Polynomial	Monomial	Binomial	Trinomial	Other Polynomial
$2x+9$		✓		
$-4x^2$	✓			
$12p-1$		✓		
$3x^4-6x^2+9x+1$				✓

3.
a degree 1
b. degree 2
c. degree 3
d. degree 0

4.

Polynomial	Constant Term	Leading Term	Leading Coefficient
$-3x^7+9$	9	$-3x^7$	-3
x^5	0	x^5	1
x^4-7x-1	-1	x^4	1
$3x+5x^3+20$	20	$5x^3$	5

5.
a. $9x^5-7x^4+9x^2-8x-6$
b. $7x^5+x^3-3x^2+8$
6. $2x^2+3x-x^2+5x^3+3x-5x^3+20=$
$5x^3-5x^3+2x^2-x^2+3x+3x+20=$
$x^2+6x+20$
7.
a. $(2)^2-5(2)+5=4-10+5=-1$
b. $(-2)^2-5(-2)+5=4+10+5=19$
8. $500-16(3)^2=500-144=356$ *ft*

Exercises 5.3

1. Polynomial
3. Not a polynomial
5. Polynomial
7. Not a polynomial
9.

Polynomial	Monomial	Binomial	Trinomial	Other Polynomial
$5x-1$		✓		
$-5a^2$	✓			
$-6a+3$		✓		
x^3+4x^2+2			✓	

11. $-4x^3+3x^2-2x+8$; degree

13. $-3y+2$; degree 1

15. $-5x^2+7x$; degree

17. $-4y^5-y^3-2y+2$; degree 5

19. $5a^2-a$; degree 2

21. $3p^3+9p$; degree 3

23.

Polynomial	Constant Term	Leading Term	Leading Coefficient
$-x^7+2$	2	$-x^7$	-1
$2x-30$	-30	$2x$	2
$-5x+1+x^2$	1	x^2	1
$7x^3-2x-3$	-3	$7x^3$	7

25. $10x^3-7x^2+10x+6$

27. $r^3+3r^2-8r+14$

29. $0x^2$

31. $0x$

33. $7(2)-3=14-3=11$
$7(-2)-3=-14-3=-17$

35. $(7)^2-3(7)+9=49-21+9=37$
$(-7)^2-3(-7)+9=49+21+9=79$

37. $2.1(2.37)^2+3.9(2.37)-7.3=13.73849$
$2.1(-2.37)^2+3.9(-2.37)-7.3=-4.74751$

39. $4x^2-x^2-2x-3x-10+4=3x^2-5x-6$

41. $6n^3-4n^3-n^2+15n^2+20n+2+8=$
$2n^3+14n^2+20n+10$

Applications

43. A polynomial in one variable; degree 16

45. $x+\dfrac{x^2}{20}=40+\dfrac{40^2}{20}=40+80=120$ ft

47. $72x+2342=72(50)+2342=5942$
The world population in 2000 was 5,942,000,000 people (5942 million).

49. $1.68x^3+9.95x^2-11.6x+730=$
$1.68(4)^3+9.95(4)^2-11.6(4)+730=950.32$
There were about 950 U.S. radio stations with a rock music format in 2003.

5.4 Addition and Subtraction of Polynomials

Practice 5.4

1.
$(6x-3)+(9x^2-3x-40)=$
$6x-3+9x^2-3x-40=9x^2+3x-43$

2.
$(9p^2+4pq+2q^2)+(-p^2-5q^2)+(2p^2-3pq-7q^2)=$
$9p^2+4pq+2q^2-p^2-5q^2+2p^2-3pq-7q^2=$
$10p^2+pq-10q^2$

3.

$$\begin{array}{l} 8n^2+2n-1 \\ \underline{3n^2\qquad\;\; -2} \\ 11n^2+2n-3 \end{array}$$

4.

$$\begin{array}{llll} 7p^3 & -8p^2q-3pq^2 & & +20 \\ & 10p^2q+pq^2 & -q^2 & +5 \\ \underline{p^3} & & \underline{-q^3} & \\ 8p^3 & +2p^2q-2pq^2 & -2q^3 & +25 \end{array}$$

5.

a. $-(4r-3s)+7r=-4r+3s+7r=3r+3s$

b. $(2p+5q)+(p-6q)-(3p+2q)=$
$2p+5q+p-6q-3p-2q=-3q$

6.
$(2x-1)-(3x^2+15x-1)=$
$2x-1-3x^2-15x+1=-3x^2-13x$

7.

$$20x-13$$
$$-(5x^2-12x+13)$$

$$20x-13$$
$$-5x^2+12x-13$$

$$-5x^2+32x-26$$

8.

$$2p^2-7pq+\ 5q^2$$
$$-(3p^2+4pq-12q^2)$$

$$2p^2-7pq+\ 5q^2$$
$$-3p^2-4pq+12q^2$$

$$-p^2-11pq+17q^2$$

9.

$(0.3x+74.8)-(0.3x+67.2)=$

$0.3x+74.8-0.3x-67.2=7.6$

The polynomial 7.6 approximates how much greater the life expectancy is for females than for males.

Exercises 5.4

1. $(3x^2+6x-5)+(-x^2+2x+7)=$

$3x^2+6x-5+-x^2+2x+7=2x^2+8x+2$

3.

$(2n^3+n)+(3n^3+8n)=2n^3+n+3n^3+8n=$

$5n^3+9n$

5. $(10p+3+p^2)+(p^2-7p-4)=$

$10p+3+p^2+p^2-7p-4=2p^2+3p-1$

7. $(8x^2+7xy-y^2)+(3x^2-10xy+3y^2)=$

$8x^2+7xy-y^2+3x^2-10xy+3y^2=$

$11x^2-3xy+2y^2$

9.

$(2p^3-p^2q-5pq^2+1)+(3p^2q+2pq^2-4q^3+4)+$

$(p^3+q^3)=3p^3+2p^2q-3pq^2-3q^3+5$

11.

$$10x^2-3x-8$$
$$20x+3$$

$$10x^2+17x-5$$

13.

$$5x^3\quad+7x-1$$
$$x^2+2x+3$$

$$5x^3+x^2+9x+2$$

15.

$$5ab^2-3a^2+a^3$$
$$2ab^2+9a^2-4a^3$$

$$7ab^2+6a^2-3a^3=$$
$$-3a^3+6a^2+7ab^2$$

17.

$(x^2+x+4)-(2x^2+3x-7)=$

$x^2+x+4-2x^2-3x=7=-x^2-2x+11$

19.

$(x^3+10x^2-8x+3)-(3x^3+x^2+5x-8)=$

$x^3+10x^2-8x+3-3x^3-x^2-5x+8=$

$-2x^3+9x^2-13x+11$

21. $(x^2+3x)-(5x+9)=$

$x^2+3x-5x-9=$

x^2-2x-9

23. $(5x^2-y^2-6xy+1)-(4y^2-6xy-3)=$

$5x^2-y^2-6xy+1-4y^2=6xy+3=$

$5x^2-5y^2+4$

25.

$$2p^2-3p+\ 5$$
$$-(7p^2-10p-1)$$

$$2p^2-3p+\ 5$$
$$-7p^2+10p+1$$

$$-5p^2+7p+6$$

27.

$$\begin{array}{l} 9t^3 - 12t^2 + 3 \\ -(8t^3 \quad\quad -5) \\ \hline 9t^3 - 12t^2 + 3 \\ -8t^3 \quad\quad\quad +5 \\ \hline t^3 - 12t^2 + 8 \end{array}$$

29.

$$\begin{array}{l} 4r^3 - 20r^2s - 7 \\ -(r^3 - 3r^2s \quad -5) \\ \hline 4r^3 - 20r^2s - 7 \\ -r^3 + 3r^2s \quad +5 \\ \hline -3r^3 - 17r^2s - 2 \end{array}$$

31. $7x-(8x+r)=7x-8x-r=-x-r$

33. $2p-(3q+r)=2p-3q-r$

35.

$(4y-1)+(3y^2-y+5)=$

$4y-1+3y^2-y+5=3y^2+3y+4$

37. $(m^3-6m+7)-(-9+6m)=$

$m^3-6m+7+9-6m=m^3-12m+16$

39. $(2x^3-7x+8)-(5x^2+3x-1)=$

$2x^3-7x+8-5x^2-3x+1=$

$2x^3-5x^2-10x+9$

41. $(8x^2+3x)+(x-2)+(x^2+9)=$

$8x^2+3x+x-2+x^2+9=9x^2+4x+7$

43. $(3x-7)+(2x+9)-(7x-10)=$

$3x-7+2x+9-7x+10=-2x+12$

45.

$(7x^2y^2-10xy+4)-(2xy+8)+(x^2y^2-10)=$

$7x^2y^2-10xy+4-2xy-8+x^2y^2-10=$

$8x^2y^2-12xy-14$

Applications

47.

a. $3.14r^2+3.14r^2+6.28rh=6.28r^2+6.28rh$

b.

$SA=2(3.14r^2)+6.28r^2=12.56r^2$ *or* $12.56h^2$

49. $(31x^3-522x^2+2083x+6051)-$

$(-x^3+16x^2+22x+189)=$

$32x^3-538x^2+2061x+5862$

In a given year, $32x^3-538x^2+2061x+5862$

51. $(2x^3-27x^2+172x+391)-$

$(2x^3-22x^2+18x+360)=-5x^2+154x+31$

In a given year, $-5x^2+154x+31$ million more CDs than cassettes were sold in a given year.

5.5 Multiplication of Polynomials

Practice 5.5

1.

$(-10x^2)(-4x^3)=(-10\bullet-4)(x^2\bullet x^3)=$

$(-10\bullet-4)(x^{2+3})=40x^5$

2.

$(7ab^2)(10a^2b^3)(-5a)=$

$(7\bullet10\bullet-5)(ab^2\bullet a^2b^3\bullet a)=(-350)(a^{1+2+1}b^{2+3})=$

$-350a^4b^5$

3. $(-5xy^2)^2=(-5)^2(x)^2(y^2)^2=25x^2y^4$

4.

$(10s^2-3)(7s)=(7s)(10s^2-3)$

$(7s)(10s^2)+(7s)(-3)=70s^2-21s$

5.

$(-2m^3n^2)(-6m^3n^5+2mn^2+n)=$

$(-2m^3n^2)(-6m^3n^5)+(-2m^3n^2)(2mn^2)+(-2m^3n^2)(n)=$

$12m^6n^7-4m^4n^4-2m^3n^3$

6.

$7s^3(-2s^2+5s+4)-s^2(s^2+6s-1)=$

$(7s^3\left(2s^2+5s+4\right)-s^2(s^2+6s-1)=$

$-14s^5+35s^4+28s^3-s^4-6s^3+s^2)=$

7.
$(a-1)(2a+3)=a(2a+3)+(-1)(2a+3)=$
$2a^2+3a-2a-3=2a^2+a-3$
8.
$(8x+3)(2x-1)=16x^2-8x+6x-3=$
$16x^2-2x-3$
9.
$(7m-n)(2m+n)=14m^2+7mn-2mn-n^2=$
$14m^2+5mn-n^2$

10. $8n^2-n+3$
$\underline{\quad n+2}$
$16n^2-2n+6$
$\underline{8n^3-n^2+3n}$
$8n^3+15n^2+n+6$

11. $8x^3+0x^2+9x-1$
$\underline{\quad 3x+7}$
$56x^3+0x^2+63x-7$
$\underline{24x^4+0x^3+27x^2-3x}$
$24x^4+56x^3+27x^2+60x-7$

12. $p^2-2pq+q^2$
$\underline{\quad p-q}$
$-p^2q+2pq^2-q^3$
$\underline{p^3-2p^2q+pq^2}$
$p^3-3p^2q+3pq^2-q^3$

13.
$(P+\mathrm{Pr})+(P+\mathrm{Pr})r=$
Remove parentheses: $P+\mathrm{Pr}+\mathrm{Pr}+\mathrm{Pr}^2=$
Combine like terms: $P+2\mathrm{Pr}+\mathrm{Pr}^2$

Exercises 5.5

1. $(6x)(-4x)=-24x^2$
3. $(9t^2)(-t^3)=-9t^5$
5. $(-5x^2)(-4x^4)=20x^6$
7. $(10x^3)(-7x^5)=-70x^8$
9. $(-4pq^2)(-4p^2qr^2)=16p^3q^3r^2$
11. $(-8x)^2=(-8)^2(x)^2=64x^2$
13. $(\frac{1}{2}t^4)^3=(\frac{1}{2})^3(t^4)^3=\frac{1}{8}t^{12}$
15.
$(7a)(10a^2)(-5a)=(70a^3)(-5a)=-350a^4$
17.
$(2ab^2)(-3abc)(4a^2)=(-6a^2b^3c)(4a^2)=$
$-24a^4b^3c$
19.
$(7x-5)x=(7x)x+(-5)x=7x^2-5x$
21.
$(9t+t^2)(5t)=(9t)(5t)+(t^2)(5t)=$
$45t^2+5t^3=5t^3+45t^2$
23.
$6a^3(4a^2-7a)=6a^3(4a^2)+(6a^3)(-7a)=$
$24a^5-42a^4$
25.
$4x^2(3x-2)=4x^2(3x)+(4x^2)(-2)=$
$12x^3-8x^2$
27.
$x^3(x^2-2x+4)=$
$x^3(x^2)+(x^3)(-2x)+x^3(4)=$
$x^5-2x^4+4x^3$
29.
$5x(3x^2+5x+6)=$
$5x(3x^2)+(5x)(5x)+5x(6)=$
$15x^3+25x^2+30x$
31.
$(5x^2-3x-7)(-9x)=$
$(5x^2)(-9x)+(-3x)(-9x)+(-7)(-9x)=$
$-45x^3+27x^2+63x$
33.
$6x^2(x^3+4x^2-x-1)=$
$6x^2(x^3)+6x^2(4x^2)+6x^2(-x)+6x^2(-1)=$
$6x^5+24x^4-6x^3-6x^2$
35.
$4p(7q-p^2)=4p(7q)+4p(-p^2)=$
$28pq-4p^3=-4p^3+28pq$
37.
$(v+3w^2)(-7v)=v(-7v)+(3w^2)(-7v)=$
$-7v^2-21vw^2$

39.
$2a^2b^3(3a^4b^2+10ab^5)=$
$2a^2b^3(3a^4b^2)+2a^2b^3(10ab^5)=$
$6a^6b^5+20a^3b^8$

41.
$10x+2x(-3x+8)=10x-6x^2+16x=$
$-6x^2+26x$

43.
$-x+8x(x^2-2x+1)=-x+8x^3-16x^2+8x=$
$8x^3-16x^2+7x$

45.
$9x(x^2+3x-5)+8x(-4x^2+x)=$
$9x^3+27x^2-45x-32x^3+8x^2=$
$-23x^3+35x^2-45x$

47.
$-4xy(2x^2+4xy)+x^2y(7x^2-2y)=$
$-8x^3y-16x^2y^2+7x^4y-2x^2y^2=$
$7x^4y-8x^3y-18x^2y^2$

49.
$5a^2b^2(3ab^4-a^3b^2)+4a^2b^2(9ab^4-10a^3b^2)=$

$15a^3b^6-5a^5b^4+36a^3b^6-40a^5b^4=$
$-45a^5b^4+51a^3b^6$

51.
$(y+2)(y+3)=y^2+3y+2y+6=$
y^2+5y+6

53.
$(x-3)(x-5)=x^2-5x-3x+15=$
$x^2-8x+15$

55. $(a-2)(a+2)=a^2+2a-2a-4=a^2-4$

57.
$(w+3)(2w-7)=w^2-7w+6w-21=$
w^2-w-21

59.
$(3-2y)(5y-1)=15y-3-10y^2+2y=$
$-10y^2+17y-3$

61.
$(10p-4)(2p-1)=20p^2-10p-8p+4=$
$20p^2-18p+4$

63. $(u+v)(u-v)=u^2-uv+uv-v^2=u^2-v^2$

65.
$(2p-q)(q-p)=2pq-2p^2-q^2+pq=$
$-2p^2+3pq-q^2$

67.
$(3a-b)(a-2b)=3a^2-6ab-ab-2b^2=$
$3a^2-7ab-2b^2$

69. $(p-8)(4q+3)=4pq+3p-32q-24$

71.
$(x-3)(x^2-3x+1)=$
$x(x^2-3x+1)+(-3)(x^2-3x+1)=$
$x^3-3x^2+x-3x^2+9x-3=x^3-6x^2+10x-3$

73.
$(2x-1)(x^2+3x-5)=$
$2x(x^2+3x-5)+(-1)(x^2+3x-5)=$
$2x^3+6x^2-10x-x^2-3x+5=2x^3+5x^2-13x+5$

75.
$(a-b)(a^2+ab+b^2)=$
$a(a^2+ab+b^2)+(-b)(a^2+ab+b^2)=$
$a^3+a^2b+ab^2-a^2b-ab^2-b^3=a^3-b^3$

77.
$(3x)(x+5)(x-7)=(3x^2+15x)(x-7)=$
$3x^3-21x^2+15x^2-105x=3x^3-6x^2-105x$

Applications

79. $x(x+30)=x(x)+x(30)=x^2+30x\ mm^2$

81.
$(1500+100x)(1000-30x)=$
$1500(1000-30x)+100x(1000-30x)=$
$1{,}500{,}000-45{,}000x+100{,}000x-3000x^2=$
$-3000x^2+55{,}000x+1{,}500{,}000$ dollars

83.

a. $A = 5000(1+r)^3 =$

$5000(1+r)(1+r)(1+r) = 5000(1+2r+r^2)(1+r) =$

$5000(1+3r+3r^2+r^3) =$

$5000+15{,}000r+15{,}000r^2+5000r^3$ dollars

b. 2 periods: $A = 5000(1+r)^2 =$

$5000(1+r)(1+r) = 5000(1+2r+r^2) =$

$5000+10{,}000r+5000r^2$

difference:

$(5000+15{,}000r+15{,}000r^2+5000r^3)$

$-(5000+10{,}000r+5000r^2)$

$5000r+10{,}000r^2+5000r^3$ dollars

c.

$5000(0.10)+10{,}000(0.10)^2+5000(0.10)^3 =$

$500+100+5 = \$605$

5.6 Special Products

Practice

1.

$(p+10)^2 = p^2+(2)(p)(10)+10^2 =$

$p^2+20p+100$

2. $(s+t)^2 = s^2+(2)(s)(t)+t^2 = s^2+2st+t^2$

3.

$(4p+5q)^2 = (4p)^2+(2)(4p)(5q)+(5q)^2 =$

$16p^2+40pq+25q^2$

4.

$(5x-2)^2 = (5x)^2+(2)(5x)(-2)+(-2)^2 =$

$25x^2-20x+4$

5.

$(u-v)^2 = u^2+(2)(u)(-v)+(-v)^2 =$

$u^2-2uv+v^2$

6.

$(2x-9y)^2 = (2x)^2+(2)(2x)(-9y)+(-9y)^2 =$

$4x^2-36xy+81y^2$

7. $(t+10)(t-10) = (t)^2-(10)^2 = t^2-100$

8.

a. $(r-s)(r+s) = (r)^2-(s)^2 = r^2-s^2$

b.

$(8s-3t)(8s+3t) = (8s)^2-(3t)^2 = 64s^2-9t^2$

9.

$(10-7k^2)(10+7k^2) = (10)^2-(7k^2)^2 = 100-49k^4$

10. $(S+s)(S-s) = (S)^2-(s)^2 = S^2-s^2$

Exercises 5.6

1. $(y+2)^2 = (y)^2+(2)(y)(2)+2^2 = y^2+4y+4$

3. $(x+4)^2 = (x)^2+(2)(x)(4)+4^2 = x^2+8x+16$

5.

$(x-11)^2 = x^2+(2)(x)(-11)+(-11)^2 =$

$x^2-22x+121$

7.

$(6-n)^2 = 6^2+(2)(6)(-n)+(-n)^2 = n^2-12n+64$

9.

$(x+y)^2 = x^2+(2)(x)(y)+y^2 = x^2+2xy+y^2$

11.

$(3x+1)^2 = (3x)^2+(2)(3x)(1)+1^2 = 9x^2+6x+1$

13.

$(4n-5)^2 = (4n)^2+2(4n)(-5)+(-5)^2 =$

$16n^2-40n+25$

15.

$(9x+2)^2 = (9x)^2+(2)(9x)(2)+2^2 =$

$81x^2+36n+4$

17.

$(a+\frac{1}{2})^2 = a^2+(2)(a)(\frac{1}{2})+(\frac{1}{2})^2 =$

$a^2+a+\frac{1}{4}$

19.

$(8b+c)^2 = (8b)^2+(2)(8b)(c)+c^2 =$

$64b^2+16bc+c^2$

21.

$(5x-2y)^2 = (5x)^2+(2)(5x)(-2y)+(-2y)^2 =$

$25x^2-20xy+4y^2$

23.

$(-x=3y)^2 = (-x)^2+(2)(-x)(3y)+(3y)^2 =$

$x^2-6xy+9y^2$

25.
$(4x^3+y^4)^2=(4x^3)^2+(2)(4x^3)(y^4)+(y^4)^2=$
$16x^6+8x^3y^4+y^8$

27. $(a+1)(a-1)=(a)^2-(1)^2=a^2-1$

29. $(4x-3)(4x+3)=(4x)^2-(3)^2=16x^2-9$

31.
$(10+3y)(3y-10)=(3y+10)(3y-10)=$
$9y^2-100$

33. $(m-\frac{1}{2})(m+\frac{1}{2})=(m)^2-(\frac{1}{2})^2=m^2-\frac{1}{4}$

35. $(4a+b)(4a-b)=(4a)^2-b^2=16a^2-b^2$

37.
$(3x-2y)(3x+2y)=(3x)^2-(2y)^2=9x^2-4y^2$

39.
$(1-5n)(5n+1)=(1-5n)(1+5n)=$
$1^2-(5n)^2=1-25n^2$

41. $x(x+5)(x-5)=x(x^2-25)=x^3-25x$

43.
$5n^2(n+7)^2=5n^2(n^2+14n+49)=$
$5n^4+70n^3+245n^2$

45.
$(n^2-m^4)(n^2+m^4)=(n^2)^2-(m^4)^2=$
n^4-m^8

47.
$(a-b)(a+b)(a^2+b^2)=(a^2-b^2)(a^2+b^2)=$
$(a^2)^2-(b^2)^2=a^4-b^4$

Applications

49.
$(x-5)^2+(y-1)^2=$
$(x^2-10x+25)+(y^2-2y+1)=$
$x^2+y^2-10x-2y+26$

51.
$$A\left(1+\frac{P}{100}\right)\left(1+\frac{P}{100}\right)=A\left(1+\frac{2P}{100}+\frac{P^2}{10,000}\right)=$$
$$A\left(1+\frac{P}{50}+\frac{P^2}{10,000}\right)=A+\frac{AP}{50}+\frac{P^2}{10.000}$$

53.
$$\frac{(a-m)^2+(b-m)^2+(c-m)^2}{2}=$$
$$\frac{a^2-2am+m^2+b^2-2bm+m^2+c^2-2cm+m^2}{2}=$$
$$\frac{3m^2-2am-2bm-2cm+a^2+b^2+c^2}{2}$$

5.7 Division of Polynomials

Practice

1.
$-12n^6\div 3n=\frac{-12n^6}{3n}=\frac{-12}{3}\bullet\frac{n^6}{n}=$
$-4n^{6-1}=-4n^5$

2.
$(20p^3q^2r^4)\div(-5p^2q^2r)=$
$\frac{20p^3q^2r^4}{-5p^2q^2r}=\frac{20}{-5}\bullet\frac{p^3q^2r^4}{p^2q^2r}=$
$-4p^{3-2}q^{2-2}r^{4-1}=-4pq^0r^3=-4pr^3$

3.
$(21x^3-14x^2)\div(7x)=$
$\frac{21x^3-14x^2}{7x}=\frac{21x^3}{7x}-\frac{14x^2}{7x}=$
$3x^{3-1}-2x^{2-1}=3x^2-2x$

4.
$\frac{14x^8+10x^5-8x^3}{-2x^3}=\frac{14x^8}{-2x^3}+\frac{10x^5}{-2x^3}-\frac{8x^3}{-2x^3}=$
$-7x^{8-3}-5x^{5-3}+4x^{3-3}=$
$-7x^5-5x^2+4x^0=$
$-7x^5-5x^2+4$

5.

$$\frac{-5a^7b^6 + a^2b^4 - 15ab^3}{5ab^3} =$$

$$\frac{-5a^7b^7}{5ab^3} + \frac{a^2b^4}{5ab^3} - \frac{15ab^3}{5ab^3} =$$

$$-a^{7-1}b^{6-3} + \frac{a^{2-1}b^{4-3}}{5} - 3a^{1-1}b^{3-3} =$$

$$-a^6b^3 + \frac{ab}{5} - 3a^0b^0 =$$

$$-a^6b^3 + \frac{ab}{5} - 3$$

6.

$$\begin{array}{r}
2x+3 \\
5x+1\overline{)10x^2+17x+3} \\
\underline{10x+2x} \\
15x+3 \\
\underline{15x+3}
\end{array}$$

The quotient is 2x + 3.

7.

$$\begin{array}{r}
x^2+2x+3 \\
3x+1\overline{)3x^3+7x^2+11x+5} \\
\underline{3x^3+x^2} \\
6x^2+11x+5 \\
\underline{6x^2+2x} \\
9x+5 \\
9x+3 \\
2
\end{array}$$

The quotient is $x^2 + 2x + 3 + \frac{2}{3x+1}$

8.

$$\begin{array}{r}
3s-5 \\
3s-2\overline{)9s^2-21s+10} \\
\underline{9s^2-6s} \\
-15s+10 \\
\underline{-15s+10} \\
0
\end{array}$$

The quotient is 3s – 5.

9.

$$\begin{array}{r}
n^2-4n-3 \\
4n-3\overline{)4n^3-19n^2+0n-4} \\
\underline{4n^3-\ \ 3n^2} \\
-16n^2+0n-4 \\
\underline{-16n^2+12n} \\
-12n-4 \\
\underline{-12n+9} \\
-13
\end{array}$$

The quotient is $n^2 - 4n - 3 + \frac{-13}{4n-3}$

10.

a. The future value of the investment after 1 year is $10(1+r)^1$. The future value of the investment after 2 years is $10(1+r)^2$.

b.

$$10(1+r)^1 = 10 + 10r = 10r + 10$$

$$10(1+r)^2 = 10(r^2 + 2r + 1) =$$

$$10r^2 + 20r + 10$$

c.

$$\begin{array}{r}
r+1 \\
10r+10\overline{)10r^2+20r+10} \\
\underline{10r^2+10r} \\
10r+10 \\
\underline{10r+10} \\
0
\end{array}$$

The future value of the investment after 2 years is $r+1$ times as great as the future value of the investment after 1 year.

Exercises 5.7

1. $\frac{10x^4}{5x^2} = \frac{10}{5} \bullet \frac{x^4}{x^2} = 2x^{4-2} = 2x^2$

3. $\frac{16a^8}{-4a} = \frac{16}{-4} \bullet \frac{a^8}{a} = -4a^{8-1} = -4a^7$

5. $\frac{8x^5}{-6x^4} = \frac{8}{-6} \bullet \frac{x^5}{x^4} = -\frac{4}{3}x^{5-4} = -\frac{4}{3}x$

7. $\frac{12p^2q^3}{3p^2q} = \frac{12}{3} \bullet \frac{p^2}{p^2} \bullet \frac{q^3}{q} = 4p^{2-2}q^{3-1} = 4q^2$

9.

$$\frac{-24u^6v^4}{-8u^4v^2} = \frac{-24}{-8} \bullet \frac{u^6}{u^4} \bullet \frac{v^4}{v^2} =$$

$$3u^{6-4}v^{4-2} = 3u^2v^2$$

11.

$$\frac{-15a^2b^5}{7ab^3} = \frac{-15}{7} \bullet \frac{a^2}{a} \bullet \frac{b^5}{b^3} =$$

$$-\frac{15}{7}a^{2-1}b^{5-3} = -\frac{15}{7}ab^2$$

13.

$$\frac{-6u^5v^3w^3}{4u^2vw^3} = \frac{-6}{4} \bullet \frac{u^5}{u^2} \bullet \frac{v^3}{v} \bullet \frac{w^3}{w^3} =$$

$$-\frac{3}{2}u^{5-2}v^{3-1}w^{3-3} = -\frac{3}{2}u^3v^2$$

15. $\dfrac{6n^2+10n}{2n} = \dfrac{6n^2}{2n} + \dfrac{10n}{2n} = 3n+5$

17. $\dfrac{20b^4-10b}{10b} = \dfrac{20b^4}{10b} - \dfrac{10b}{10b} = 2b^3-1$

19. $\dfrac{18a^2+12a}{-3a} = \dfrac{18a^2}{-3a} + \dfrac{12a}{-3a} = -6a-4$

21.

$$\frac{9x^5-6x^7}{3x^5} = \frac{9x^5}{3x^5} - \frac{6x^7}{3x^5} = 3x^0 - 2x^2 = 3-2x^2$$

23.

$$\frac{12a^4-18a^3+30a^2}{6a^2} = \frac{12a^4}{6a^2} - \frac{18a^3}{6a^2} + \frac{30a^2}{6a^2} =$$

$$2a^2-3a+5$$

25.

$$\frac{n^5-10n^4-5n^3}{-5n^3} = \frac{n^5}{-5n^3} - \frac{10n^4}{-5n^3} - \frac{5n^3}{-5n^3} =$$

$$-\frac{n^2}{5}+2n+1$$

27.

$$\frac{20a^2b+4ab^3}{8ab} = \frac{20a^2b}{8ab} + \frac{4ab^3}{8ab} = \frac{5}{2}a + \frac{1}{2}b^2$$

29.

$$\frac{12x^2y^3-9xy-3xy^2}{-3xy} =$$

$$\frac{12x^2y^3}{-3xy} - \frac{9xy}{-3xy} - \frac{3xy^2}{-3xy} =$$

$$-4xy^2+3+y$$

31.

$$\frac{8p^2q^3-4p^3q^3+6p^4q}{4p^2q} =$$

$$\frac{8p^2q^3}{4p^2q} - \frac{4p^3q^3}{4p^2q} + \frac{6p^4q}{4p^2q} =$$

$$2q^2 - pq^2 + \frac{3}{2}p^2$$

33.

$$\begin{array}{r} x-7 \\ x+3\overline{)\,x^2-4x-21} \\ \underline{x^2+3x} \\ -7x-21 \\ \underline{-7x-21} \\ 0 \end{array}$$

35.

$$\begin{array}{r} 7x-2 \\ 8x-1\overline{)\,56x^2-23x+2} \\ \underline{56x^2-7x} \\ -16x+2 \\ \underline{-16x+2} \\ 0 \end{array}$$

37.

$$\begin{array}{r} 3x-1 \\ 2x+5\overline{)\,6x^2+13x-5} \\ \underline{6x^2+15x} \\ -2x-5 \\ \underline{-2x-5} \\ 0 \end{array}$$

39.

$$\begin{array}{r} 5x+3 \\ x-1\overline{)\,5x^2-2x-3} \\ \underline{5x^2-5x} \\ 3x-3 \\ \underline{3x-3} \\ 0 \end{array}$$

41.

$$\begin{array}{r} 7x+2 \\ 3x+2\overline{)\,21x^2+20x+4} \\ \underline{21x^2+14x} \\ 6x+4 \\ \underline{6x+4} \\ 0 \end{array}$$

43.
$$\begin{array}{r} x \\ x+2\overline{)x^2+2x-5} \\ \underline{x^2+2x} \\ -5 \end{array}$$

The quotient is $x+\dfrac{-5}{x+2}$.

45.
$$\begin{array}{r} 2x+1 \\ x-3\overline{)2x^2-5x-3} \\ \underline{2x^2-6x} \\ x-3 \\ \underline{x-3} \\ 0 \end{array}$$

47.
$$\begin{array}{r} 2x-3 \\ 4x+3\overline{)8x^2-6x-7} \\ \underline{8x^2+6x} \\ -12x-7 \\ \underline{-12x-9} \\ 2 \end{array}$$

The quotient is $2x-3+\dfrac{2}{4x+3}$.

49.
$$\begin{array}{r} x^2-6x+5 \\ x+1\overline{)x^3-5x^2-x+5} \\ \underline{x^3+x^2} \\ -6x^2-x+5 \\ \underline{-6x^2-6x} \\ 5x+5 \\ \underline{5x+5} \\ 0 \end{array}$$

51.
$$\begin{array}{r} 2x^2-x-3 \\ 3x-4\overline{)6x^3-11x^2-5x+19} \\ \underline{6x^3-8x^2} \\ -3x^2-5x+19 \\ \underline{-3x^2+4x} \\ -9x+19 \\ \underline{-9x+12} \\ 7 \end{array}$$

The quotient is $2x^2-x-3+\dfrac{7}{3x-4}$.

53.
$$\begin{array}{r} 5x+20 \\ x-4\overline{)5x^2+0x-2} \\ \underline{5x^2-20x} \\ 20x-2 \\ \underline{20x-80} \\ 78 \end{array}$$

The quotient is $5x+20+\dfrac{78}{x-4}$.

55.
$$\begin{array}{r} 2x^2+3x+4 \\ 2x-3\overline{)4x^3+0x^2-x+3} \\ \underline{4x^3-6x^2} \\ 6x^2-x+3 \\ \underline{6x^2-9x} \\ 8x+3 \\ \underline{8x-12} \\ 15 \end{array}$$

The quotient is $2x^2+3x+4+\dfrac{15}{2x-3}$.

57.
$$\begin{array}{r} x^2-3x+9 \\ x+3\overline{)x^3+0x^2+0x+27} \\ \underline{x^3+3x^2} \\ -3x^2+0x+27 \\ \underline{-3x^2-9x} \\ 9x+27 \\ \underline{9x+27} \\ 0 \end{array}$$

59.

a. $d = rt$

$\frac{d}{r} = \frac{rt}{r}$

$\frac{d}{r} = t$

$t = \frac{d}{r}$

b. $t = \frac{t^3 - 6t^2 + 7t + 14}{t+1}$

$$\begin{array}{r} t^2 - 7t + 14\,hr. \\ t+1 \overline{)\,t^3 - 6t^2 + 7t + 14} \\ \underline{t^3 + t^2} \quad\quad\quad\quad \\ -7t^2 + 7t + 14 \\ \underline{-7t^2 - 7x} \quad\quad \\ 14t + 14 \\ \underline{14t + 14} \\ 0 \end{array}$$

It takes $t^2 - 7t + 14$ hours.

61. $$\begin{array}{r} 3x - 14 \\ 200x - 300 \overline{)\,600x^2 - 3700x + 4400} \\ \underline{600x^2 - 900x} \quad\quad\quad \\ -2800x + 4400 \\ \underline{-2800x + 4200} \\ 200 \end{array}$$

There are 3x – 14 thousand subscribers per cell system.

Chapter 5 Review Exercises

1. $(-x)^3 = (-x)(-x)(-x) = -x^3$

2. $-31^0 = 1$

3. $n^4 \bullet n^7 = n^{4+7} = n^{11}$

4. $x^6 \bullet x = x^{6+1} = x^7$

5. $\frac{n^8}{n^5} = n^{8-5} = n^3$

6. $p^{10} \div p^7 = p^{10-7} = p^3$

7. $y^4 \bullet y^2 \bullet y = y^{4+2+1} = y^7$

8. $(a^2b)(ab^2) = a^{2+1}b^{1+2} = a^3b^3$

9. $x^0 y = (1)(y) = y$

10.

$\frac{n^4 \bullet n^7}{n^9} = \frac{n^{4+7}}{n^9} = \frac{n^{11}}{n^9} =$

$n^{11-9} = n^2$

11. $(5x)^{-1} = \frac{1}{(5x)^1} = \frac{1}{5x}$

12. $-3n^{-2} = (-3)\frac{1}{n^2} = \frac{-3}{n^2}$

13. $8^{-2}v^4 = \frac{1}{8^2}v^4 = \frac{v^4}{64}$

14. $\frac{1}{y^{-4}} = y^4$

15. $x^{-8} \bullet x^7 = x^{-8+7} = x^{-1} = \frac{1}{x}$

16. $5^{-1} \bullet y^6 \bullet y^{-3} = \frac{1}{5^1} y^{6+-3} = \frac{1}{5}y^3$

17. $\frac{a^5}{a^{-5}} = a^{5-(-5)} = a^{10}$

18. $\frac{t^{-2}}{t^4} = t^{-2-4} = t^{-6} = \frac{1}{t^6}$

19. $\frac{x^{-2}}{y} = \frac{1}{x^2 y}$

20. $\frac{x^2}{y^{-1}} = x^2 y$

21.

$(10^2)^4 = 10^{2\bullet 4} =$

$10^8 = 100{,}000{,}000$

22. $-(x^3)^3 = -(x^{3\bullet 3}) = -x^9$

23. $(2x^3)^2 = 2^2 x^{3\bullet 2} = 4x^6$

24. $(-4m^5n)^3 = (-4)^3 m^{5\bullet 3} n^{1\bullet 3} = -64m^{15}n^3$

25.

$3(x^{-2})^6 = 3x^{-2\bullet 6} =$

$3x^{-12} = 3\bullet\frac{1}{x^{12}} = \frac{3}{x^{12}}$

26.

$(a^3b^{-4})^{-2} = a^{3\bullet -2}b^{-4\bullet -2} =$

$a^{-6}b^8 = \frac{b^8}{a^6}$

27. $(\frac{x}{3})^4 = \frac{x^4}{3^4} = \frac{x^4}{81}$

28. $(\frac{-a}{b^3})^2 = \frac{(-a)^2}{b^{3\cdot2}} = \frac{a^2}{b^6}$

29. $(\frac{x}{y})^{-6} = (\frac{y}{x})^6 = \frac{y^6}{x^6}$

30. $(\frac{x^2}{y^{-1}})^5 = (x^2y^1)^5 = x^{10}y^5$

31. $(\frac{4a^3}{b^4c})^2 = \frac{4^2a^{3\cdot2}}{b^{4\cdot2}c^2} = \frac{16a^6}{b^8c^2}$

32.

$(\frac{-u^{-5}v^2}{7w})^2 = \frac{(-u)^{-5\cdot2}v^{2\cdot2}}{7^2w^2} =$

$\frac{v^4}{49(-u)^{10}w^2} = \frac{v^4}{49u^{10}w^2}$

33. 37,000,000,000
34. 1,630,000,000
35. 0.00005022
36. 0.00000000006
37. 1.2×10^{12}
38. 4.27×10^8
39. 4×10^{-14}
40. 5.6×10^{-7}
41.

$(1.4\times10^6)(4.2\times10^3) = (1.4\times4.2)(10^{6+3}) =$

5.88×10^9

42.

$(3\times10^{-2})(2.1\times10^5) = (3\times2.1)(10^{-2+5}) =$

6.3×10^3

43.

$(1.8\times10^4)\div(3\times10^{-3}) =$

$(\frac{1.8}{3})(10^{4-(-3)}) = 0.6\times10^7 =$

6×10^6

44.

$(9.6\times10^{-4})\div(1.6\times10^6) =$

$(\frac{9.6}{1.6})(10^{-4-6}) = 6\times10^{-10}$

45. Polynomial
46. Not a polynomial
47. Trinomial
48. Binomial
49. $-3y^3 + y^2 + 8y - 1$
degree 3; leading term: $-3y^3$;
leading coefficient: -3
50. $n^4 - 7n^3 - 6n^2 + n$
degree 4; leading term: n^4;
leading coefficient: 1
51. $-x^3 + x^2 + 2x + 13$
52. $3n^3 + 4n^2 - 6n + 4$
53.

$2(-1)^2 - 7(-1) + 3 = 2 + 7 + 3 = 12$

$2(3)^2 - 7(3) + 3 = 18 - 21 + 3 = 0$

54.

$2^3 - 8 = 8 - 8 = 0$

$(-2)^3 - 8 = -8 - 8 = -16$

55.

$(4x^2 - x + 4) + (-3x^2 + 9) =$

$4x^2 - x + 4 - 3x^2 + 9 =$

$x^2 - x + 13$

56.

$(5y^4 - 2y^3 + 7y - 11) + (6 - 8y - y^2 - 5y^4) =$

$5y^4 - 2y^3 + 7y - 11 + 6 - 8y - y^2 - 5y^4 =$

$-2y^3 - y^2 - y - 5$

57.

$(a^2 + 5ab + 6b^2) + (3a^2 - 9b^2) + (-7ab - 3a^2) =$

$a^2 + 5ab + 6b^2 + 3a^2 - 9b^2 - 7ab - 3a^2 =$

$a^2 - 2ab - 3b^2$

58.

$(5s^3t - 2st + t^2) + (s^2t - 5t^2) + (t^2 - 4st + 9s^2) =$

$5s^3t - 2st + t^2 + s^2t - 5t^2 + t^2 - 4st + 9s^2 =$

$5s^3t + s^2t + 9s^2 - 6st - 3t^2$

59.

$(x^2 - 5x + 2) - (-x^2 + 3x + 10) =$

$x^2 - 5x + 2 + x^2 - 3x - 10 = 2x^2 - 8x - 8$

60.

$(10n^3 + n^2 - 4n + 1) - (11n^3 - 2n^2 - 5n + 1) =$

$10n^3 + n^2 - 4n + 1 - 11n^3 + 2n^2 + 5n - 1 =$

$-n^3 + 3n^2 + n$

61.

$$\begin{array}{r} 5y^4 - 4y^3 \qquad\quad + y - 6 \\ -(\qquad\quad y^3 - 2y^2 + 7y - 3) \\ \hline 5y^4 - 5y^3 + 2y^2 - 6y - 3 \end{array}$$

62.

$$\begin{array}{r} -9x^3 + 8x^2 - 11x - 12 \\ +11x^3 \qquad\quad - x + 15 \\ \hline 2x^3 + 8x^2 - 12x + 3 \end{array}$$

63.

$14t^2 - (10t^2 - 4t) = 14t^2 - 10t^2 + 4t =$

$4t^2 + 4t$

64.

$-(5x - 6y) + (3x - 7y) = -5x + 6y + 3x - 7y =$

$-2x - y$

65.

$(3y^2 - 1) - (y^2 + 3y + 2) + (-2y + 5) =$

$3y^2 - 1 - y^2 - 3y - 2 - 2y + 5 =$

$2y^2 - 5y + 2$

66.

$(1 - 4x - 6x^2) - (7x - 8) - (-11x - x^2) =$

$1 - 4x - 6x^2 - 7x + 8 + 11x + x^2 =$

$-5x^2 + 9$

67. $-3x^4 \bullet 2x = (-3)(2)(x^{4+1}) = -6x^5$

68.

$(3ab)(8a^2b^3)(-6b) =$

$(3)(8)(-6)(a^{1+2})(b^{1+3+1}) =$

$-144a^3b^5$

69.

$2xy^2(4x - 5y) = 2xy^2(4x) + 2xy^2(-5xy) =$

$8x^2y^2 - 10x^2y^3$

70.

$(x^2 - 3x + 1)(-5x^2) =$

$(x^2)(-5x^2) + (-3x)(-5x^2) + (1)(-5x^2) =$

$-5x^4 + 15x^3 - 5x^2$

71.

$(n + 3)(n + 7) = n(n + 7) + 3(n + 7) =$

$n^2 + 7n + 3n + 21 = n^2 + 10n + 21$

72.

$(3x - 9)(x + 6) = 3x(x + 6) + (-9)(x + 6) =$

$3x^2 + 18x - 9x - 54 = 3x^2 + 9x - 54$

73.

$(2x - 1)(4x - 1) = 2x(4x - 1) + (-1)(4x - 1) =$

$8x^2 - 2x - 4x + 1 = 8x^2 - 6x + 1$

74.

$(3a - b)(3a + 2b) = 3a(3a + 2b) + (-b)(3a + 2b) =$

$9a^2 + 6ab - 3ab - 2b^2 = 9a^2 + 3ab - 2b^2$

75.

$(2x^3 - 5x + 2)(x + 3) =$

$(2x^3)(x + 3) - 5x(x + 3) + 2(x + 3) =$

$2x^4 + 6x^3 - 5x^2 - 15x + 2x + 6 =$

$2x^4 + 6x^3 - 5x^2 - 13x + 6$

76.

$(y - 2)(y^2 - 7y + 1) =$

$(y)(y^2 - 7y + 1) - 2(y^2 - 7y + 1) =$

$y^3 - 7y^2 + y - 2y^2 + 14y - 2 =$

$y^3 - 9y^2 + 15y - 2$

77.

$-y + 2y(-3y + 7) =$

$-y + 2y(-3y) + 2y(7) =$

$-y - 6y^2 + 14y =$

$-6y^2 + 13y$

78.

$4x^2(2x - 6) - 3x(3x^2 - 10x + 2) =$

$8x^3 - 24x^2 - 9x^3 + 30x^2 - 6x =$

$-x^3 + 6x^2 - 6x$

79.

$(a - 1)^2 = a^2 + (2)(a)(-1) + (-1)^2 =$

$a^2 - 2a + 1$

80.

$(s + 4)^2 = s^2 + (2)(s)(4) + (4)^2 =$

$s^2 + 8a + 16$

81.

$(2x + 5)^2 = (2x)^2 + (2)(2x)(5) + (5)^2 =$

$4x^2 + 20x + 25$

82.
$(3-4t)^2 = 3^2 + (2)(3)(-4t) + (-4t)^2 =$
$9 - 24t + 16t^2$

83.
$(5a-2b)^2 = (5a)^2 + (2)(5a)(-2b) + (-2b)^2 =$
$25a^2 - 20ab + 4b^2$

84.
$(u^2+v^2)^2 = (u^2)^2 + (2)(u^2)(v^2) + (v^2)^2 =$
$u^4 + 2u^2v^2 + v^4$

85. $(m+4)(m-4) = m^2 - 4^2 = m^2 - 16$

86. $(6-n)(6+n) = 6^2 - n^2 = 36 - n^2$

87. $(7n-1)(7n+1) = (7n)^2 - 1^2 = 49n^2 - 1$

88. $(2x+y)(2x-y) = (2x)^2 - y^2 = 4x^2 - y^2$

89
$(4a-3b)(4a+3b) = (4a)^2 - (3b)^2 = 16a^2 - 9b^2$

90.
$x(x+10)(x-10) = x(x^2-100) =$
$x^3 - 100x$

91.
$-3t^2(4t-5)^2 =$
$-3t^2(16t^2 - 40t + 25) =$
$-48t^4 + 120t^3 - 75t^2$

92.
$(p^2-q^2)(p+q)(p-q) =$
$(p^2-q^2)(p^2-q^2) =$
$(p^2-q^2)^2 = (p^2)^2 + (2)(p^2)(-q^2) + (-q)^2 =$
$p^4 - 2p^2q^2 + q^4$

93.
$$12x^4 \div 4x^2 = \frac{12}{4} \bullet \frac{x^4}{x^2} = 3x^{4-2} = 3x^2$$

94.
$$\frac{-20a^3b^5c}{10ab^2} = -2a^{3-1}b^{5-2}c = -2a^2b^3c$$

95.
$(18x^3 - 6x) \div (3x) =$
$$\frac{18x^3 - 6x}{3x} = \frac{18x^3}{3x} - \frac{6x}{3x} = 6x^2 - 2$$

96.
$$\frac{10x^5 + 6x^4 - 4x^3 - 2x^2}{2x^2} =$$
$$\frac{10x^5}{2x^2} + \frac{6x^4}{2x^2} - \frac{4x^3}{2x^2} - \frac{2x^2}{2x^2} =$$
$5x^3 + 3x^2 - 2x - 1$

97.
$$\begin{array}{r} 3x-7 \\ x+5\overline{)3x^2+8x-35} \\ \underline{3x^2+15x} \\ -7x-35 \\ \underline{-7x-35} \\ 0 \end{array}$$

98.
$$\begin{array}{r} x^2-2x-1 \\ 2x-1\overline{)2x^3-5x^2+0x+13} \\ \underline{2x^3-x^2} \\ -4x^2+0x+13 \\ \underline{-4x^2+2x} \\ -2x+13 \\ \underline{-2x+1} \\ 12 \end{array}$$

The quotient is $x^2 - 2x - 1 + \dfrac{12}{2x-1}$

99. 1.39×10^{10} years

100.
6,240,000,000,000,000,000 eV

101. 3×10^{-5} m

102. 0.00000000011 m

103.
$$\frac{9^2}{2} - \frac{9}{2} = \frac{81}{2} - \frac{9}{2} = \frac{72}{2} = 36$$
There will be 36 handshakes.

104. $-4.9(2)^2 + 500 = -19.6 + 500 = 480.4$
The object is 480.4 m above the ground.

105.
$-0.3(1)^2 + 36.7(1) + 213 = 249.4$ thousand
There were about 249,000 divorces in 1951.

106.
$-0.8(0)^2 + 41.5(0) + 898.6 = 898.6$
There were about 899 two-year colleges in 1970.

107.
a. $A = w(3w - 10) = (3w^2 - 10w)$ square feet
b.
$A = (w + 12)(3w - 10 + 12) - (3w^2 - 10w) =$
$(w + 12)(3w + 2) - (3w^2 - 10w) =$
$3w^2 + 38w + 24 - 3w^2 + 10w = 48w + 24$
The area of the concrete walk is 48w +24 square feet.
c. $A = 48(12) + 24 = 600$ square feet

108.

$$\begin{array}{r|l} & x^2 + 4x + 4 \\ \hline 2x - 3 & 2x^3 + 5x^2 - 4x - 12 \\ & \underline{2x^3 - 3x^2} \\ & 8x^2 - 4x - 12 \\ & \underline{8x^2 - 12x} \\ & 8x - 12 \\ & \underline{8x - 12} \\ & 0 \end{array}$$

The area of the base $x^2 + 4x + 4$

Chapter 5 Posttest

1. $x^6 \bullet x = x^{6+1} = x^7$
2. $n^{10} \div n^4 = n^{10-4} = n^6$
3. $7a^{-1}b^0 = (7)(\frac{1}{a^1})(1) = \frac{7}{a}$
4. $(-3x^3y)^3 = -27x^9y^3$
5. $(\frac{x^2}{y^3})^4 = \frac{x^8}{y^{12}}$
6. $(\frac{3x^2}{y})^{-3} = (\frac{y}{3x^2})^3 = \frac{y^3}{27x^6}$

7.
a. the terms: $-x^3, 2x^2, 9x, -1$
b. the coefficients -1, 2, 9, and -1
c. the degree: 3
d. the constant term: -1

8. $(y^2 - 1) + (y^2 - y + 6) = 2y^2 - y + 5$

9.
$(x^2 - 7x - 4) - (2x^2 - 8x + 5) =$
$x^2 - 7x - 4 - 2x^2 + 8x - 5 =$
$-x^2 + x - 9$

10.
$(4x^2y^2 - 6xy - y^2) - (3x^2 + x^2y^2 - 2y^2) -$
$(x^2 - 6xy + y^2) =$
$4x^2y^2 - 6xy - y^2 - 3x^2 - x^2y^2 + 2y^2 -$
$x^2 + 6xy - y^2 =$
$3x^2y^2 - 4x^2$

11.
$(2mn^2)(5m^2n - 10mn + mn^2) =$
$(2mn^2)(5m^2n) + (2mn^2)(-10mn) + (2mn^2)(mn^2) =$
$10m^3n^3 - 20m^2n^3 + 2m^2n^4$

12.
$(y^3 - 2y^2 + 4)(y - 1) =$
$(y^3 - 2y^2 + 4)(y) + (y^3 - 2y^2 + 4)(-1) =$
$y^4 - 2y^3 + 4y - y^3 + 2y^2 - 4 =$
$y^4 - 3y^3 + 2y^2 + 4y - 4$

13.
$(3x + 1)(2x + 7) = 3x(2x + 7) + 1(2x + 7) =$
$6x^2 + 21x + 2x + 7 = 6x^2 + 23x + 7$

14.
$(7 - 2n)(7 + 2n) = 7^2 - (2n)^2 = 49 - 4n^2$

15.
$(2m - 3)^2 = (2m)^2 + (2)(2m)(-3) + (-3)^2 =$
$4m^2 - 12m + 9$

16.
$\frac{12s^3 + 15s^2 - 27s}{-3s} = \frac{12s^3}{-3s} + \frac{15s^2}{-3s} - \frac{27s}{-3s} =$
$-4s^2 - 5s + 9$

17.

$$\begin{array}{r} t^2 - t - 1 \\ 3t-2\overline{)3t^3 - 5t^2 - t + 6} \\ \underline{3t^3 - 2t^2} \\ -3t^2 - t + 6 \\ \underline{-3t^2 + 2t} \\ -3t + 6 \\ \underline{-3t + 2} \\ 4 \end{array}$$

The quotient is $t^2 - t - 1 + \dfrac{4}{3t-2}$.

18.

$(1000)(10^{-10}) = (1\times 10^3)(1\times 10^{-10}) =$

$1\times 10^{-7}\ m$

19.

a. First house: 1500x + 140,000 dollars
Second house: 800x + 90,000 dollars

b. $(1500x + 140{,}000) + (800x + 90{,}000) =$

$2300x + 230{,}000$ dollars

20. $1000(1 + 0.03)^2 = 1000(1.03)^2 =$

$1000(1.0609) = \$1060.90$

The account balance is $1060.90.

Cumulative Review Exercises

1.

$y = mx + b$

$y - b = mx + b - b$

$y - b = mx$

$(\frac{1}{x})(y - b) = (\frac{1}{x})(mx)$

$\frac{y-b}{x} = m$

$m = \frac{y-b}{x}$

2.

$3(-3)^2 - 5(-3)(2) + 2^2 =$

$27 + 30 + 4 = 61$

3.

$2x + 12 - 9x = 5(4 - 3x) + 6x$

$2x + 12 - 9x = 20 - 15x + 6x$

$-7x + 12 = 20 - 9x$

$-7x + 9x + 12 = 20 - 9x + 9x$

$2x + 12 = 20$

$2x + 12 - 12 = 20 - 12$

$2x = 8$

$\frac{2x}{2} = \frac{8}{2}$

$x = 4$

check:

$2(4) + 12 - 9(4) \stackrel{?}{=} 5(4 - 3(4)) + 6(4)$

$8 + 12 - 36 \stackrel{?}{=} 5(4 - 12) + 24$

$8 + 12 - 36 \stackrel{?}{=} 5(-8) + 24$

$20 - 36 \stackrel{?}{=} -40 + 24$

$-16 = -16$

4.

$2x - 3y = 6$

$2x - 2x - 3y = -2x + 6$

$-3y = -2x + 6$

$\frac{-3y}{-3} = \frac{-2x}{-3} + \frac{6}{-3}$

$y = \frac{2}{3}x - 2$

Slope: $\frac{2}{3}$; y-intercept: (0,-2)

5.

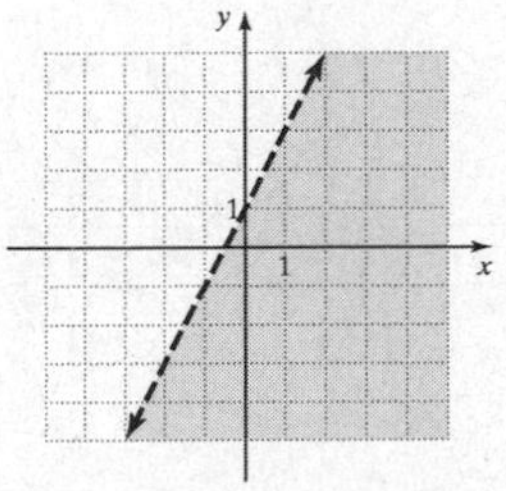

6.

$$\begin{aligned} 3x - 2y &= 10 \\ 2x + 3y &= -2 \end{aligned}$$

$$\begin{aligned} 9x - 6y &= 30 \\ 4x + 6y &= -4 \end{aligned}$$

$$13x \quad = 26$$

$$\frac{13x}{13} = \frac{26}{13}$$

$$x = 2$$

$$3(2) - 2y = 10$$

$$6 - 2y = 10$$

$$6 - 6 - 2y = 10 - 6$$

$$-2y = 4$$

$$\frac{-2y}{-2} = \frac{4}{-2}$$

$$y = -2$$

$$(x, y) = (2, -2)$$

7.

$$(3m^2 - 8m + 7) - (2m^2 + 8m - 9) =$$

$$3m^2 - 8m + 7 - 2m^2 - 8m + 9 =$$

$$m^2 - 16m + 16$$

8.

$$E = (1\times10^{-3})(3\times10^{8})^2$$

$$E = (1\times10^{-3})(3\times10^{8})(3\times10^{8})$$

$$E = (1\times10^{-3})(9\times10^{16})$$

$$E = 9\times10^{13}\, kg \bullet m^2 / \sec^2$$

9.

a. x + 4, x + 8, x + 12

b. Yes: 1980 – 1972 = 8, which is a multiple of 4.

10.

a. $0.20b$

b. $b =$ amount of original bill

$c = b + 0.20b = 1.20b$

Chapter 6 Factoring Polynomials

Chapter 6 Pretest

1.

$18ab = 2{\bullet}3^2{\bullet}a{\bullet}b$

$36a^4 = 2^2{\bullet}3^2{\bullet}a{\bullet}a^3$

$GCF = 2{\bullet}3^2{\bullet}a = 18a$

2.

$GCF = 4p$

$4pq + 16p = 4p(q+4)$

3.

$GCF = 5xy$

$10x^2y - 5x^3y^3 + 5xy^2 = 5xy(2x - x^2y^2 + y)$

4.

$3x^2 + 6x + 2x + 4 = (3x^2 + 6x) + (2x + 4)$

$= 3x(x+2) + 2(x+2)$

$= (x+2)(3x+2)$

5. $n^2 - 11n + 24 = (n-3)(n-8)$

6.

$4a + a^2 - 21 = a^2 + 4a - 21$

$= (a-3)(a+7)$

7.

$9y - 12y^2 + 3y^3 - 3y^3 - 12y^2 + 9y$

$= 3y(y^2 - 4y + 3)$

$= 3y(y-1)(y-3)$

8. $5a^2 + 6ab - 8b^2 = (5a - 4b)(a + 2b)$

9.

$-12n^2 + 38n + 14 = -2(6n^2 - 19n - 7)$

$= -2(3n+1)(2n-7)$

10. $4x^2 - 28x + 49 = (2x-7)^2$

11. $25n^2 - 9 = (5n+3)(5n-3)$

12.

$x^2y - 4y^3 = y(x^2 - 4y^2)$

$= y(x+2y)(x-2y)$

13. $y^6 - 9y^3 + 20 = (y^3 - 4)(y^3 - 5)$

14.

$n(n-6) = 0$

$n = 0 \quad n - 6 = 0$

$n - 6 + 6 = 0 + 6$

$n = 6$

$0, 6$

15.

$3x^2 + x = 2$

$3x^2 + x - 2 = 2 - 2$

$3x^2 + x - 2 = 0$

$(3x-2)(x+1) = 0$

$3x - 2 = 0 \quad x + 1 = 0$

$3x = 2 \quad x = -1$

$x = \frac{2}{3}$

$\frac{2}{3}, \quad -1$

16.

$(y+4)(y-2) = 7$

$y^2 + 2y - 8 = 7$

$y^2 + 2y - 8 - 7 = 7 - 7$

$y^2 + 2y - 15 = 0$

$(y-3)(y+5) = 0$

$y - 3 = 0 \quad y + 5 = 0$

$y = 3 \quad y = -5$

3, -5

17.

$A = 2lw + 2lh + 2wh$

$A - 2lw = 2lh + 2wh$

$A - 2lw = h(2l + 2w)$

$\frac{A - 2lw}{2l + 2w} = h$

$h = \frac{A - 2lw}{2l + 2w}$

18.

$-16t^2 + 63t + 4 = -(16t^2 - 63t - 4)$

$= -(16t+1)(t-4)\text{ft}$

19.

$A = s^2 - 15^2 = (s+15)(s-15)$

20.

$h^2 + (h+8)^2 = 40^2$

$h^2 + h^2 + 16h + 64 = 1600$

$2h^2 + 16h - 1536 = 0$

$h^2 + 8h - 768 = 0$

$(h+32)(h-24) = 0$

$h+32 = 0 \quad h-24 = 0$

$h = -32 \qquad h = 24$

$h+8 = 32$

The height is 24 feet and the length is 24 + 8 or 32 feet.

Practice 6.1

1.

$24 = 2^3 \cdot 3$

$72 = 2^3 \cdot 3^2$

$96 = 2^5 \cdot 3$

$GCF = 2^3 \cdot 3 = 24$

2.

$a^3 = a \cdot a \cdot a$

$a^2 = a \cdot a$

$a = a$

$GCF = a$

3.

$-18x^3y^4 = -1 \cdot 2 \cdot 3^2 \cdot x \cdot x^2 \cdot y^2 \cdot y^2$

$12xy^2 = 2^2 \cdot 3 \cdot x \cdot y^2$

$GCF = 2 \cdot 3 \cdot x \cdot y^2 = 6xy^2$

4.

$GCF = 2y^2$

$10y^2 + 8y^5 = 2y^2(5 + 4y^3)$

5.

$GCF = 7a$

$21a^2b - 14a = 7a(3ab - 2)$

6.

$GCF = 2ab^2$

$8a^2b^2 - 6ab^3 = 2ab^2(4a - 3b)$

7.

$GCF = 12$

$24a^2 - 48a + 12 = 12(2a^2 - 4a + 1)$

8.

$ab = s^2 - ac$

$ab + ac = s^2 - ac + ac$

$ab + ac = s^2$

$a(b+c) = s^2$

$$\frac{a(b+c)}{(b+c)} = \frac{s^2}{(b+c)}$$

$$a = \frac{s^2}{(b+c)}$$

9.

$4(y-3) + y(y-3) = (y-3)(4+y)$
or $(y-3)(y+4)$

10.

$$\begin{aligned} 3y(x-1) + 2(1-x) &= 3y(x-1) + 2[-1(x-1)] \\ &= 3y(x-1) - 2(x-1) \\ &= (x-1)(3y-2) \end{aligned}$$

11.

$$\begin{aligned} (4-3x) + 2x(4-3x) &= 1(4-3x) + 2x(4-3x) \\ &= (4-3x)(1+2x) \end{aligned}$$

12.

$$\begin{aligned} a^2 + 2a - 2ab - 4b &= (a^2 + 2a) + (-2ab - 4b) \\ &= a(a+2) - 2b(a+2) \\ &= (a+2)(a-2b) \end{aligned}$$

13.

$$\begin{aligned} 5y - 5z - y^5 + y^4z &= (5y - 5z) + (-y^5 + y^4z) \\ &= 5(y-z) - y^4(y-z) \\ &= (y-z)(5-y^4) \end{aligned}$$

14. $v_0t + \frac{1}{2}at^2 = t(v_0 + \frac{1}{2}at)$

6.1 Common Factoring and Factoring by Grouping

Exercises 6.1

1.
$27 = 3^3$

$54 = 2 \cdot 3^3$

$81 = 3^4$

$GCF = 3^3 = 27$

3.
$x^4 = x \cdot x \cdot x \cdot x$

$x^6 = x \cdot x \cdot x \cdot x \cdot x \cdot x$

$x^3 = x \cdot x \cdot x$

$GCF = x^3$

5.
$16b = 2^4 \cdot b$

$8b^3 = 2^3 \cdot b^3$

$12b^2 = 2^2 \cdot 3 \cdot b^2$

$GCF = 2^2 \cdot b = 4b$

7.
$-12x^5y^7 = -1 \cdot 2^2 \cdot 3 \cdot x^5 \cdot y^7$

$4y^3 = 2^2 \cdot y^3$

$GCF = 2^2 \cdot y^3 = 4y^3$

9.
$18a^5b^4 = 2 \cdot 3^2 \cdot a^5 \cdot b^4$

$-6a^4b^3 = -1 \cdot 2 \cdot 3 \cdot a^4 \cdot b^3$

$9a^2b^2 = 3^2 \cdot a^2 \cdot b^2$

$3a^2b^2 = 3 \cdot a^2 \cdot b^2$

$GCF = 3 \cdot a^2 \cdot b^2 = 3a^2b^2$

11.
$x(3x-1) = x \cdot (3x-1)$

$8(3x-1) = 8 \cdot (3x-1)$

$GCF = 3x-1$

13.
$4x(x+7) = 4 \cdot x \cdot (x+7)$

$9x(x+7) = 9 \cdot x \cdot (x+7)$

$GCF = x \cdot (x+7) = x(x+7)$

15.
$GCF = 3$

$3x+6 = 3(x+2)$

17.
$GCF = 8$

$24x^2+8 = 8(3x^2+1)$

19.
$GCF = 9$

$27m+9n = 9(3m+n)$

21.
$GCF = x$

$2x-7x^2 = x(2-7x)$

23.
$GCF = 4z^2$

$4z^5+12z^2 = 4z^2(z^3+3)$

25.
$GCF = 5x$

$10x^3-15x = 5x(2x^2-3)$

27.
$GCF = ab$

$a^2b^2-ab = ab(ab-1)$

30.
$GCF = xy$

$6xy^2+7x^2y = xy(6y+7x)$

31.
$GCF = 9pq$

$27pq^2+18p^2q = 9pq(3q+2p)$

33.
$GCF = 2x^3y$

$2x^3y-12x^3y^4 = 2x^3y(1-6y^3)$

35.
$GCF = 3$

$3c^3+6c^2+12 = 3(c^3+2c^2+4)$

37.
$GCF = b^2$

$9b^4-3b^3+b^2 = b^2(9b^3-3b+1)$

39.
$GCF = 2m^2$

$2m^4+10m^3-6m^2 = 2m^2(m^2+5m-3)$

41.

$GCF = b^2$

$5b^5 - 3b^3 + 2b^2 = b^2(5b^3 - 3b + 2)$

43.

$GCF = 5x$

$15x^4 - 10x^3 - 25x = 5x(3x^3 - 2x^2 - 5)$

45.

$GCF = 4ab$

$4ab^2 + 8a^2b^2 - 12ab = 4ab(a + 2ab - 3)$

47.

$GCF = 3cd$

$9c^2d^4 + 12c^3d + 3cd^3 = 3cd(3cd^3 = 4c^2 + d^2)$

49. $x(x-1) + 3(x-1) = (x-1)(x+3)$

51. $5a(a-1) - 3(a-1) = (a-1)(5a-3)$

53.

$$\begin{aligned} r(s+7) - 2(7+s) &= r(s+7) - 2(s+7) \\ &= (s+7)(r-2) \end{aligned}$$

55. $a(x-y) - b(x-y) = (x-y)(a-b)$

57. $3x(y+2) - (y+2) = (y+2)(3x-1)$

59.

$$\begin{aligned} b(b-3) + 5(1-b) &= b(b-1) + 5[-1(b-1)] \\ &= b(b-1) - 5(b-1) \\ &= (b-1)(b-5) \end{aligned}$$

61.

$$\begin{aligned} y(y-1) - 5(1-y) &= y(y-1) - 5[-1(y-1)] \\ &= y(y-1) + 5(y-1) \\ &= (y-1)(y+5) \end{aligned}$$

63.

$$\begin{aligned} (t-3) - t(3-t) &= 1(t-3) - t[-1(t-3)] \\ &= 1(t-3) + t(t-3) \\ &= (t-3)(1+t) \end{aligned}$$

65.

$$\begin{aligned} 9a(b-7) + 2(7-b) &= 9a(b-7) + 2[-1(b-7)] \\ &= 9a(b-7) - 2(b-7) \\ &= (b-7)(9a-2) \end{aligned}$$

67.

$$\begin{aligned} rs + 3s + rt + 3t &= (rs + 3s) + (rt + 3t) \\ &= s(r+3) + t(r+3) \\ &= (r+3)(s+t) \end{aligned}$$

69.

$$\begin{aligned} xy + 6y - 4x - 24 &= (xy + 6y) + (-4x - 24) \\ &= y(x+6) - 4(x+6) \\ &= (x+6)(y-4) \end{aligned}$$

71.

$$\begin{aligned} 15xy - 9yz + 20xz - 12z^2 &= (15xy - 9yz) + (20xz - 12z^2) \\ &= 3y(5x - 3z) + 4z(5x - 3z) \\ &= (5x - 3z)(3y + 4z) \end{aligned}$$

73.

$$\begin{aligned} 2xz + 8x + 5yz + 20y &= (2xz + 8x) + (5yz + 20y) \\ &= 2x(z+4) + 5y(z+4) \\ &= (z+4)(2x+5y) \end{aligned}$$

75.

$TM = PC + PL$

$TM = P(C + L)$

$\frac{TM}{C+L} = \frac{P(C+L)}{C+L}$

$\frac{TM}{C+L} = P$

$P = \frac{TM}{C+L}$

77.

$S = 2lw + 2lh + 2wh$

$S - 2wh = 2lw + 2lh$

$S - 2wh = l(2w + 2h)$

$\frac{S - 2wh}{2w + 2h} = l$

$l = \frac{S - 2wh}{2w + 2h}$

79. $mv_2 - mv_1 = m(v_2 - v_1)$

81. $0.5n^2 - 0.5n = 0.5n(n-1)$

83. $\frac{1}{2}n^2 - \frac{3}{2}n = \frac{1}{2}n(n-3)$

85.

$P = nC + nT + D$

$P - D = nC + nT$

$P - D = n(C + T)$

$\frac{P - D}{C + T} = n$

$n = \frac{P - D}{C + T}$

6.2 Factoring Trinomials Whose Leading Coefficient is 1

Practice 6.2

1.

$x^2 + 5x + 4 = (x + ?)(x + ?)$

factors of 4	sum of factors
1, 4	5
2, 2	4

$x^2 + 5x + 4 = (x + 1)(x + 4)$

or $(x + 4)(x + 1)$

2.

$y^2 - 9y + 20 = (y - ?)(y - ?)$

factors of 20	sum of factors
−1, −20	21
−2, −10	12
−4, −5	9

$y^2 - 9y + 20 = (y - 4)(y - 5)$

or $(y - 5)(y - 4)$

3.

$x^2 + 3x + 5 = (x + ?)(x + ?)$

factors of 5	sum of factors
1, 5	6

Prime polynomial: cannot be factored

4.

$32 - 12y + y^2 = y^2 - 12y + 32$

$y^2 - 12y + 32 = (y - ?)(y - ?)$

factors of 32	sum of factors
−1, −32	−33
−2, −16	−18
−4, −8	−12

$y^2 - 12y + 32 = (y - 4)(y - 8)$

or $(y - 8)(y - 4)$

5.

$p^2 - 4pq + 3q^2 = (p - ?q)(p - ?q)$

factors of 3	sum of factors
−1, −3	−4

$p^2 - 4pq + 3q^2 = (p - q)(p - 3q)$

or $(p - 3q)(p - q)$

6.

$x^2 + x - 6 = (x + ?)(x - ?)$

factors of −6	sum of factors
−1, 6	5
1, −6	−5
2, −3	−1
−2, 3	1

$x^2 + x - 6 = (x + 3)(x - 2)$

7.

$x^2 - 21x - 46 = (x + ?)(x + ?)$

factors of −46	sum of factors
1, −46	−45
−1, 46	45
2, −23	−21
−2, 23	21

$x^2 - 21x - 46 = (x + 2)(x - 23)$

or $(x - 23)(x + 2)$

8.

$y^2-24+2y=y^2+2y-24$

$y^2+2y-24=(y+?)(y-?)$

factors of -24	sum of factors
$1,-24$	-23
$-1,\ 24$	23
$2,-12$	-10
$-2,\ 12$	10
$3,\ -8$	-5
$-3,\ 8$	5
$4,\ -6$	-2
$-4,\ 6$	2

$y^2+2y-24=(y+6)(x-4)$

or $(x-4)(x+6)$

9.

$a^2-5ab-24b^2=(a+?b)(a-?b)$

factors of -24	sum of factors
$-1,\ 24$	23
$1,\ -24$	-23
$-2,\ 12$	10
$2,\ -12$	-10
$3,\ -8$	-5
$-3,\ 8$	5
$-4,\ 6$	2
$4,\ -6$	-2

$a^2-5ab-24b^2=(a+3b)(a-8b)$

or $(a-8b)(a+3b)$

10.

$y^3-9y^2-10y=y(y^2-9y-10)$

factors of -10	sum of factors
$-1,\ 10$	9
$1,\ -10$	-9
$-2,\ 5$	3
$2,\ -5$	-3

$y^3-9y^2-10y=y(y^2-9y-10)=$

$y(y+1)(y-10)$

11.

$8x^3-24x^2+16x=8x\left(x^2-3x+2\right)$

factors of 2	sum of factors
1,2	3
-1,-2	-3

$8x^3-24x^2+16x=$

$8x\left(x^2-3x+2\right)=$

$8x(x-1)(x-2)$

12.

$-x^2-10x+11=(-1)(x^2+10x-11)$

factors of -11	sum of factors
$-1,\ 11$	10
$1,\ -11$	-10

$-x^2-10x+11=(-1)(x^2+10x-11)$

$-(x-1)(x+11)$ or $(-x+1)(x+11)$ or

$(x-1)(-x-11)$

13.

$-16t^2+32t+48=(-16)(t^2-2t-3)$

factors of -3	sum of factors
-1,3	2
1,-3	-2

$-16t^2+32t+48=(-16)(t^2-2t-3)$

$-16(t+1)(t-3)$

Exercises 6.2

1. f
3. e
5. b

7. $x^2-3x-10=(x+2)(x-5)$

9. $x^2+5x+4=(x+1)(x+4)$

11. $x^2+5x-6=(x+6)(x-1)$

13. $x^2+6x+8=(x+2)(x+4)$

15. $x^2+5x-6=(x-1)(x+6)$

17. Prime polynomial

19. $x^2+5x+4=(x+1)(x+4)$

21. $x^2-4x+3=(x-1)(x-3)$

23. $y^2-12y+32=(y-4)(y-8)$

25. $t^2-4t-5=(t+1)(m-5)$

27. $n^2-9n-36=(n+3)(n-12)$

29. $x^2+4x-45=(x-5)(x+9)$

31. $y^2-9y+20=(y-4)(y-5)$

33. $b^2+11b+28=(b+4)(b+7)$

35. $m^2-15m+44=(m-4)(m-11)$

37.

$-y^2+5y+50=-(y^2-5y-50)=$
$-(y+5)(y-10)$

39.

$$\begin{aligned} x^2+64-16x &= x^2-16x+64 \\ &= (x-8)(x-8) \\ &= (x-8)^2 \end{aligned}$$

41.

$$\begin{aligned} 16-10x+x^2 &= x^2-10x+16 \\ &= (x-2)(x-8) \end{aligned}$$

43.

$$\begin{aligned} 81-30w+w^2 &= w^2-30w+81 \\ &= (w-3)(w-27) \end{aligned}$$

45. $p^2-8pq+7q^2=(p-q)(p-7q)$

47. $p^2-4pq-5q^2=(p+q)(p-5q)$

49. $m^2-12mn+35n^2=(m-5n)(m-7n)$

51. $x^2+9xy+8y^2=(x+y)(x+8y)$

53.

$$\begin{aligned} 5x^2-5x-30 &= 5(x^2-x-6) \\ &= 5(x+2)(x-3) \end{aligned}$$

55.

$$\begin{aligned} 2x^2+10x-28 &= 2(x^2+5x-14) \\ &= 2(x-2)(x+7) \end{aligned}$$

57.

$$\begin{aligned} 12-18t+6t^2 &= 6t^2-18t+12 \\ &= 6(t^2-3t+2) \\ &= 6(t-1)(t-2) \end{aligned}$$

59.

$$\begin{aligned} 3x^2+24+18x &= 3x^2+18x+24 \\ &= 3(x^2+6x+8) \\ &= 3(x+2)(x+4) \end{aligned}$$

61.

$$\begin{aligned} y^3+3y^2-10y &= y(y^2+3y-10) \\ &= y(y-2)(y+5) \end{aligned}$$

63.

$$\begin{aligned} a^3+8a^2+15a &= a(a^2+8a+15) \\ &= a(a+3)(a+5) \end{aligned}$$

65.

$$\begin{aligned} t^4-14t^3+24t^2 &= t^2(t^2-14t+24) \\ &= t^2(t-2)(t-12) \end{aligned}$$

67.

$$\begin{aligned} 4a^3-12a^2+8a &= 4a(a^2-3a+2) \\ &= 4a(a-1)(a-2) \end{aligned}$$

69.

$$\begin{aligned} 2x^3+30x+16x^2 &= 2x^3+16x^2+30x \\ &= 2x(x^2+8x+15) \\ &= 2x(x+3)(x+5) \end{aligned}$$

71.

$$\begin{aligned} 4x^3+48x-28x^2 &= 4x^3-28x^2+48x \\ &= 4x(x^2-7x+12) \\ &= 4x(x-3)(x-4) \end{aligned}$$

73.

$$\begin{aligned} -56s+6s^2+2s^3 &= 2s^3+6s^2-56s \\ &= 2s(s^2+3s-28) \\ &= 2s(s-4)(s+7) \end{aligned}$$

75.

$$\begin{aligned} 2c^4+4c^3-70c^2 &= 2c^2(c^2+2c-35) \\ &= 2c^2(c-5)(c+7) \end{aligned}$$

77.

$$\begin{aligned} ax^3-18ax^2+32ax &= ax(x^2-18x+32) \\ &= ax(x-2)(x-16) \end{aligned}$$

79. $n^2+11n+30=(n+5)(n+6)$;
The factors represent two whole numbers that differ by 1.

81. $C=x^2-14x+45=(x-5)(x-9)$

6.3 Factoring Trinomials Whose Leading Coefficient is Not 1

Practice 6.3

1. $5x^2+14x+8=(5x+4)(x+2)$

2.
$21-25x+6x^2=6x^2-25x+21=$
$(6x-7)(x-3)$

3. $7y^2+47y-14=(7y-2)(y+7)$

4. $2x^2-x-10=(2x-5)(x+2)$

5.
$18x^3-21x^2-9x=3x(6x^2-7x-3)$
$=3x(3x+1)(2x-3)$

6.
$36c^2-12cd-15d^2=3(12c^2-4cd-5d^2)$
$=3(6c-5d)(2c+d)$

7. $2x^2-7x-4=(2x+1)(x-4)$

8.
$4x^3-24x^2+35x=x(4x^2-24x+35)$
$=x(2x-5)(2x-7)$

Exercises 6.3

1. e
3. b
5. c
7. $3x^2+16x+5=(x+5)(3x+1)$
9. $5x^2-13x-6=(5x+2)(x-3)$
11. $3x^2-11x+6=(3x-2)(x-3)$
13. $3x^2+8x+5=(3x+5)(x+1)$
15. $2y^2-11y+5=(2y-1)(y-5)$
17. $3x^2+14x+8=(3x+2)(x+4)$
19. Prime polynomial
21. $6y^2-y-5=(6y+5)(y-1)$
23. $2y^2-11y+14=(2y-7)(y-2)$
25. $9a^2-18a-16=(3a+2)(3a-8)$
27. $4x^2-13x+3=(4x-1)(x-3)$

29.
$6+17y+12y^2=12y^2+17y+6$
$=(3y+2)(4y+3)$

31.
$-17m+21+2m^2=2m^2-17m+21$
$=(2m-3)(m-7)$

33.
$-6a^2-7a+3=(-1)(6a^2+7a-3)$
$=-(3a-1)(2a+3)$

35. $8y^2+5y-22=(8y-11)(y+2)$
37. Prime polynomial
39. $8a^2+65a+8=(8a+1)(a+8)$
41. $6x^2+25x-9=(3x-1)(2x+9)$
43. $8y^2-26y+15=(4y-3)(2y-5)$

45.
$14y^2-38y+20=2(7y^2-19y+10)$
$=2(7y-5)(y-2)$

47.
$28a^2+24a-4=4(7a^2+6a-1)$
$=4(7a-1)(a+1)$

49.
$-6b^2+40b+14=-2(3b^2-20b-7)$
$=-2(3b+1)(b-7)$

51.
$12y^3+50y^2+28y=2y(6y^2+25y+14)$
$=2y(3y+2)(2y+7)$

53.
$14a^4-38a^3+20a^2=2a^2(7a^2-19a+10)$
$=2a^2(7a-5)(a-2)$

55.
$2x^3y+13x^2y+15xy=xy(2x^2+13x+15)$
$=xy(2x+3)(x+5)$

57.
$6ab^3-44ab^2+14ab=2ab(3b^2-22b+7)$
$=2ab(3b-1)(b-7)$

59. $20c^2 - 9cd + d^2 = (5c - d)(4c - d)$

61. $2x^2 - 5xy - 3y^2 = (2x + y)(x - 3y)$

63. $8a^2 - 6ab + b^2 = (4a - b)(2a - b)$

65.

$$18x^2 + 3xy - 6y^2 = 3(6x^2 + xy - 2y^2) = 3(3x + 2y)(2x - y)$$

67.

$$16c^2 - 44cd + 30d^2 = 2(8c^2 - 22cd + 15d^2) = 2(4c - 5d)(2c - 3d)$$

69.

$$27u^2 + 18uv + 3v^2 = 3(9u^2 + 6uv + v^2) = 3(3u + v)(3u + v)$$

71.

$$42x^3 + 45x^2y - 27xy^2 = 3x(14x^2 + 15xy - 9y^2) = 3x(7x - 3y)(2x + 3y)$$

$-30x^4y + 35x^3y^2 + 15x^2y^3 =$

73. $-5x^2y(6x^2 - 7xy - 3y^2) =$

$-5x^2y(3x + y)(2x - 3y)$

75.

$$5ax^2 - 28axy - 12ay^2 = a(5x^2 - 28xy - 12y^2) = a(5x + 2y)(x - 6y)$$

Applications

77.

$$-5t^3 - 21t^2 + 20 = (-1)(5t^2 + 21t - 20) = -(5t - 4)(t + 5)$$

79. $4n^2 - 12n + 5 = (2n - 5)(2n - 1)$ The difference between the two integers is $(2n - 1) - (2n - 5) = 2n - 1 - 2n + 5 = 4$. The factors represent two integers that differ by 4 no matter what integer n represents.

6.4 Factoring Perfect Square Trinomials and the Difference of Squares

Practice 6.4

1.a. $x^2 + 6x + 9 = x^2 + 2 \cdot x \cdot 3 + 3^2 = (x + 3)^2$
The trinomial is a perfect square.

b. In the polynomial $-4t^2 - 4t + 1$, the coefficient of the first term is negative. Therefore the polynomial is not a perfect square.

c. $y^2 - 14y + 49 = y^2 + 2 \cdot y \cdot 7 + 7^2 = (y + 7)^2$
The trinomial is a perfect square.

d. $x^2 - 2x - 1$ The trinomial is not a perfect square. The constant term is negative.

e. $4p^2 - 4pq + q^2 = (2p)^2 + 2 \cdot 2p \cdot q + (-q)^2$
The trinomial is a perfect square.

2. $n^2 + 20n + 100 = n^2 + 2 \cdot n + 10^2 = (n + 10)^2$

3.

$t^2 + 4 - 4t = t^2 - 4t + 4 =$

$t^2 + 2 \cdot t \cdot (-2)^2 = (t - 2)^2$

4.

$25c^2 - 40cd + 16d^2 = (5c)^2 + 2 \cdot 5c + (-4d)^2$

$= (5c - 4d)^2$

5.

$x^4 + 8x^2 + 16 = (x^2)^2 + 2 \cdot x^2 \cdot 4 + 4^2 = (x^2 + 4)^2$

6.a. $x^2 - 64 = x^2 - 8^2$
The binomial is a difference of squares.

b. $x^2 + 49 = x^2 + 7^2$
The binomial is not a difference of squares.

c. $x^3 - 16 = x^3 - 4^2$
The binomial is not a difference of squares.

d. $r^4 - 9s^6 = (r^2)^2 - (3s^3)^2$
The binomial is a difference of squares.

7. $y^2 - 121 = y^2 - 11^2 = (y + 11)(y - 11)$

8.

$9x^2 - 25y^2 = (3x)^2 - (5y)^2 = (3x + 5y)(3x - 5y)$

9.

$64x^8 - 81y^2 = (8x^4)^2 - (9y)^2 = (8x^4 + 9y)(8x^4 - 9y)$

10. $256 - 16t^2 = 16(16 - t^2) = 16(4 + t)(4 - t)$

Exercises 6.4

1. Perfect square trinomial
3. Neither
5. Perfect square trinomial
7. Difference of squares
9. Difference of squares
11. Perfect square trinomial
13. Neither

15. Neither

17. Neither

19.

$x^2-12x+36=x^2+2\bullet y\bullet(-6)+(-6)^2$

$=(y-6)^2$

21.

$y^2+20y+100=y^2+2\bullet y+10^2=(y+10)^2$

23.

$a^2-4a+4=a^2+2(-2)a+(-2)^2=$

$(a-2)^2$

25. Prime polynomial

27. $m^2-64=m^2-8^2=(m+8)(m-8)$

29. $y^2-81=y^2-9^2=(y+9)(y-9)$

31. $144-x^2=12^2-x^2=(12+x)(12-x)$

33.

$4a^2-36a+81=(2a)^2+2(2a)(-9)+(-9)^2$

$=(2a-9)^2$

35.

$49x^2+28x+4=(7x)^2+2\bullet 7x\bullet 2+2^2=$

$(7x+2)^2$

37.

$36-60x+25x^2=6^2+2(-5x)(6)+(5x)^2=$

$(6-5x)^2$

39.

$100m^2-81=(10m)^2-9^2=(10m+9)(10m-9)$

41. Prime polynomial

43.

$1-9x^2=1^2-(3x)^2=(1+3x)(1-3x)$

45.

$m^2+26mn+169n^2=m^2+2\bullet m\bullet 13n+(13n)^2$

$=(m+13n)^2$

47.

$4a^2+36ab+81b^2=(2a)^2+2\bullet 2a\bullet(9b)+(9b)^2=$

$(2a+9b)^2$

49. $x^2-4y^2=x^2-(2y)^2=(x+2y)(x-2y)$

51.

$100x^2-9y^2=(10x)^2-(3y)^2=$

$(10x+3y)(10x-3y)$

53.

$y^6+2y^3+1=(y^3)^2+2\bullet y^3\bullet 1+1^2=(y^3+1)^2$

55.

$64x^8+16x^4+1=(8x^4)^2+2\bullet 8x^4\bullet 1+1^2$

$=(8x^4+1)^2$

57.

$100x^{10}-20x^5y^5+y^{10}=$

$(10x^5)^2+2\bullet 10x^5\bullet(-y)^5+(y^5)^2$

$=(10x^5-y^5)^2$

59.

$25m^6-36=(5m^3)^2-(6)^2=(5m^3+6)(5m^3-6)$

61.

$x^4-144y^2=(x^2)^2-(12y)^2=(x^2+12y)(x^2-12y)$

63.

$6x^2+12x+6=6(x^2+2x+1)=$

$6(x^2+2\bullet x\bullet 1+1^2)=6(x+1)^2$

65.

$27m^3-36m^2+12m=3m(9m^2-12m+4)=$

$3m[(3m)^2+2\bullet 3m\bullet(-2)+(-2)^2]=$

$3m(3m-2)^2$

67.

$4s^2t^3+80s^2t^2+400s^2t=$

$4s^2t(t^2+20t+100)=$

$4s^2t[t^2+2\bullet t\bullet 10+10^2]=$

$4s^2t(t+10)^2$

69.

$12x^6y^4-36x^3y^2+27=$

$3(4x^6y^4-12x^3y^2+9)=$

$3[(2x^3y^2)^2-2(2x^3y^2)(-3)+(-3)^2]=$

$3(2x^3y^2-3)^2$

71.

$3k^3-147k=3k(k^2-49)=$

$3k[k^2-(7)^2]=3k(k+7)(k-7)$

73.

$4y^4-36y^2=4y^2(y^2-9)=$

$4y^2[y^2-3^2]=4y^2(y+3)(y-3)$

75.

$27x^2y-3x^2y^3=3x^2y(9-y^2)=$

$3x^2y[3^2-y^2]=3x^2y(3+y)(3-y)$

77.
$2a^2b^2 - 98 = 2(a^2b^2 - 49) =$
$2[(ab)^2 - 7^2] = 2(ab+7)(ab-7)$

79.
$16b^4 - 121 = (4b^2)^2 - 11^2 =$
$(4b^2+11)(4b^2-11)$

81.
$256 - r^4 = 16^2 - (r^2)^2 = (16+r^2)(16-r^2) =$
$(16+r^2)(4^2 - r^2) = (16+r^2)(4+r)(4-r)$

83.
$5x^4 - 80y^8 = 5(x^4 - 16y^8) =$
$5[(x^2)^2 - (4y^4)^2] = 5(x^2+4y^4)(x^2-4y^4) =$
$5(x^2+4y^4)(x-2y^2)(x+2y^2)$

85.
$x^2(c-d) - 4(c-d) = (c-d)(x^2-4) =$
$(c-d)(x+2)(x-2)$

87.
$16(x-y) - a^2(x-y) = (x-y)(16-a^2) =$
$(x-y)(4+a)(4-a)$

Applications

89.
$4\pi r_1^2 - 4\pi r_2^2 = 4\pi\left(r_1^2 - r_2^2\right) =$
$4\pi\left(r_1 - r_2\right)\left(r_1 + r_2\right)$

91.
$16{,}000 + 32{,}000r + 16{,}000r^2 =$
$16{,}000(1+2r+r^2) = 16{,}000(1+r)^2$

93.
$kv_2^2 - kv_1^2 = k(v_2^2 - v_1^2) =$
$k(v_2+v_1)(v_2-v_1)$

6.5 Solving Quadratic Equations by Factoring

Practice 6.5

1.
$(3x-1)(x+5) = 0$

$3x - 1 = 0 \qquad x + 5 = 0$

$3x = 1 \qquad x = -5$

$x = \frac{1}{3}$

$check:$

$\left(3(\frac{1}{3}) - 1\right)\left(\frac{1}{3} + 5\right) \stackrel{?}{=} 0 \qquad (3(-5) - 1)(-5+5) \stackrel{?}{=} 0$

$(0)\left(\frac{16}{3}\right) \stackrel{?}{=} 0 \qquad (-16)(0) \stackrel{?}{=} 0$

$0 = 0 \qquad 0 = 0$

2.
$y^2 + 6y = 0$

$y(y+6) = 0$

$y = 0 \qquad y + 6 = 0$

$y = -6$

$check:$

$0^2 + 6(0) \stackrel{?}{=} 0 \qquad (-6)^2 + 6(-6) \stackrel{?}{=} 0$

$0 + 0 \stackrel{?}{=} 0 \qquad 36 - 36 \stackrel{?}{=} 0$

$0 = 0 \qquad 0 = 0$

3.

$4y^2-11y=3$

$4y^2-11y-3=0$

$(4y+1)(y-3)=0$

$4y+1=0 \qquad y-3=0$

$4y=-1 \qquad y=3$

$y=-\frac{1}{4}$

check :

$4\left(-\frac{1}{4}\right)^2-11\left(-\frac{1}{4}\right)\overset{?}{=}3 \qquad 4(3)^2-11(3)\overset{?}{=}3$

$\frac{1}{4}+\frac{11}{4}\overset{?}{=}3 \qquad 36-33\overset{?}{=}3$

$3=3 \qquad 3=3$

4.

$3t(t+4)=15$

$3t^2+12t-15=0$

$(3t-3)(t+5)$

$3t-3=0 \qquad t+5=0$

$3t=3 \qquad t=-5$

$t=1$

check :

$3(1)(1+4)\overset{?}{=}15 \qquad 3(-5)(-5+4)\overset{?}{=}15$

$(3)(5)\overset{?}{=}15 \qquad (-15)(-1)\overset{?}{=}15$

$15=15 \qquad 15=15$

5.

$P=3ft=36in$

$2l+2w=36$

$l=18-w$

$A=w(18-w)=18w-w^2$

$80=18w-w^2$

$w^2-18w+80=0$

$(w-8)(w-10)=0$

$w-8=0 \qquad w-10=0$

$w=8 \qquad w=10$

$l=10 \qquad l=8$

The dimensions of the frame should be 8 inches by 10 inches.

6.

$a^2+b^2=c^2$

$x^2+5^2=13^2$

$x^2+25=169$

$x^2=144$

$x=12$

The scooter going north has traveled 12 miles.

Exercises 6.5

1. Quadratic

3. Linear

5. Quadratic

7.

$(x+3)(x-4)=0$

$x+3=0 \qquad x-4=0$

$x=-3 \qquad x=4$

check :

$(-3+3)(-3-4)\overset{?}{=}0 \qquad (4+3)(4-4)\overset{?}{=}0$

$(0)(-7)\overset{?}{=}0 \qquad (7)(0)\overset{?}{=}0$

$0=0 \qquad 0=0$

9.

$y(3y+5)=0$

$y=0 \qquad 3y+5=0$

$3y=-5$

$y=-5/3$

check :

$(0)(3(0)+5)\overset{?}{=}0 \qquad (-\frac{5}{3})(3(-\frac{5}{3})+5)\overset{?}{=}0$

$(0)(5)\overset{?}{=}0 \qquad -\frac{5}{3}(0)\overset{?}{=}0$

$0=0 \qquad 0=0$

11.

$(2t+1)(t-5)=0$

$2t+1=0 \quad t-5=0$

$2t=-1 \quad t=5$

$t=-\frac{1}{2}$

check :

$\left(2\left(-\frac{1}{2}\right)+1\right)\left(-\frac{1}{2}-5\right)\stackrel{?}{=}0 \quad (2(5)+1)(5-5)\stackrel{?}{=}0$

$(0)\left(-\frac{11}{2}\right)\stackrel{?}{=}0 \quad (11)(0)\stackrel{?}{=}0$

$0=0 \quad 0=0$

13.

$(2x+3)(2x-3)=0$

$2x+3=0 \quad 2x-3=0$

$2x=-3 \quad 2x=3$

$x=-\frac{3}{2} \quad x=\frac{3}{2}$

check :

$\left(2\left(-\frac{3}{2}\right)+3\right)\left(2\left(-\frac{3}{2}\right)-3\right)\stackrel{?}{=}0$

$(0)(-6)\stackrel{?}{=}0$

$0=0$

$\left(2\left(\frac{3}{2}\right)+3\right)\left(2\left(\frac{3}{2}\right)-3\right)\stackrel{?}{=}0$

$(6)(0)\stackrel{?}{=}0$

$0=0$

15.

$t(2-3t)=0$

$t=0 \quad 2-3t=0$

$-3t=-2$

$t=\frac{2}{3}$

check :

$(0)(2-3(0))\stackrel{?}{=}0 \quad (\frac{2}{3})(2-3(\frac{2}{3}))\stackrel{?}{=}0$

$(0)(2)\stackrel{?}{=}0 \quad (\frac{2}{3})(0)\stackrel{?}{=}0$

$0=0 \quad 0=0$

17.

$y^2-2y=0$

$y(y-2)=0$

$y=0 \quad y-2=0$

$y=2$

check :

$0^2-2(0)\stackrel{?}{=}0 \quad (2)^2-2(2)\stackrel{?}{=}0$

$0-0\stackrel{?}{=}0 \quad 4-4\stackrel{?}{=}0$

$0=0 \quad 0=0$

19.

$5x-25x^2=0$

$5x(1-5x)=0$

$5x=0 \quad 1-5x=0$

$t=0 \quad -5x=-1$

$x=\frac{1}{5}$

check :

$5(0)-25(0)^2\stackrel{?}{=}0 \quad 5(\frac{1}{5})-25(\frac{1}{5})^2\stackrel{?}{=}0$

$0-0\stackrel{?}{=}0 \quad 1-1\stackrel{?}{=}0$

$0=0 \quad 0=0$

21.

$x^2+5x+6=0$

$(x+2)(x+3)=0$

$x+2=0 \qquad x+3=0$

$x=-2 \qquad x=-3$

check :

$(-2)^2+5(-2)+6\overset{?}{=}0 \quad (-3)^2+5(-3)+6\overset{?}{=}0$

$4-10+6\overset{?}{=}0 \qquad 9-15+6\overset{?}{=}0$

$0=0 \qquad 0=0$

23.

$x^2+x-56=0$

$(x+8)(x-7)=0$

$y+8=0 \qquad y-7=0$

$y=-8 \qquad y=7$

check :

$(-8)^2+(-8)-56\overset{?}{=}0 \quad 7^2+7-56\overset{?}{=}0$

$64-8-56\overset{?}{=}0 \qquad 49+7-56\overset{?}{=}0$

$0=0 \qquad 0=0$

25.

$2x^2-5x-3=0$

$(2x+1)(x-3)=0$

$2x+1=0 \qquad x-3=0$

$2x=-1 \qquad x=3$

$x=-\frac{1}{2}$

check :

$2(-\frac{1}{2})^2-5(-\frac{1}{2})-3\overset{?}{=}0 \quad 2(3)^2-5(3)-3\overset{?}{=}0$

$\frac{1}{2}+\frac{5}{2}-3\overset{?}{=}0 \qquad 18-15-3\overset{?}{=}0$

$0=0 \qquad 0=0$

27.

$6x^2-x-2=0$

$(3x-2)(2x+1)=0$

$3x-2=0 \qquad 2x+1=0$

$3x=2 \qquad 2x=-1$

$x=\frac{2}{3} \qquad x=-\frac{1}{2}$

check :

$6(\frac{2}{3})^2-\frac{2}{3}-2\overset{?}{=}0 \quad 6(-\frac{1}{2})^2-(-\frac{1}{2})-2\overset{?}{=}0$

$\frac{8}{3}-\frac{2}{3}-2\overset{?}{=}0 \qquad \frac{3}{2}+\frac{1}{2}-2\overset{?}{=}0$

$0=0 \qquad 0=0$

29.

$0=36x^2-12x+1$

$0=(6x-1)(6x-1)$

$6x-1=0 \qquad 6x-1=0$

$6x=1 \qquad 6x=1$

$x=\frac{1}{6} \qquad x=\frac{1}{6}$

check :

$0\overset{?}{=}36(\frac{1}{6})^2-12(\frac{1}{6})+1$

$0\overset{?}{=}1-2+1$

$0=0$

31.

$r^2-121=0$

$(r-11)(r+11)=0$

$r-11=0 \qquad r+11=0$

$r=11 \qquad r=-11$

check :

$11^2-121\overset{?}{=}0 \quad (-11)^2-121\overset{?}{=}0$

$121-121\overset{?}{=}0 \quad 121-121\overset{?}{=}0$

$0=0 \qquad 0=0$

33.

$0=(2x-3)^2$

$0=2x-3$

$3=2x$

$2x=3$

$x=\dfrac{3}{2}$

check :

$0\overset{?}{=}(2(\frac{3}{2})-3)^2$

$0=0$

35.

$16x^2-16x+4=0$

$4x^2-4x+1=0$

$(2x-1)^2=0$

$2x-1=0$

$2x=1$

$x=\dfrac{1}{2}$

check :

$16(\frac{1}{2})^2-16(\frac{1}{2})+4\overset{?}{=}0$

$4-8+4\overset{?}{=}0$

$0=0$

37.

$9m^2+15m-6=0$

$3m^2+5m-2=0$

$(3m-1)(m+2)=0$

$3m-1=0 \quad m+2=0$

$3m=1 \qquad m=-2$

$m=\dfrac{1}{3}$

check :

$9(\frac{1}{3})^2+15(\frac{1}{3})-6\overset{?}{=}0 \quad 9(-2)^2+15(-2)-6\overset{?}{=}0$

$1+5-6\overset{?}{=}0 \qquad 36-30-6\overset{?}{=}0$

$0=0 \qquad 0=0$

39.

$r^2-r=6$

$r^2-r-6=0$

$(r+2)(r-3)=0$

$r+2=0 \qquad r-3=0$

$r=-2 \qquad r=3$

check :

$(-2)^2-(-2)\overset{?}{=}6 \quad (3)^2-3\overset{?}{=}6$

$6=6 \qquad 6=6$

41.

$y^2-7y=-12$

$y^2-7y+12=0$

$(y-4)(y-3)=0$

$r-4=0 \qquad r-3=0$

$r=4 \qquad r=3$

check :

$4^2-7(4)\overset{?}{=}-12 \quad (3)^2-7(3)\overset{?}{=}-12$

$16-28\overset{?}{=}-12 \qquad 9-21\overset{?}{=}-12$

$-12=-12 \qquad -12=-12$

43.

$n^2+2n=8$

$n^2+2n-8=0$

$(n-2)(n+4)=0$

$n-2=0 \qquad n+4=0$

$n=2 \qquad n=-4$

check :

$2^2+2(2)\overset{?}{=}8 \quad (-4)^2+2(-4)\overset{?}{=}8$

$4+4\overset{?}{=}8 \qquad 16-8\overset{?}{=}8$

$8=8 \qquad 0=0$

45.

$3y^2+4y=-1$

$3y^2+4y+1=0$

$(3y+1)(y+1)=0$

$3y+1=0 \quad y+1=0$

$3y=-1$

$y=-\frac{1}{3} \quad y=-1$

check :

$3(-\frac{1}{3})^2+4(-\frac{1}{3})\stackrel{?}{=}-1 \quad 3(-1)^2+4(-1)\stackrel{?}{=}-1$

$\frac{1}{3}+(-\frac{4}{3})\stackrel{?}{=}-1 \quad 3-4\stackrel{?}{=}-1$

$-1=-1 \quad -1=-1$

47.

$4x^2+6x=-2$

$4x^2+6x+2=0$

$2x^2+3x+1=0$

$(2x+1)(x+1)=0$

$2x+1=0 \quad x+1=0$

$x=-\frac{1}{2} \quad x=-1$

check :

$4(-\frac{1}{2})^2+6(-\frac{1}{2})\stackrel{?}{=}-2 \quad 4(-1)^2+6(-1)=-2$

$1-3\stackrel{?}{=}-2 \quad 4-6\stackrel{?}{=}-2$

$-2=-2 \quad -2=-2$

49.

$2n^2=-10n$

$2n^2+10n=0$

$2n(n+5)=0$

$2n=0 \quad n+5=0$

$n=0 \quad n=-5$

check :

$2(0)^2\stackrel{?}{=}-10(0) \quad 2(-5)^2\stackrel{?}{=}-10(-5)$

$0=0 \quad 50=50$

51.

$4x^2=1$

$4x^2-1=0$

$(2x+1)(2x-1)=0$

$2x+1=0 \quad 2x-1=0$

$2x=-1 \quad 2x=1$

$x=-\frac{1}{2} \quad x=\frac{1}{2}$

check :

$4(-\frac{1}{2})^2\stackrel{?}{=}1 \quad 4(\frac{1}{2})^2\stackrel{?}{=}1$

$1=1 \quad 1=1$

53.

$8y^2=2$

$8y^2-2=0$

$2(4y^2-1)=0$

$4y^2-1=0$

$(2y+1)(2y-1)=0$

$2y+1=0 \quad 2y-1=0$

$2y=-1 \quad 2y=1$

$y=-\frac{1}{2} \quad y=\frac{1}{2}$

check :

$8(-\frac{1}{2})^2\stackrel{?}{=}2 \quad 8(\frac{1}{2})^2\stackrel{?}{=}2$

$2=2 \quad 2=2$

55.

$3r^2+6r=2r^2-9$

$r^2+6r+9=0$

$(r+3)^2=0$

$r+3=0$

$r=-3$

check :

$3(-3)^2+6(-3)\stackrel{?}{=}2(-3)^2-9$

$27-18\stackrel{?}{=}18-9$

$9=9$

57.

$x(x-1)=12$

$x^2-x=12$

$x^2-x-12=0$

$(x-4)(x+3)=0$

$x-4=0 \quad x+3=0$

$x=4 \qquad x=-3$

check :

$4(4-1)\stackrel{?}{=}12 \quad -3(-3-1)\stackrel{?}{=}12$

$12=12 \qquad 12=12$

59.

$4t(t-1)=24$

$4t^2-4t=24$

$4t^2-4t-24=0$

$t^2-t-6=0$

$(t+2)(t-3)=0$

$t+2=0 \quad t-3=0$

$t=-2 \qquad t=3$

check :

$4(-2)(-2-1)\stackrel{?}{=}24 \quad 4(3)(3-1)\stackrel{?}{=}24$

$24=24 \qquad 24=24$

61.

$(y+3)(y-2)=14$

$y^2+y-6=14$

$y^2+y-20=0$

$(y-4)(y+5)=0$

$y-4=0 \qquad y+5=0$

$y=4 \qquad y=-5$

check :

$(4+3)(4-2)\stackrel{?}{=}14 \quad (-5+3)(-5-2)\stackrel{?}{=}14$

$14=14 \qquad 14=14$

63.

$(3n-2)(n+5)=-14$

$3n^2+13n-10=-14$

$3n^2+13n+4=0$

$(3n+1)(n+4)=0$

$3n+1=0 \qquad n+4=0$

$n=-\frac{1}{3} \qquad n=-4$

check :

$(3(-\frac{1}{3})-2)(-\frac{1}{3}+5)\stackrel{?}{=}-14 \quad (3(-4)-2)(-4+5)\stackrel{?}{=}-14$

$(-1-2)(\frac{14}{3})\stackrel{?}{=}-14 \qquad (-12-2)(1)\stackrel{?}{=}-14$

$(-3)(\frac{14}{3})\stackrel{?}{=}-14 \qquad (-14)(1)\stackrel{?}{=}-14$

$-14=-14 \qquad -14=-14$

65.

$(n+2)(n+4)=12n$

$n^2+6n+8=12n$

$n^2-6n+8=0$

$(n-2)(n-4)=0$

$n-2=0 \qquad n-4=0$

$n=2 \qquad n=4$

check :

$(2+2)(2+4)\stackrel{?}{=}12(2) \quad (4+2)(4+4)\stackrel{?}{=}12(4)$

$24=24 \qquad 48=48$

67.

$3x(2x-5)=x^2-10$

$6x^2-15x=x^2-10$

$5x^2-15x+10=0$

$x^2-3x+2=0$

$(x-2)(x-1)=0$

$x-2=0 \qquad x-1=0$

$x=2 \qquad x=1$

check :

$3(2)(2(2)-5)\stackrel{?}{=}2^2-10 \quad 3(1)(2(1)-5)\stackrel{?}{=}1^2-10$

$6(-1)\stackrel{?}{=}4-10 \qquad 3(-3)\stackrel{?}{=}1-10$

$-6=-6 \qquad -9=-9$

Applications

69.

$n^2 - n = 210$

$n^2 - n - 210 = 0$

$(n-15)(n+14) = 0$

$n - 15 = 0 \quad n + 14 = 0$

$n = 15 \qquad n = -14$

There were 15 teams in the league.

71.

$x^2 + (x+2)^2 = 10^2$

$x^2 + x^2 + 4x + 4 = 100$

$2x^2 + 4x - 96 = 0$

$x^2 + 2x - 48 = 0$

$(x-6)(x+8) = 0$

$x - 6 = 0 \quad x + 8 = 0$

$x = 6 \qquad x = -8$

$x + 2 = 8 \quad x + 2 = -6$

One car traveled 6 miles and the other traveled 8 miles.

73. w=width of room;(w+4)=length of room

$w(w+4) = 192$

$w^2 + 4w - 192 = 0$

$(w-12)(w+16) = 0$

$w - 12 = 0 \quad w + 16 = 0$

$w = 12 \qquad w = -16$

$w + 4 = 16 \quad w + 4 = -12$

The length of the room is 16 feet and the width is 12 feet.

75.

$-16t^2 + 8t + 24 = 0$

$-8(2t^2 - t - 3) = 0$

$2t^2 - t - 3 = 0$

$(2t-3)(t+1) = 0$

$2t - 3 = 0 \qquad t + 1 = 0$

$t = \frac{3}{2} \qquad t = -1$

The diver will hit the water in $\frac{3}{2}$, or 1.5 seconds.

Chapter 6 Review Exercises

1.

$48 = 2^4 \cdot 3$

$36 = 2^2 \cdot 3^2$

$60 = 2^2 \cdot 3 \cdot 5$

$GCF = 2^2 \cdot 3 = 12$

2.

$9m^3n = 3^2 \cdot m^3 \cdot n$

$24m^4 = 2^3 \cdot 3 \cdot m^4$

$15m^2n^2 = 3 \cdot 5 \cdot m^2 \cdot n^2$

$GCF = 3 \cdot m^2 = 3m^2$

3. $3x - 6y = 3(x - 2y)$

4.

$16p^3q^2 + 18p^2q - 4pq^2 = 2pq(8p^2q + 9p - 2q)$

5.

$(n-1) + n(n-1) = 1(n-1) + n(n-1) = (n-1)(1+n)$

6.

$xb - 5b - 2x + 10 = (xb - 5b) + (-2x + 10) =$

$b(x-5) + (-2)(x-5) = (x-5)(b-2)$

7.

$d = rt_1 + rt_2$

$d = r(t_1 + t_2)$

$\frac{d}{t_1 + t_2} = r \quad or \quad r = \frac{d}{t_1 + t_2}$

8.

$ax + y = bx + c$

$ax + y - y = bx + c - y$

$ax = bx + c - y$

$ax - bx = bx - bx + c - y$

$ax - bx = c - y$

$x(a-b) = c - y$

$x = \frac{c-y}{a-b}$

9. Prime polynomial

10. Prime polynomial

11.

$y^2 + 42 + 13y = y^2 + 13y + 42 =$

$(y+6)(y+7)$

12. $m^2 - 7mn + 10n^2 = (m-2n)(m-5n)$

13.

$24 - 8x - 2x^2 = -2x^2 - 8x + 24 =$

$-2(x^2 + 4x - 12) = -2(x-2)(x+6)$

14.

$-15xy^2 + 3x^3 - 12x^2y = 3x(-5y^2 + x^2 - 4xy) =$

$3x(x^2 - 4xy - 5y^2) = 3x(x+y)(x-5y)$

15. $3x^2 + 5x - 2 = (3x-1)(x+2)$

16. $5n^2 + 13n + 6 = (5n+3)(n+2)$

17. Prime polynomial

18. $6x^2 - x - 12 = (3x+4)(2x-3)$

19. $2a^2 + 3ab - 35b^2 = (2a-7b)(a+5b)$

20.

$16a - 4a^2 - 15 = -4a^2 + 16a - 15 =$

$-(4a^2 - 16a + 15) = -(2a-3)(2a-5)$

21.

$9y^3 - 21y + 60y^2 = 9y^3 + 60y^2 - 21y =$

$3y\left(3y^2 + 20y - 7\right) = 3y(3y-1)(y+7)$

22.

$2p^2q - 3pq^2 - 2q^3 = q\left(2p^2 - 3pq - 2q^2\right) =$

$q(2p+q)(p-2q)$

23.

$b^2 - 6b + 9 = b^2 + 2 \bullet b \bullet (-3) + (-3)^2$

$= (b-3)^2$

24. $64 - x^2 = 8^2 - x^2 = (8-x)(8+x)$

25.

$25y^2 - 20y + 4 = (5y)^2 + 2(5y)(-2) + (-2)^2$

$= (5y-2)^2$

26.

$9a^2 + 24ab + 16b^2 = (3a)^2 + 2 \bullet 3a \bullet 4b + (4b)^2 =$

$(3a+4b)^2$

27.

$81p^2 - 100q^2 = (9p)^2 - (10q)^2 =$

$(9p-10q)(9p+10q)$

28.

$4x^8 - 28x^4 + 49 = (2x^4)^2 + 2(2x^4)(-7) + (-7)^2$

$= (2x^4 - 7)^2$

29.

$48x^4 - 3y^4 = 3(16x^4 - y^4) =$

$3[(4x^2)^2 - (y^2)^2] = 3(4x^2 + y^2)(4x^2 - y^2) =$

$3(4x^2 + y^2)(2x+y)(2x-y)$

30.

$x^2(x-1) - 9(x-1) = (x-1)(x^2 - 9) =$

$(x-1)(x+3)(x-3)$

31.

$(x+2)(x-1) = 0$

$x+2=0 \quad x-1=0$

$x=-2 \qquad x=1$

check :

$(-2+2)(-2-1) \stackrel{?}{=} 0 \quad (1+2)(1-1) \stackrel{?}{=} 0$

$(0)(-3) \stackrel{?}{=} 0 \qquad (3)(0) \stackrel{?}{=} 0$

$0=0 \qquad 0=0$

32.

$t(t-4) = 0$

$t=0 \quad t-4=0$

$t-4$

check :

$0(0-4) \stackrel{?}{=} 0 \quad 4(4-4) \stackrel{?}{=} 0$

$0(-4) \stackrel{?}{=} 0 \qquad 4(0) \stackrel{?}{=} 0$

$0=0 \qquad 0=0$

33.

$3x^2 + 18x = 0$

$3x(x+6) = 0$

$3x=0 \quad x+6=0$

$x=0 \quad x=-6$

check :

$3(0)^2 + 18(0) \stackrel{?}{=} 0 \quad 3(-6)^2 + 18(-6) \stackrel{?}{=} 0$

$0+0 \stackrel{?}{=} 0 \qquad 108 + (-108) \stackrel{?}{=} 0$

$0=0 \qquad 0=0$

34.

$4x^2+4x+1=0$

$(2x+1)^2=0$

$2x+1=0$

$2x=-1$

$x=-\frac{1}{2}$

check :

$4(-\frac{1}{2})^2+4(-\frac{1}{2})+1\overset{?}{=}0$

$1-2+1\overset{?}{=}0$

$0=0$

35.

$y^2-10y=-16$

$y^2-10y+16=0$

$(y-2)(y-8)=0$

$y-2=0 \quad y-8=0$

$y=2 \qquad y=8$

check :

$2^2-10(2)\overset{?}{=}-16 \quad 8^2-10(8)\overset{?}{=}-16$

$4-20\overset{?}{=}-16 \qquad 64-80\overset{?}{=}-16$

$-16=-16 \qquad -16=-16$

36.

$3k^2-k=2$

$3k^2-k-2=0$

$(3k+2)(k-1)=0$

$3k+2=0 \quad k-1=0$

$3k=-2 \qquad k=1$

$k=-\frac{2}{3}$

check :

$3\left(-\frac{2}{3}\right)^2-\left(-\frac{2}{3}\right)\overset{?}{=}2 \quad 3(1)^2-1\overset{?}{=}2$

$\frac{4}{3}+\frac{4}{3}\overset{?}{=}2 \qquad 3-1\overset{?}{=}2$

$2=2 \qquad 2=2$

37.

$4n(2n+3)=20$

$8n^2+12n-20=0$

$4(2n^2+3n-5)=0$

$2n^2+3n-5=0$

$(2n+5)(n-1)=0$

$2n+5=0 \quad n-1=0$

$2n=-5 \qquad n=1$

$n=-\frac{5}{2}$

check :

$4\left(-\frac{5}{2}\right)\left(2\left(-\frac{5}{2}\right)+3\right)\overset{?}{=}20$

$(-10)(-2)\overset{?}{=}20$

$20=20$

$4(1)(2(1)+3)\overset{?}{=}20$

$(4)(5)\overset{?}{=}20$

$20=20$

38.

$(y-1)(y+2)=10$

$y^2+y-2=10$

$y^2+y-12=0$

$(y-3)(y+4)=0$

$y-3=0 \quad y+4=0$

$y=3 \qquad y=-4$

check :

$(3-1)(3+2)\overset{?}{=}10 \quad (-4-1)(-4+2)\overset{?}{=}10$

$10=10 \qquad 10=10$

Mixed Applications

39. $aLt_2-aLt_1=aL(t_2-t_2)$

40.

$at_2 - 16t_2^2 - (at_1 - 16t_1^2) =$

$at_2 - 16t_2^2 - at_1 + 16t_1^2 =$

$(at_2 - at_1) + (-16t_2^2 + 16t_1^2) =$

$a(t_2 - t_1) - 16(t_2^2 - t_1^2) =$

$(t_2 - t_1)[a - 16(t_2 + t_1)]$

41.

$c =$ number of blocks

1 block = 500 feet

$a^2 + b^2 = c^2$

$4^2 + 3^2 = c^2$

$16 + 9 = c^2$

$25 = c^2$

$c = 5$ blocks

5 blocks = 2500 ft

The distance between the two intersections is 2500 feet.

42.

2x = length of the horizontal diagonal.

$12^2 + x^2 = 20^2$

$144 + x^2 = 400$

$x^2 = 256$

$x = 16$

$2x = 32$

The length of the horizontal diagonal of the kite is 32 inches.

43.

$76t - 16t^2 = 18$

$-16t^2 + 76t - 18 = 0$

$-2(8t^2 - 38t + 9) = 0$

$8t^2 - 38t + 9 = 0$

$(4t - 1)(2t - 9) = 0$

$4t - 1 = 0 \quad 2t - 9 = 0$

$4t = 1 \quad 2t = 9$

$t = \frac{1}{4} \quad t = \frac{9}{2}$

The rocket will reach a height of 18 feet above the launch in $\frac{1}{4}$ second and $\frac{9}{2}$ seconds, or in 0.25 second and 4.5 seconds.

Chapter 6 Posttest

1.

$12x^3 = 2^2 \cdot 3 \cdot x^3$

$15x^2 = 3 \cdot 5 \cdot x^2$

$GCF = 3 \cdot x^2 = 3x^2$

2.

$2xy - 14y = 2y(x - 7)$

3.

$6pq^2 + 8p^3 - 16p^2q = 8p^3 - 16p^2q + 6pq^2 =$

$2p(4p^2 - 8pq + 3q^2) = 2p(2p - 3q)(2p - q)$

4.

$ax - bx + by - ay = (ax - bx) + (-1)(ay - by) =$

$x(a - b) - y(a - b) = (a - b)(x - y)$

5. $n^2 - 13n - 48 = (n + 3)(n - 16)$

6. $-8 + x^2 - 2x = x^2 - 2x - 8 = (x + 2)(x - 4)$

7.

$15x^2 - 5x^3 + 20x = -5x^3 + 15x^2 + 20x =$

$-5x(x^2 - 3x - 4) = -5x(x + 1)(x - 4)$

8. $4x^2 + 13xy - 12y^2 = (4x - 3y)(x + 4y)$

9.

$-12x^2 + 36x - 27 = -3(4x^2 - 12x + 9) =$

$-3(2x - 3)^2$

10.
$9x^2+30xy+25y^2=(3x)^2+2\bullet 3x+(5y)^2=$
$(3x+5y)^2$

11.
$121-4x^2=11^2-(2x)^2=$
$(11+2x)(11-2x)$

12.
$p^2q^2-1=(pq)^2-1^2=$
$(pq+1)(pq-1)$

13.
$y^4-8y^2+16=(y^2-4)(y^2-4)=$
$(y+2)(y-2)(y+2)(y-2)=$
$(y+2)^2(y-2)^2$

14.
$(n+8)(n-1)=0$
$n+8=0 \quad n-1=0$
$n=-8 \qquad n=1$
check:
$(-8+8)(-8-1)\overset{?}{=}0 \quad (1+8)(1-1)\overset{?}{=}0$
$(0)(-9)\overset{?}{=}0 \qquad (9)(0)\overset{?}{=}0$
$0=0 \qquad 0=0$

15.
$6x^2+10x=4$
$6x^2+10x-4=0$
$(6x-2)(x+2)=0$
$6x-2=0 \quad x+2=0$
$6x=2 \qquad x=-2$
$x=\frac{1}{3}$
check:
$6(\frac{1}{3})^2+10(\frac{1}{3})\overset{?}{=}4 \quad 6(-2)^2+10(-2)\overset{?}{=}4$
$\frac{2}{3}+\frac{10}{3}\overset{?}{=}4 \qquad 24+(-20)\overset{?}{=}4$
$4=4 \qquad 4=4$

16.
$(2n+1)(n-1)=5$
$2n^2-n-6=0$
$(2n+3)(n-2)=0$
$2n+3=0 \quad n-2=0$
$2n=-3 \qquad n=2$
$n=-\frac{3}{2}$
check:
$(2(-\frac{3}{2})+1)(-\frac{3}{2}-1)\overset{?}{=}5 \quad (2(2)+1)(2-1)\overset{?}{=}5$
$(-2)(-\frac{5}{2})\overset{?}{=}5 \qquad (5)(1)\overset{?}{=}5$
$5=5 \qquad 5=5$

17.
$x^2+15^2=17^2$
$x^2+225=289$
$x^2=64$
$x=8$
The missing side is 8 m in length.

18. $mgy_2-mgy_1=mg(y_2-y_1)$

19.
$9-16t^2=0$
$(3-4t)(3+4t)=0$
$3-4t=0 \quad 3+4t=0$
$-4t=-3 \quad 4t=-3$
$t=\frac{3}{4} \qquad t=-\frac{3}{4}$
The weight reaches the ground in $\frac{3}{4}$, or 0.75 second.

20.
$A=(2x+25)(2x+30)-(25)(30)$
$=4x^2+110x+750-750$
$=4x^2+110x=2x(2x+55)$
The area of the mulch border is 2x(2x + 55) square feet.

Cumulative Review Exercises

1. $2(-4)+0-(-5)=-8+0+5=-3$

2. $5a+2-2b+3a+9b=8a+7b+2$

3.

$$z+4=5z-2(z+6)$$
$$z+4=5z-2z-12$$
$$z+4=3z-12$$
$$z+4-4=3z-12-4$$
$$z=3z-16$$
$$z-3z=3z-3z-16$$
$$-2z=-16$$
$$\frac{-2z}{-2}=\frac{-16}{-2}$$
$$z=8$$

check:

$$8+4\overset{?}{=}5(8)-2(8+6)$$
$$12\overset{?}{=}40-28$$
$$12=12$$

4.

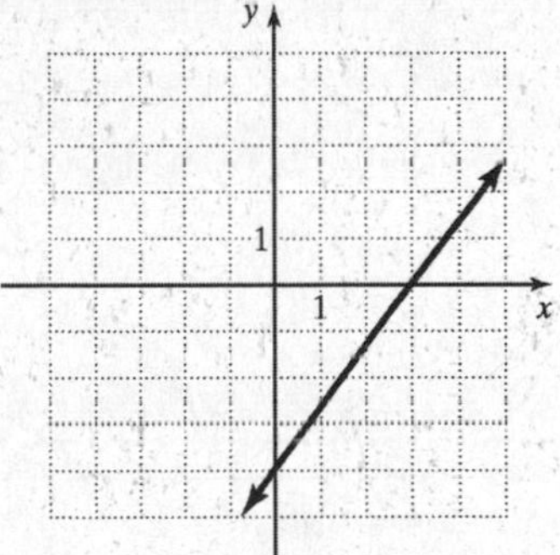

5.

$$x+y=6$$
$$y=2x+9$$
$$x+(2x+9)=6$$
$$3x+9=6$$
$$3x+9-9=6-9$$
$$3x=-3$$
$$\frac{3x}{3}=\frac{-3}{3}$$
$$x=-1$$
$$y=2(-1)+9=7$$

check:

$$-1+7\overset{?}{=}6 \quad 7\overset{?}{=}2(-1)+9$$
$$6=6 \qquad 7=7$$

6.

$$\frac{n^2 \bullet n^3 \bullet n}{n^5}=\frac{n^6}{n^5}=n^{6-5}=n$$

7. $(3x+1)(7x-2)=21x^2+x-2$

8. The charge is $f+8c$ dollars.

9.

x = number of shirts produced and sold

$$15x+500=25x$$
$$500=10x$$
$$10x=500$$
$$x=50$$

The company must sell 50 shirts per day in order to break even.

10. $2\pi r^2+2\pi rh=2\pi r(r+h)$

Chapter 7 Rational Expressions and Equations

Pretest Chapter 7

1.

$x+6=0$

$x=-6$

The expression is undefined when x = -6.

2. $\frac{n^2-2n}{3n}=\frac{n(n-2)}{3n}=\frac{n-2}{3}$

3. $\frac{24x^2y^3}{6xy^5}=\frac{4\cdot 6x\cdot x\cdot y^3}{6\cdot x\cdot y^3\cdot y^2}=\frac{4x}{y^2}$

4. $\frac{4a^2-8a}{a-2}=\frac{4a(a-2)}{(a-2)}=4a$

5. $\frac{w^2-6w}{36-w^2}=\frac{w(w-6)}{(-1)(w+6)(w-6)}=-\frac{w}{w+6}$

6. $\frac{\frac{5n}{8}}{\frac{n^3}{16}}=\frac{5n}{8}\div\frac{n^3}{16}=\frac{5n}{8}\cdot\frac{16}{n^3}=\frac{10}{n^2}$

7.

$3a=3\cdot a$

$9a-18=3^2\cdot(a-2)$

$LCD=3^2\cdot a(a-2)=9a(a-2)$

$\frac{2}{3a}=\frac{2\cdot 3(a-2)}{3a\cdot 3(a-2)}=\frac{6(a-2)}{9a(a-2)}$

$\frac{1}{9a-18}=\frac{1\cdot a}{(9a-18)\cdot a}=\frac{a}{9a(a-2)}$

8.

$\frac{12y}{y+1}-\frac{7y+2}{y+1}=\frac{12y-7y-2}{y+1}=\frac{5y-2}{y+1}$

9.

$\frac{1}{6x^2}+\frac{5}{4x}=\frac{2}{12x^2}+\frac{15x}{12x^2}=\frac{15x+2}{12x^2}$

10.

$\frac{3}{c-3}-\frac{1}{c+3}=\frac{3(c+3)}{(c-3)(c+3)}-\frac{1(c-3)}{(c-3)(c+3)}=$

$\frac{3c+9-c+3}{(c-3)(c+3)}=\frac{2c+12}{(c-3)(c+3)}$

11.

$\frac{1}{x^2-2x+1}-\frac{2}{1-x^2}=$

$\frac{1(-1)(x+1)}{(-1)(x+1)(x-1)^2}-\frac{2(x-1)}{(-1)(x+1)(x-1)^2}=$

$\frac{-x-1-2x+2}{(-1)(x+1)(x-1)^2}=\frac{-3x+1}{(-1)(x+1)(x-1)^2}=$

$\frac{3x-1}{(x+1)(x-1)^2}$

12.

$\frac{15a^3b}{20n^4}\cdot\frac{16n^2}{9ab}=\frac{15\cdot a^3\cdot b}{20\cdot n^4}\cdot\frac{16\cdot n^2}{9\cdot a\cdot b}=\frac{20a^2}{15n^2}=\frac{4a^2}{3n^2}$

13.

$\frac{y-4}{5y^2+10y}\cdot\frac{y^2-2y-8}{y^2-16}=\frac{(y-4)}{5y(y+2)}\cdot\frac{(y+2)(y-4)}{(y+4)\cdot(y-4)}=$

$\frac{y-4}{5y(y+4)}$

14.

$\frac{x^2-x-2}{x^2+5x+4}\div\frac{x^2-7x+10}{x-5}=\frac{(x-2)(x+1)}{(x+4)(x+1)}\cdot\frac{x-5}{(x-5)(x-2)}=$

$\frac{(x-2)(x+1)}{(x+4)(x+1)}\cdot\frac{(x-5)}{(x-5)\cdot(x-2)}=$

$\frac{1}{(x+4)}$

15.

$$\frac{3x-8}{x^2-4}+\frac{2}{x-2}=\frac{7}{x+2}$$

$$\left(x^2-4\right)\left(\frac{3x-8}{x^2-4}+\frac{2}{x-2}\right)=\left(\frac{7}{x+2}\right)\left(x^2-4\right)$$

$$3x-8+2(x+2)=7(x-2)$$

$$5x-4=7x-14$$

$$-2x=-10$$

$$x=5$$

check :

$$\frac{3(5)-8}{5^2-4}+\frac{2}{5-2}\stackrel{?}{=}\frac{7}{5+2}$$

$$\frac{7}{21}+\frac{2}{3}=\frac{7}{7}$$

$$1=1$$

16.

$$\frac{x}{x+4}-1=\frac{2}{x-1}$$

$$(x+4)(x-1)\left(\frac{x}{x+4}-1\right)=\left(\frac{2}{x-1}\right)(x+4)(x-1)$$

$$x(x-1)-(x+4)(x-1)=2(x+4)$$

$$x^2-x-x^2-3x+4=2x+8$$

$$-4x+4=2x+8$$

$$-6x=4$$

$$x=-\frac{2}{3}$$

check :

$$\frac{-\frac{2}{3}}{-\frac{2}{3}+4}-1\stackrel{?}{=}\frac{2}{-\frac{2}{3}-1}$$

$$\frac{-\frac{2}{3}}{\frac{10}{3}}-1\stackrel{?}{=}\frac{2}{-\frac{5}{3}}$$

$$-\frac{2}{10}-1\stackrel{?}{=}-\frac{6}{5}$$

$$-\frac{6}{5}=-\frac{6}{5}$$

17.

$$\frac{9}{2x}=\frac{x}{x-1}$$

$$9(x-1)=2x(x)$$

$$9x-9=2x^2$$

$$2x^2-9x+9=0$$

$$(2x-3)(x-3)=0$$

$$2x-3=0 \quad x-3=0$$

$$x=\frac{3}{2} \qquad x=3$$

check :

$$\frac{9}{2\left(\frac{3}{2}\right)}\stackrel{?}{=}\frac{\frac{3}{2}}{\frac{3}{2}-1} \qquad \frac{9}{2(3)}\stackrel{?}{=}\frac{3}{3-1}$$

$$\frac{9}{3}\stackrel{?}{=}\frac{\frac{3}{2}}{\frac{1}{2}} \qquad \frac{9}{6}\stackrel{?}{=}\frac{3}{2}$$

$$3=3 \qquad \frac{3}{2}=\frac{3}{2}$$

18.

$$2+\frac{1500}{x}=\frac{2\cdot x}{x}+\frac{1500}{x}=\frac{2x+1500}{x}$$

19.

	r	t	d
Part 1	r	$\frac{195}{r}$	195
Part 2	r-10	$\frac{165}{r-10}$	165

$$\frac{195}{r}+\frac{165}{r-10}=6$$

$$195(r-10)+165r=6r^2-60r$$

$$195r-1950+165r=6r^2-60r$$

$$6r^2-420r+1950=0$$

$$r^2+70r+325=0$$

$$(r-65)(r-5)=0$$

$$r-65=0 \quad r-5=0$$

$$r=65 \qquad r=5$$

The average speed during the first part of the trip was 65 mph and the average speed during the second part was 55 mph.

20.

$$\frac{30}{t}=\frac{90}{t+5}$$

$$30t+150=90t$$

$$150=60t$$

$$2.5=t$$

It takes the photocopier $2\frac{1}{2}$ (or 2.5) minutes to make 30 copies.

7.1 Rational Expressions and Equations

Practice 7.1

1. a. The expression is undefined when n is equal to 3.

$$n-3=0$$

$$n=3$$

b. The expression is undefined when n is equal to -3 or 3.

$$n^2-9=0$$

$$n^2=9$$

$$n=\pm 3$$

2. a. The expressions are equivalent.

$$\frac{b\bullet 3b}{3\bullet 3b}=\frac{3b^2}{9b}$$

b. The expressions are equivalent.

$$\frac{5\bullet a}{(a+7)\bullet a}=\frac{5a}{a^2+7a}$$

3.

a. $$\frac{-12n^3}{-6mn}=\frac{2\bullet(\cancel{-6})\bullet \cancel{n^3}}{\cancel{-6}\bullet m\bullet \cancel{n}}=\frac{2n^2}{m}$$

b. $$\frac{-3x^4y}{2x^2y}=\frac{(-1)\bullet 3\bullet(\cancel{x^4})\bullet \cancel{y}}{2\bullet \cancel{x^2}\bullet \cancel{y}}=-\frac{3x^2}{2}$$

4.

a. $$\frac{2x-8}{4y+6}=\frac{\cancel{2}\bullet(x-4)}{\cancel{2}\bullet(2y+3)}=\frac{x-4}{2y+3}$$

b. The expression cannot be simplified.

c. $$\frac{x+2}{3x+6}=\frac{\cancel{x+2}}{3\bullet(\cancel{x+2})}=\frac{1}{3}$$

5.

a. $$\frac{wt-wx}{wz-3wg}=\frac{\cancel{w}\bullet(t-x)}{\cancel{w}\bullet(z-g)}=\frac{t-x}{z-3g}$$

b. $$\frac{4n^2-4}{n^2+3n+2}=\frac{4\bullet(\cancel{n+1})\bullet(n-1)}{(\cancel{n+1})\bullet(n+2)}=\frac{4(n-1)}{n+2}$$

c. $$\frac{2y^2-y-1}{2y^2+7y+3}=\frac{(\cancel{2y+1})\bullet(y-1)}{(\cancel{2y+1})\bullet(y+3)}=\frac{y-1}{y+3}$$

6.

a. $\frac{-y-1}{1+y}=\frac{(-1)\cdot(1+y)}{(1+y)}=-1$

b. $\frac{3x-12}{-x+4}=\frac{(-3)\cdot(-x+4)}{(-x+4)}=-3$

c. $\frac{3n-15}{25-n^2}=\frac{3\cdot(n-5)}{(-1)(n-5)(n+5)}=-\frac{3}{n+5}$

d.

$$\frac{4-9s^2}{3s^2+s-2}=\frac{(-1)\cdot(3s-2)(3s+2)}{(3s-2)(s+1)}=-\frac{3s+2}{s+1}$$

7.

a. $\frac{\pi r^2}{2\pi r}=\frac{(\pi)\cdot r^2}{2\cdot\pi\cdot r}=\frac{r}{2}$

b. The expression is undefined when $r=0$.

$2\pi r=0$

$\frac{2\pi r}{2\pi}=\frac{0}{2\pi}$

$r=0$

Exercises 7.1

1. $x=0$

3.

$y-2=0$

$y=2$

5.

$x+5=0$

$x=-5$

7.

$2n-1=0$

$2n=1$

$n=\frac{1}{2}$

9.

$x^2-1=0$

$(x+1)(x-1)=0$

$x+1=0 \quad x-1=0$

$x=-1 \quad x=1$

11.

$x^2-x-20=0$

$(x+4)(x-5)=0$

$x+4=0 \quad x-5=0$

$x=-4 \quad x=5$

13.

$\frac{p\cdot r}{q\cdot r}=\frac{pr}{qr}$

$\frac{p}{q}=\frac{pr}{qr}$

The expressions are equivalent.

15.

$\frac{(3t+5)\cdot t}{(t+1)\cdot t}=\frac{3t^2+5t}{t^2+t}$

$\frac{3t+5}{t+1}=\frac{3t^2+5t}{t^2+t}$

The expressions are equivalent.

17. The expressions are not equivalent.

19. $\frac{x-2}{2-x}=\frac{(x-2)(-1)}{(2-x)(-1)}=\frac{2-x}{x-2}$

The expressions are equivalent.

21. The expressions are not equivalent.

23. $\frac{10a^4}{12a}=\frac{2\cdot5\cdot a^4}{2\cdot6\cdot a}=\frac{5a^3}{6}$

25. $\frac{3x^2}{12x^5}=\frac{-1\cdot3\cdot x^2}{4\cdot3\cdot x^5}=\frac{1}{4x^3}$

27. $\frac{9s^3t^2}{6s^5t}=\frac{3\cdot3\cdot s^3\cdot t^2}{3\cdot2\cdot s^5\cdot t}=\frac{3t}{2s^2}$

29. $\frac{24a^4b^5}{3ab^2}=\frac{3\cdot8\cdot a^4\cdot b^5}{3\cdot a\cdot b^2}=8a^3b^3$

31. $\dfrac{-2p^2q^3}{-10pq^4}=\dfrac{-2\cdot p^2\cdot q^3}{5\cdot -2\cdot p\cdot q^4}=\dfrac{p}{5q}$

33. $\dfrac{5x(x+8)}{4x(x+8)}=\dfrac{5\cdot x\cdot (x+8)}{4\cdot x\cdot (x+8)}=\dfrac{5}{4}$

35. $\dfrac{8x^2(5-2x)}{3x(2x-5)}=\dfrac{8\cdot x^2\cdot(-1)\,(2x-5)}{3\cdot x\cdot(2x-5)}=-\dfrac{8x}{3}$

37. $\dfrac{5x-10}{5}=\dfrac{5\cdot(x-2)}{5}=x-2$

39. $\dfrac{2x^2+2x}{4x^2+6x}=\dfrac{2\cdot x\cdot(x+1)}{2\cdot x\cdot(2x+3)}=\dfrac{x+1}{2x+3}$

41. $\dfrac{a^2-4a}{ab-4b}=\dfrac{a\cdot(a-4)}{b\cdot(a-4)}=\dfrac{a}{b}$

43. $\dfrac{6x-4y}{9x-6y}=\dfrac{2\cdot(3x-2y)}{3\cdot(3x-2y)}=\dfrac{2}{3}$

45. $\dfrac{t^2-1}{t+1}=\dfrac{(t-1)\,(t+1)}{(t+1)}=t-1$

47, $\dfrac{p^2-q^2}{q^2-p^2}=\dfrac{-1\,(q^2-p^2)}{(q^2-p^2)}=\dfrac{-1}{1}=-1$

49. $\dfrac{n-1}{2-2n}=\dfrac{n-1}{-2\,(n-1)}=-\dfrac{1}{2}$

51. $\dfrac{(b-4)^2}{b^2-16}=\dfrac{(b-4)\,(b-4)}{(b-4)\,(b+4)}=\dfrac{b-4}{b+4}$

53. $\dfrac{2x^2+5x-3}{10x-5}=\dfrac{(2x-1)\,(x+3)}{5\cdot(2x-1)}=\dfrac{x+3}{5}$

55.

$\dfrac{a^3+9a^2+14a}{a^2-10a-24}=\dfrac{a\,(a+2)\,(a+7)}{(a-12)\,(a+2)}=\dfrac{a(a+7)}{a-12}$

57. $\dfrac{t^2-4t-5}{t^2-3t-10}=\dfrac{(t-5)\,(t+1)}{(t-5)\,(t+2)}=\dfrac{t+1}{t+2}$

59.

$\dfrac{9-16d^2}{16d^2-24d+9}=\dfrac{(-1)\,(4d-3)\,(4d+3)}{(4d-3)\,(4d-3)}=$

$-\dfrac{4d+3}{4d-3}$

61 $\dfrac{8s-2x^2}{2s^2-11s+12}=\dfrac{(-2s)\,(s-4)}{(2s-3)\,(s-4)}=-\dfrac{2s}{2s-3}$

63. $\dfrac{6x^2+5x+1}{6x^2-x-1}=\dfrac{(3x+1)\,(2x+1)}{(3x+1)\,(2x-1)}=\dfrac{2x+1}{2x-1}$

65.

$\dfrac{6y^2-7y+2}{6y^3+5y^2-6y}=\dfrac{(2y-1)(3y-2)}{y(6y^2+5y-6)}=$

$\dfrac{(3y-2)\,(2y-1)}{y\,(3y-2)\,(2y+3)}=\dfrac{2y-1}{y(2y+3)}$

67.

$\dfrac{2ab^2+4a^2b}{2b^2+5ab+2a^2}=\dfrac{2ab\cdot(b+2a)}{(b+2a)\,(2b+a)}=\dfrac{2ab}{2b+a}$

69.

$\dfrac{m^2+3mn-28n^2}{2m^2+4mn-48n^2}=\dfrac{(m+7n)\,(m-4n)}{2\cdot(m-4n)\,(m+6n)}=$

$\dfrac{m+7n}{2(m+6n)}$

Applications

71. $\dfrac{mu^2-mv^2}{mu-mv}=\dfrac{\cancel{m}\,\cancel{(u-v)}(u+v)}{\cancel{m}\,\cancel{(u-v)}}=u+v$

73.

a. $A=(x+5)(x+2)=x^2+7x+10$

b. $A=x(x+2)=x^2+2x$

c.

$$\frac{x^2+7x+10}{x^2+2x}=\frac{(x+5)\cancel{(x+2)}}{x\cancel{(x+2)}}=\frac{(x+5)}{x}$$

d. $\dfrac{x+5}{x}=\dfrac{8+5}{8}=\dfrac{13}{8}$

75.

a. $\dfrac{\pi r_1^2}{\pi r_2^3}=\dfrac{r_1^2}{r_2^2}$

b. $\dfrac{\pi r_1^2}{\pi r_3^3}=\dfrac{r_1^2}{r_3^2}$

7.2 Multiplication and Division of Rational Expressions

Practice 7.2

1.

a. $\dfrac{n}{7}\bullet\dfrac{2}{m}=\dfrac{n\bullet 2}{7\bullet m}=\dfrac{2n}{7m}$

b. $\dfrac{p^2}{6q^2}\bullet\dfrac{2q}{5p}=\dfrac{\cancelto{p}{p^2}\bullet\cancelto{1}{2q}}{\cancelto{3q}{6q^2}\bullet 5\cancelto{1}{p}}=\dfrac{p}{15q}$

2.

a.

$$\frac{3t}{t^2+5t}\bullet\frac{3t+15}{6t}=\frac{3t}{t(t+5)}\bullet\frac{3(t+5)}{6t}=$$

$$\frac{\cancelto{1}{3t}}{t\,\cancelto{1}{(t+5)}}\bullet\frac{3\,\cancelto{1}{(t+5)}}{\cancelto{2}{6t}}=\frac{3}{2t}$$

b.

$$\frac{8}{36g^2-1}\bullet\frac{1+6g}{4}=\frac{8}{(6g+1)(6g-1)}\bullet\frac{1+6g}{4}=$$

$$\frac{\cancelto{2}{8}}{\cancelto{1}{(6g+1)}(6g-1)}\bullet\frac{\cancelto{1}{(1+6g)}}{\cancelto{1}{4}}=\frac{2}{6g-1}$$

3.

a.

$$\frac{y^2+4y-21}{3y-9}\bullet\frac{6y^2-24}{y+2}=\frac{(y+7)(y-3)}{3(y-3)}\bullet\frac{6(y+2)(y-2)}{y+2}=$$

$$\frac{(y+7)\cancelto{1}{(y-3)}}{\cancelto{1}{3}\cancelto{1}{(y-3)}}\bullet\frac{\cancelto{2}{6}\cancelto{1}{(y+2)}(y-2)}{\cancelto{1}{(y+2)}}-2(y+7)(y-2)$$

b.

$$\frac{x^2-x-30}{x^2+10x+9}\bullet\frac{18-7x-x^2}{x^2-8x+12}=\frac{(x+5)(x-6)}{(x+1)(x+9)}\bullet\frac{(-1)(x+9)(x-2)}{(x-6)(x-2)}=$$

$$\frac{(x+5)\cancelto{1}{(x-6)}}{(x+1)\cancelto{1}{(x+9)}}\bullet\frac{(-1)\cancelto{1}{(x+9)}\cancelto{1}{(x-2)}}{\cancelto{1}{(x-6)}\cancelto{1}{(x-2)}}=-\frac{x+5}{x+1}$$

4.

a. $\dfrac{a}{4}\div\dfrac{b}{6}=\dfrac{a}{4}\bullet\dfrac{6}{b}=\dfrac{a}{\cancelto{2}{4}}\bullet\dfrac{\cancelto{3}{6}}{b}=\dfrac{3a}{2b}$

b.

$$\frac{5pq}{7p^2q^4}\div\frac{p^3}{3q^2}=\frac{5pq}{7p^2q^4}\bullet\frac{3q^2}{p^3}=$$

$$\frac{5\cancelto{1}{p}\cancelto{1}{q}}{7p^2\cancelto{q}{q^4}}\bullet\frac{3\cancelto{1}{q^2}}{\cancelto{p^2}{p^3}}=\frac{15}{7p^4q}$$

5.

a.

$$\frac{x+3}{x-10} \div \frac{x+1}{x+3} = \frac{x+3}{x-10} \bullet \frac{x+3}{x+1} =$$

$$\frac{(x+3)^2}{(x-10)(x+1)} \text{ or } \frac{x^2+6x+9}{x^2-9x-10}$$

b.

$$\frac{p+4q}{3p-6q} \div \frac{2p+8q}{2p-4q} = \frac{p+4q}{3(p-2q)} \bullet \frac{2(p-2q)}{2(p+4q)} =$$

$$\frac{\cancel{(p+4q)}}{3\cancel{(p-2q)}} \bullet \frac{\cancel{2}\cancel{(p-2q)}}{\cancel{2}\cancel{(p+4q)}} = \frac{1}{3}$$

c.

$$\frac{y^2+4y+3}{5y+10} \div \frac{y^2-1}{y} = \frac{(y+3)(y+1)}{5(y+2)} \bullet \frac{y}{(y+1)(y-1)} =$$

$$\frac{(y+3)\cancel{(y+1)}}{5(y+2)} \bullet \frac{y}{\cancel{(y+1)}(y-1)} = \frac{y(y+3)}{5(y+2)(y-1)}$$

6. $\frac{W}{g} \bullet \frac{v^2}{r} = \frac{W \bullet v^2}{g \circ r} = \frac{Wv^2}{gr}$

Exercises 7.2

1. $\frac{1}{t^2} \bullet \frac{t}{4} = \frac{1}{\cancel{t^2}_{t}} \bullet \frac{\cancel{t}^{1}}{4} = \frac{1}{4t}$

3. $\frac{2}{a} \bullet \frac{3}{b} = \frac{6}{ab}$

5. $\frac{2x^4}{3x^5} \bullet \frac{5}{x^8} = \frac{2\cancel{x^4}^{1}}{3x^5} \bullet \frac{5}{\cancel{x^8}_{x^4}} = \frac{10}{3x^9}$

7. $-\frac{7x^2y}{3} \bullet \frac{6}{x^3y} = -\frac{7\cancel{x^2}\cancel{y}}{\cancel{3}} \bullet \frac{\cancel{6}^{2}}{\cancel{x^3}_{x}\cancel{y}} = -\frac{14}{x}$

9. $\frac{x}{x-2} \bullet \frac{5x-10}{x^4} = \frac{\cancel{x}}{\cancel{x-2}} \bullet \frac{5\cancel{(x-2)}}{\cancel{x^4}_{x^3}} = \frac{5}{x^3}$

11. $\frac{8n-3}{n^2} \bullet n = \frac{8n-3}{\cancel{n^2}_{n}} \bullet \frac{\cancel{n}^{1}}{1} = \frac{8n-3}{n}$

13.

$$\frac{8x-6}{5x+20} \bullet \frac{2x+8}{4x-3} = \frac{2(4x-3)}{5(x+4)} \bullet \frac{2(x+4)}{4x-3} =$$

$$\frac{2\cancel{(4x-3)}}{5\cancel{(x+4)}} \bullet \frac{2\cancel{(x+4)}}{\cancel{4x-3}} = \frac{4}{5}$$

15.

$$\frac{x^2-4y^2}{x+y} \bullet \frac{3x+3y}{4x-8y} = \frac{(x+2y)(x-2y)}{x+y} \bullet \frac{3(x+y)}{4(x-2y)} =$$

$$\frac{(x+2y)\cancel{(x-2y)}}{\cancel{(x+y)}} \bullet \frac{3\cancel{(x+y)}}{4\cancel{(x-2y)}} = \frac{3(x+2y)}{4}$$

17.

$$\frac{p^4-1}{p^4-16} \bullet \frac{p^2+4}{p^2+1} =$$

$$\frac{\cancel{(p^2+1)}(p+1)(p-1)}{\cancel{(p^2+4)}(p+2)(p-2)} \bullet \frac{\cancel{(p^2+4)}}{\cancel{(p^2+1)}} =$$

$$\frac{p^2-1}{p^2-4}$$

19.

$$\frac{n^2-2n-24}{n^2+6n+8} \bullet \frac{n^2+5n+6}{n^2-5n-6} =$$

$$\frac{\cancel{(n-6)}\cancel{(n+4)}}{\cancel{(n+2)}\cancel{(n+4)}} \bullet \frac{\cancel{(n+2)}(n+3)}{\cancel{(n-6)}(n+1)} = \frac{n+3}{n+1}$$

21.

$$\frac{2y^2-y-6}{2y^2+y-3} \bullet \frac{2y^2-3y+1}{2y^2-9y+10} =$$

$$\frac{\cancel{(2y+3)}\cancel{(y-2)}}{\cancel{(2y+3)}\cancel{(y-1)}} \bullet \frac{(2y-1)\cancel{(y-1)}}{(2y-5)\cancel{(y-2)}} = \frac{2y-1}{2y-5}$$

23.

$$\frac{2}{x^3}\bullet\frac{4x}{5}\bullet\frac{10}{x^2}=$$

$$\frac{2}{\cancel{x^3}_{x^2}}\bullet\frac{4\cancel{x}}{\cancel{5}}\bullet\frac{\cancel{10}^{2}}{x^2}=\frac{16}{x^4}$$

25.

$$\frac{x^2-7x+10}{2x-2}\bullet\frac{6x}{x^2-2x-15}\bullet\frac{x^2+2x-3}{x-2}=$$

$$\frac{\cancel{(x-5)}\cancel{(x-2)}}{\cancel{2}\cancel{(x-1)}}\bullet\frac{\cancel{6}^{3}x}{\cancel{(x-5)}\cancel{(x+3)}}\bullet\frac{\cancel{(x+3)}\cancel{(x-1)}}{\cancel{(x-2)}}=$$

$3x$

27. $\frac{7}{a}\div\frac{14}{a}=\frac{7}{a}\bullet\frac{a}{14}=\frac{\cancel{7}}{\cancel{a}}\bullet\frac{\cancel{a}}{\cancel{14}_{2}}=\frac{1}{2}$

29. $\frac{p^3}{10}\div\frac{p^3}{20}=\frac{p^3}{10}\bullet\frac{20}{p^3}=\frac{\cancel{p^3}}{\cancel{10}}\bullet\frac{\cancel{20}^{2}}{\cancel{p^3}}=2$

31.

$$\frac{12}{x^3}\div\frac{6}{5x^2}=\frac{12}{x^3}\bullet\frac{5x^2}{6}=$$

$$\frac{\cancel{12}^{2}}{\cancel{x^3}_{x}}\bullet\frac{5\cancel{x^2}}{\cancel{6}}=\frac{10}{x}$$

33. $\frac{3}{t}\div t=\frac{3}{t}\bullet\frac{1}{t}=\frac{3}{t^2}$

35.

$$\frac{9xy^2}{2x^3}\div\frac{3x^2y}{4y}=\frac{9xy^2}{2x^3}\bullet\frac{4y}{3x^2y}=$$

$$\frac{\cancel{9}^{3}\cancel{x}\cancel{y^2}^{y}}{\cancel{2}x^3}\bullet\frac{\cancel{4}^{2}y}{\cancel{3}\cancel{x^2}_{x}\cancel{y}}=\frac{6y^2}{x^4}$$

37. $\frac{c+3}{c-5}\div\frac{c+9}{c-7}=\frac{c+3}{c-5}\bullet\frac{c-7}{c+9}=\frac{(c+3)(c-7)}{(c-5)(c+9)}$

39.

$$\frac{6a-12}{8a+32}\div\frac{9a-18}{5a+20}=\frac{6(a-2)}{8(a+4)}\bullet\frac{5(a+4)}{9(a-2)}=$$

$$\frac{\cancel{6}\cancel{(a-2)}}{\cancel{8}_{4}\cancel{(a+4)}}\bullet\frac{5\cancel{(a+4)}}{\cancel{9}_{3}\cancel{(a-2)}}=\frac{5}{12}$$

41.

$$\frac{x+1}{10}\div\frac{1-x^2}{5}=\frac{x+1}{5\bullet2}\bullet\frac{5}{(-1)(x+1)(x-1)}=$$

$$\frac{\cancel{(x+1)}}{\cancel{5}\bullet2}\bullet\frac{\cancel{5}}{(-1)(x-1)\cancel{(x+1)}}=\frac{1}{2(1-x)}$$

43.

$$\frac{p^2-1}{1-p}\div\frac{p+1}{p}=\frac{(p-1)(p+1)}{-1(p-1)}\bullet\frac{p}{p+1}=$$

$$\frac{\cancel{(p+1)}\cancel{(p-1)}}{-1\cancel{(p-1)}}\bullet\frac{p}{\cancel{(p+1)}}=-p$$

45.

$$\frac{x^2y+3xy^2}{x^2-9y^2}\div\frac{5x^2y}{x^2-2xy-3y^2}=$$

$$\frac{xy(x+3y)}{(x+3y)(x-3y)}\bullet\frac{(x+y)(x-3y)}{5x^2y}=$$

$$\frac{\cancel{(xy)}\cancel{(x+3y)}}{\cancel{(x+3y)}\cancel{(x-3y)}}\bullet\frac{(x+y)\cancel{(x-3y)}}{5\cancel{(x^2y)}_{x}}=\frac{x+y}{5x}$$

47.

$$\frac{2t^2-3t-2}{2t+1}\div\left(4-t^2\right)=$$

$$\frac{(2t+1)(t-2)}{2t+1}\bullet\frac{1}{(-1)(t+2)(t-2)}=$$

$$\frac{\cancel{(2t+1)}\cancel{(t-2)}}{\cancel{2t+1}}\bullet\frac{1}{(-1)(t+2)\cancel{(t-2)}}=-\frac{1}{t+2}$$

49.

$$\frac{x^2-11x+28}{x^2-x-42}\div\frac{x^2-2x-8}{x^2+7x+10}=$$

$$\frac{(x-4)(x-7)}{(x+6)(x-7)}\bullet\frac{(x+5)(x+2)}{(x-4)(x+2)}=$$

$$\frac{\cancel{(x-4)}\,\cancel{(x-7)}}{(x+6)\,\cancel{(x-7)}}\bullet\frac{(x+5)\,\cancel{(x+2)}}{\cancel{(x-4)}\,\cancel{(x+2)}}=\frac{x+5}{x+6}$$

51.

$$\frac{3p^2-3p-18}{p^2+2p-15}\div\frac{2p^2+6p-20}{2p^2-12p+16}=$$

$$\frac{3(p-3)(p+2)}{(p-3)(p+5)}\bullet\frac{2(p-4)(p-2)}{2(p+5)(p-2)}=$$

$$\frac{3\cancel{(p-3)}(p+2)}{\cancel{(p-3)}(p+5)}\bullet\frac{\cancel{2}(p-4)\cancel{(p-2)}}{\cancel{2}(p+5)\cancel{(p-2)}}$$

$$=\frac{3(p+2)(p-4)}{(p+5)^2}$$

Applications

53.

$$\left(\frac{A-p}{p}\right)\div r=\frac{A-p}{p}\bullet\frac{1}{r}=\frac{A-p}{pr}$$

55. $\frac{p}{100}\bullet\frac{q}{100}\bullet B=\frac{pqB}{10{,}000}$

57. $P=\frac{V^2}{R+r}\bullet\frac{r}{R+r}=\frac{V^2r}{(R+r)^2}$

59.

a. $\frac{\frac{4}{3}\pi r^3}{\pi r^2h}=\frac{\frac{4}{3}\pi r^3(3)}{\pi r^2h(3)}=\frac{4\pi r^3}{3\pi r^2h}=\frac{4r}{3h}$

b. $\frac{4\pi r^2}{2\pi rh+2\pi r^2}=\frac{\cancel{2\pi r}(2r)}{\cancel{2\pi r}(h+r)}=\frac{2r}{h+r}$

c. $\frac{4r}{3h}\div\frac{2r}{h+r}=\frac{\overset{2}{\cancel{4r}}}{3h}\bullet\frac{h+r}{\underset{1}{\cancel{2r}}}=\frac{2(h+r)}{3h}$

d. $\frac{2(2r+r)}{3(2r)}=\frac{6r}{6r}=1$

7.3 Addition and Subtraction of Rational Expressions

Practice 7.3

1.

a. $\frac{9}{y+2}+\frac{1}{y+2}=\frac{9+1}{y+2}=\frac{10}{y+2}$

b. $\frac{10r}{3s}+\frac{5r}{3s}=\frac{10r+5r}{3s}=\frac{\overset{5}{\cancel{15}}r}{\underset{1}{\cancel{3}}s}=\frac{5r}{s}$

c.

$$\frac{6t-7}{t-1}+\frac{1}{t-1}=\frac{6t-7+1}{t-1}=\frac{6t-6}{t-1}=$$

$$\frac{6\cancel{(t-1)}}{\cancel{(t-1)}}=6$$

d.

$$\frac{n^2+10n+1}{n+5}+\frac{2n^2+4n-6}{n+5}=$$

$$\frac{n^2+10n+1+2n^2+4n-6}{n+5}=\frac{3n^2+14n-5}{n+5}=$$

$$\frac{(3n-1)\cancel{(n+5)}}{\cancel{n+5}}=3n-1$$

2.

a. $\frac{12}{v}-\frac{7}{v}=\frac{12-7}{v}=\frac{5}{v}$

b. $\frac{7t}{10}-\frac{t}{10}=\frac{7t-t}{10}=\frac{6t}{10}=\frac{\overset{3}{\cancel{6}}t}{\underset{5}{\cancel{10}}}=\frac{3t}{5}$

c. $\frac{7p}{3q}-\frac{8p}{3q}=\frac{7p-8p}{3q}=-\frac{p}{3q}$

3.

a.

$$\frac{7a-4b}{3a}-\frac{a-4b}{3a}=\frac{7a-4b-a+4b}{3a}=$$

$$\frac{6a}{3a}=2$$

b.

$$\frac{9xy-5xz}{4y-z}-\frac{xy-3xz}{4y-z}=\frac{9xy-5xz-xy+3xz}{4y-z}=$$

$$\frac{8xy-2xz}{4y-z}=\frac{2x\overset{1}{\cancel{(4y-z)}}}{\underset{1}{\cancel{4y-z}}}=2x$$

c.

$$\frac{2x+13}{x^2-7x+10}-\frac{5x+7}{x^2-7x+10}=\frac{2x+13-5x-7}{x^2-7x+10}=$$

$$\frac{6-3x}{x^2-7x+10}=\frac{-3\overset{1}{\cancel{(x-2)}}}{(x-5)\underset{1}{\cancel{(x-2)}}}=-\frac{3}{x-5}$$

4.

a.

$$y=y$$

$$20=2^2\cdot 5$$

$$LCD=y\cdot 2^2\cdot 5=20y$$

b.

$$6t=2\cdot 3\cdot t$$

$$3t^2=3\cdot t^2$$

$$LCD=2\cdot 3\cdot t^2=6t^2$$

c.

$$5x=5\cdot x$$

$$15xy^3=3\cdot 5\cdot x\cdot y^3$$

$$2x^2=2\cdot x^2$$

$$LCD=2\cdot 3\cdot 5\cdot x^2\cdot y^3=30x^2y^3$$

5.

a.

$$2n^2+2n=2\cdot n\cdot(n+1)$$

$$n^2+2n+1=(n+1)^2$$

$$LCD=2\cdot n\cdot(n+1)^2=2n(n+1)^2$$

b.

$$p+2=p+2$$

$$p-1=p-1$$

$$p+5=p+5$$

$$LCD=(p+2)(p-1)(p-5)$$

c.

$$s^2+4st+4t^2=(s+2t)^2$$

$$s^2-4t^2=(s+2t)(s-2t)$$

$$s^2-4st+4t^2=(s-2t)^2$$

$$LCD=(s+2t)^2(s-2t)^2$$

6.

a.

$$7p^3=7\cdot p^3$$

$$2p=2\cdot p$$

$$LCD=2\cdot 7\cdot p^3=14p^3$$

$$\frac{2}{7p^3}=\frac{2\cdot 2}{7p^3\cdot 2}=\frac{4}{14p^3}$$

$$\frac{p+3}{2p}=\frac{(p+3)\cdot 7p^2}{2p\cdot 7p^2}=\frac{7p^2(p+3)}{14p^3}$$

b.

$$y^2-9=(y-3)(y+3)$$

$$y^2-6y+9=(y-3)^2$$

$$LCD=(y-3)^2(y+3)$$

$$\frac{3y-2}{y^2-9}=\frac{(3y-2)(y-3)}{(y-3)^2(y+3)}$$

$$\frac{y}{y^2-6y+9}=\frac{y(y+3)}{(y-3)^2(y+3)}$$

7.

a.

$$\frac{3}{4p}+\frac{1}{6p}=\frac{3\cdot 3}{4p\cdot 3}+\frac{1\cdot 2}{6p\cdot 2}=$$
$$\frac{9}{12p}+\frac{2}{12p}=\frac{11}{12p}$$

b.

$$\frac{1}{5y}-\frac{2}{15y^2}=\frac{1\cdot 3y}{5y\cdot 3y}-\frac{2}{15y^2}=$$
$$\frac{3y}{15y^2}-\frac{2}{15y^2}=\frac{3y-2}{15y^2}$$

8.

a.

$$\frac{x+2}{x}-\frac{x-4}{x+3}=\frac{(x+2)(x+3)}{x(x+3)}-\frac{(x-4)x}{(x+3)x}=$$
$$\frac{x^2+5x+6}{x(x+3)}-\frac{x^2-4x}{x(x+3)}=\frac{9x+6}{x(x+3)}=\frac{3(3x+2)}{x(x+3)}$$

b.

$$\frac{3x-4}{x-1}+\frac{x+1}{1-x}=\frac{(3x-4)(-1)}{(x-1)(-1)}+\frac{(x+1)}{(x-1)(-1)}=$$
$$\frac{-3x+4}{(x-1)(-1)}+\frac{x+1}{(x-1)(-1)}=\frac{-3x+4+x+1}{(x-1)(-1)}=$$
$$\frac{2x-5}{x-1}$$

9.

a.

$$\frac{1}{4x-16}+\frac{2x}{x^2-16}=\frac{1(x+4)}{4(x+4)(x-4)}+\frac{2x(4)}{4(x+4)(x-4)}=$$
$$\frac{x+4+8x}{4(x+4)(x-4)}=\frac{9x+4}{4(x+4)(x-4)}$$

b.

$$\frac{3x-7}{x^2-1}+\frac{2}{1-x}=\frac{(-1)(3x-7)}{(-1)(x+1)(x-1)}+\frac{2(x+1)}{(-1)(x+1)(x-1)}=$$
$$\frac{-3x+7+2x+2}{(-1)(x+1)(x-1)}=\frac{x-9}{(x+1)(x-1)}$$

10.

$$\frac{y}{y^2+5y+6}-\frac{4y+1}{y^2+3y+2}=$$
$$\frac{y(y+1)}{(y+2)(y+3)(y+1)}-\frac{(4y+1)(y+3)}{(y+1)(y+2)(y+3)}=$$
$$\frac{y^2+y-4y^2-13y-3}{(y+1)(y+2)(y+3)}=\frac{-3y^2-12y-3}{(y+1)(y+2)(y+3)}=$$
$$\frac{-3(y^2+4y+1)}{(y+1)(y+2)(y+3)}$$

11. $100\left(\frac{C_1}{C_0}-1\right)=\frac{100(C_1-C_0)}{C_0}$

Exercises 7.3

1. $\frac{5a}{12}+\frac{11a}{12}=\frac{5a+11a}{12}=\frac{\overset{4}{\cancel{16}}\,a}{\underset{3}{\cancel{12}}}=\frac{4a}{3}$

3. $\frac{5t}{3}-\frac{2t}{3}=\frac{3t}{3}=-\frac{\overset{1}{\cancel{3}}\,t}{\underset{1}{\cancel{3}}}=t$

5. $\frac{10}{x}+\frac{1}{x}=\frac{10+1}{x}=\frac{11}{x}$

7. $\frac{6}{7y}-\frac{1}{7y}=\frac{6-1}{7y}=\frac{5}{7y}.$

9. $\frac{5x}{2y}+\frac{x}{2y}=\frac{5x+x}{2y}=\frac{\overset{3}{\cancel{6}}\,x}{\underset{1}{\cancel{2}}\,y}=\frac{3x}{y}$

11. $\frac{2p}{5q}-\frac{3p}{5q}=\frac{2p-3p}{5q}=-\frac{p}{5q}$

13. $\frac{2}{x+1}+\frac{7}{x+1}=\frac{2+7}{x+1}=\frac{9}{x+1}$

15. $\frac{5}{x+2}-\frac{9}{x+2}=\frac{5-9}{x+2}=-\frac{4}{x+2}$

17. $\frac{a}{a+3}+\frac{1}{a+3}=\frac{a+1}{a+3}$

19. $\frac{3x}{x-8}+\frac{2x+1}{x-8}=\frac{3x+2x+1}{x-8}=\frac{5x+1}{x-8}$

21. $\frac{7x+1}{5x+2}-\frac{3x}{5x+2}=\frac{7x+1-3x}{5x+2}=\frac{4x+1}{5x+2}$

23.

$$\frac{9x+17}{2x+5}-\frac{3x+2}{2x+5}=\frac{9x+17-3x-2}{2x+5}=$$

$$\frac{6x+15}{2x+5}=\frac{3\cancel{(2x+5)}^{1}}{\cancel{(2x+5)}_{1}}=3$$

25.

$$\frac{-7+5n}{3n-1}+\frac{7n+3}{3n-1}=\frac{-7+5n+7n+3}{3n-1}=$$

$$\frac{12n-4}{3n-1}=\frac{4\cancel{(3n-1)}^{1}}{\cancel{3n-1}_{1}}=4$$

27.

$$\frac{x^2-1}{x^2-4x-2}-\frac{x^2-x+3}{x^2-4x-2}=$$

$$\frac{x^2-1-x^2+x-3}{x^2-4x-2}=\frac{x-4}{x^2-4x-2}$$

29.

$$\frac{x}{x^2-3x+2}+\frac{2}{x^2-3x+2}+\frac{x^2-4x}{x^2-3x+2}=$$

$$\frac{x+2+x^2-4x}{x^2-3x+2}=\frac{x^2-3x+2}{x^2-3x+2}=1$$

31.

$$\frac{2x\quad 1}{3x^2-x+2}+\frac{8}{3x^2-x+2}-\frac{3x}{3x^2-x+2}=$$

$$\frac{2x-1+8-3x}{3x^2-x+2}=\frac{-x+7}{3x^2-x+2}$$

33.

$5(x+2)=5\bullet(x+2)$

$3(x+2)=3\bullet(x+2)$

$LCD=3\bullet5\bullet(x+2)=15(x+2)$

35.

$(p-3)(p+8)=(p-3)(p+8)$

$(p-3)(p-8)=(p-3)(p-8)$

$LCD=(p-3)(p+8)(p-8)$

37.

$t=t$

$t+3=t+3$

$t-3=t-3$

$LCD=t(t+3)(t-3)$

39.

$t^2+7t+10=(t+2)(t+5)$

$t^2-25=(t-5)(t+5)$

$LCD=(t+2)(t+5)(t-5)$

41.

$3s^2-11s+6=(3s-2)(s-3)$

$3s^2+4s-4=(3s-2)(s+2)$

$LCD=(3s-2)(s-3)(s+2)$

43.

$LCD=12x^2$

$$\frac{1}{3x}=\frac{1\bullet4x}{3x\bullet4x}=\frac{4x}{12x^2}$$

$$\frac{5}{4x^2}=\frac{5\bullet3}{4x^2\bullet3}=\frac{15}{12x^2}$$

45.

$LCD=14a^2b$

$$\frac{5}{2a^2}=\frac{5\bullet7b}{2a^2\bullet7b}=\frac{35b}{14a^2b}$$

$$\frac{a-3}{7ab}=\frac{(a-3)2a}{7ab\bullet2a}=\frac{2a(a-3)}{14a^2b}$$

47.

$LCD=n(n+1)^2$

$$\frac{8}{n(n+1)}=\frac{8(n+1)}{n(n+1)^2}$$

$$\frac{5}{(n+1)^2}=\frac{5\bullet n}{(n+1)^2\bullet n}=\frac{5n}{n(n+1)^2}$$

49.

$LCD=4(n-1)(n+1)$

$$\frac{3n}{4n+4}=\frac{3n(n-1)}{4(n+1)(n-1)}=\frac{3n(n-1)}{4(n-1)(n+1)}$$

$$\frac{2n}{n^2-1}=\frac{2n(4)}{(n+1)(n-1)(4)}=\frac{8n}{4(n-1)(n+1)}$$

51.

$LCD=(n+1)(n+5)(n-3)$

$$\frac{2n}{n^2+6n+5}=\frac{2n(n-3)}{(n+1)(n+5)(n-3)}$$

$$\frac{3n}{n^2+2n-15}=\frac{3n(n+1)}{(n+1)(n+5)(n-3)}$$

53. $\frac{5}{3x}+\frac{1}{2x}=\frac{10}{6x}+\frac{3}{6x}=\frac{13}{6x}$

55.
$\frac{2}{3x^2}-\frac{5}{6x}=\frac{4}{6x^2}-\frac{5x}{6x^2}=\frac{4-5x}{6x^2}$

57.
$\frac{-2}{3x^2y}+\frac{4}{3xy^2}=\frac{-2y}{3x^2y^2}+\frac{4x}{3x^2y^2}=$
$\frac{-2y+4x}{3x^2y^2}=-\frac{2(y-2x)}{3x^2y^2}$

59.
$\frac{1}{x+1}+\frac{1}{x-1}=\frac{x-1}{x^2-1}+\frac{x+1}{x^2-1}=\frac{2x}{x^2-1}$

61.
$\frac{p+6}{3}-\frac{2p+1}{7}=\frac{7p+42}{21}-\frac{6p+3}{21}=\frac{p+39}{21}$

63.
$x-\frac{10-4x}{2}=\frac{2x}{2}-\frac{10-4x}{2}=\frac{6x-10}{2}=3x-5$

65.
$\frac{3a+1}{6a}-\frac{a^2-2}{2a^2}=\frac{a(3a+1)}{6a^2}-\frac{3(a^2-2)}{6a^2}=$
$\frac{3a^2+a-3a^2+6}{6a^2}=\frac{a+6}{6a^2}$

67. $\frac{a^2}{a-1}-\frac{1}{1-a}=\frac{a^2}{a-1}-\frac{(-1)1}{a-1}=\frac{a^2+1}{a-1}$

69.
$\frac{4}{c-4}+\frac{c}{4-c}=\frac{4}{c-4}+\frac{(-1)c}{c-4}=$
$\frac{4-c}{c-4}=\frac{(-1)(c-4)}{c-4}=-1$

71.
$\frac{x-5}{x+1}-\frac{x+2}{x}=\frac{x^2-5x}{x(x+1)}-\frac{x^2+3x+2}{x(x+1)}=$
$\frac{-8x-2}{x(x+1)}=-\frac{2(4x+1)}{x(x+1)}$

73.
$\frac{4x-5}{x-4}+\frac{1-3x}{4-x}=\frac{4x-5}{x-4}+\frac{(-1)(1-3x)}{(x-4)}=$
$\frac{4x-5-1+3x}{x-4}=\frac{7x-6}{x-4}$

75.
$\frac{5x}{x^2+x-2}+\frac{6}{x+2}=$
$\frac{5x}{(x+2)(x-1)}+\frac{6(x-1)}{(x+2)(x-1)}=$
$\frac{5x+6x-6}{(x+2)(x-1)}=\frac{11x-6}{(x+2)(x-1)}$

77.
$\frac{4}{3n-9}-\frac{n}{n^2+2n-15}=$
$\frac{4(n+5)}{3(n-3)(n+5)}-\frac{3\bullet n}{3(n-3)(n+5)}=$
$\frac{4n+20-3n}{3(n-3)(n+5)}=\frac{n+20}{3(n-3)(n+5)}$

79.
$\frac{2}{t+5}-\frac{t+6}{25-t^2}=$
$\frac{2(-1)(t-5)}{(t+5)(-1)(t-5)}-\frac{t+6}{(-1)(t-5)(t+5)}=$
$\frac{-2t+10-t-6}{(t+5)(-1)(t-5)}=\frac{4-3t}{(5+t)(5-t)}$ *or* $\frac{3t-4}{(t+5)(t-5)}$

81.
$\frac{4x}{x^2+2x+1}-\frac{2x+5}{x^2+4x+3}=$
$\frac{4x(x+3)}{(x+1)^2(x+3)}-\frac{(2x+5)(x+1)}{(x+1)^2(x+3)}=$
$\frac{4x^2+12x-2x^2-7x-5}{(x+1)^2(x+3)}=\frac{2x^2+5x-5}{(x+1)^2(x+3)}$

83.
$\frac{2t-1}{2t^2+t-3}+\frac{2}{t-1}=$
$\frac{2t-1}{(2t+3)(t-1)}+\frac{2(2t+3)}{(t-1)(2t+3)}=$
$\frac{2t-1+4t+6}{(2t+3)(t-1)}=\frac{6t+5}{(t-1)(2t+3)}$

85.

$$\frac{4x}{x-1}+\frac{2}{3x}+\frac{x}{x^2-1}=$$

$$\frac{4x(3x)(x+1)}{3x(x+1)(x-1)}+\frac{2(x+1)(x-1)}{3x(x+1)(x-1)}+\frac{x(3x)}{3x(x+1)(x-1)}=$$

$$\frac{12x^3+12x^2+2x^2-2+3x^2}{3x(x+1)(x-1)}=\frac{12x^3+17x^2-2}{3x(x+1)(x-1)}$$

87.

$$\frac{5y}{3y-1}-\frac{3}{y-4}+\frac{y+1}{3y^2-13y+4}=$$

$$\frac{5y(y-4)}{(3y-1)(y-4)}-\frac{3(3y-1)}{(3y-1)(y-4)}+\frac{y+1}{(3y-1)(y-4)}=$$

$$\frac{5y^2-20y-9y+3+y+1}{(3y-1)(y-4)}=\frac{5y^2-28y+4}{(3y-1)(y-4)}$$

89.

$$\frac{a-1}{(a+3)^2}-\frac{2a-3}{a+3}-\frac{a}{4a+12}=$$

$$\frac{4(a-1)}{4(a+3)^2}-\frac{(2a-3)(4)(a+3)}{4(a+3)^2}-\frac{a(a+3)}{4(a+3)^2}=$$

$$\frac{4a-4-8a^2-12a+36-a^2-3a}{3(a+3)^2}=$$

$$\frac{-9a^2-11a+32}{3(a+3)^2}$$

Applications

91. $vt+\frac{at^2}{2}=\frac{2vt}{2}+\frac{at^2}{2}=\frac{2vt+at^2}{2}$

93.

$$\frac{1000}{1+r}-\frac{1000}{(1+r)^2}=\frac{1000(1+r)}{(1+r)^2}-\frac{1000}{(1+r)^2}=$$

$$\frac{1000+1000r-1000}{(1+r)^2}=\frac{1000r}{(1+r)^2}$$

95. $\frac{20}{r}+\frac{20}{2r}=\frac{20}{r}+\frac{10}{r}=\frac{30}{r}$ hours

The trip took $\frac{30}{r}$ hours.

97. $\frac{3}{x}+0.1=\frac{3}{x}+\frac{0.1x}{x}=\frac{3+0.1x}{x}$ dollars

The average cost per check is $\frac{3+0.1x}{x}$ dollars.

7.4 Complex Rational Expressions

Practice 7.4

1.

a. $\frac{\frac{3}{x^4}}{\frac{5}{x}}=\frac{3}{x^4}\div\frac{5}{x}=\frac{3}{\cancel{x^4}_{x^3}}\bullet\frac{\cancel{x}^1}{5}=\frac{3}{5x^3}$

b.

$$\frac{2x}{\frac{x^2}{4}+\frac{x}{2}}=\frac{2x}{\frac{x^2}{4}+\frac{x\cdot 2}{2\cdot 2}}=\frac{2x}{\frac{x^2+2x}{4}}=$$

$$2x\div\frac{x^2+2x}{4}=\frac{2\cancel{x}^1}{1}\bullet\frac{4}{\cancel{x^2+2x}_{x+2}}=\frac{8}{x+2}$$

2.

$$\frac{2-\frac{1}{n}}{2+\frac{1}{n}}=\frac{\frac{2n-1}{n}}{\frac{2n+1}{n}}=\frac{2n-1}{n}\div\frac{2n+1}{n}=$$

$$\frac{2n-1}{\cancel{n}_1}\bullet\frac{\cancel{n}^1}{2n+1}=\frac{2n-1}{2n+1}$$

3.

a.

$$\frac{\frac{4}{x}}{\frac{2}{x^3}}=\frac{4}{x}\div\frac{2}{x^3}=\frac{\cancel{4}^2}{\cancel{x}_1}\bullet\frac{\cancel{x^3}^{x^2}}{\cancel{2}_1}=2x^2$$

b.

$$\frac{\frac{1}{y}+\frac{3}{y^2}}{2y}=\frac{\left(\frac{1}{y}+\frac{3}{y^2}\right)y^2}{2y\cdot y^2}=\frac{y+3}{2y^3}$$

4.

a.

$$\frac{4+\frac{1}{y}}{4-\frac{1}{y^2}}=\frac{\left(4+\frac{1}{y}\right)y^2}{\left(4-\frac{1}{y^2}\right)y^2}=\frac{4y^2+y}{4y^2-1}$$

b.

$$\frac{\frac{1}{2a^2}-\frac{1}{2b^2}}{\frac{5}{a}+\frac{5}{b}}=\frac{\left(\frac{1}{2a^2}-\frac{1}{2b^2}\right)2a^2b^2}{\left(\frac{5}{a}+\frac{5}{b}\right)2a^2b^2}=$$

$$\frac{b^2-a^2}{10ab^2+10a^2b}=\frac{(b-a)\cancel{(b+a)}}{10ab\,\cancel{(b+a)}}=\frac{b-a}{10ab}$$

5.

$$\frac{3}{\frac{1}{a}+\frac{1}{b}+\frac{1}{c}}=\frac{3\cdot abc}{\left(\frac{1}{a}+\frac{1}{b}+\frac{1}{c}\right)abc}=\frac{3abc}{bc+ac+ab}$$

Exercises 7.4

1.

$$\frac{\frac{x}{5}}{\frac{x^2}{10}}=\frac{\frac{x}{5}\bullet 10}{\frac{x^2}{10}\bullet 10}=\frac{2x}{x^2}=\frac{2}{x}$$

3.

$$\frac{\frac{a+1}{2}}{\frac{a-1}{2}}=\frac{\frac{a+1}{2}\bullet 2}{\frac{a-1}{2}\bullet 2}=\frac{a+1}{a-1}$$

5.

$$\frac{3+\frac{1}{x}}{3-\frac{1}{x^2}}=\frac{\frac{3x+1}{x}}{\frac{3x^2-1}{x^2}}=\frac{3x+1}{x}\div\frac{3x^2-1}{x^2}=$$

$$\frac{3x+1}{\cancel{x}}\bullet\frac{\cancel{x^2}}{3x^2-1}=\frac{x(3x+1)}{3x^2-1}$$

7.

$$\frac{\frac{1}{3d}-\frac{1}{d^2}}{d-\frac{9}{d}}=\frac{\frac{d-3}{3d^2}}{\frac{d^2-9}{d}}=\frac{d-3}{3d^2}\div\frac{d^2-9}{d}=$$

$$\frac{\cancel{(d-3)}}{3\cancel{d^2}}\bullet\frac{\cancel{d}}{\cancel{(d-3)}(d+3)}=\frac{1}{3d(d+3)}$$

9.

$$\frac{1-\frac{4y^2}{x^2}}{3+\frac{6y}{x}}=\frac{\frac{x^2-4y^2}{x^2}}{\frac{3x+6y}{x}}=\frac{\frac{x^2-4y^2}{x^2}\bullet x^2}{\frac{3x+6y}{x}\bullet x^2}=$$

$$\frac{x^2-4y^2}{x(3x+6y)}=\frac{(x-2y)\cancel{(x+2y)}}{3x\,\cancel{(x+2y)}}=$$

$$\frac{x-2y}{3x}$$

11.

$$\frac{\frac{2}{y}-\frac{1}{5}}{\frac{5}{y}-1}=\frac{\frac{10-y}{5y}\bullet 5y}{\frac{5-y}{y}\bullet 5y}=\frac{10-y}{5(5-y)}$$

13.

$$\frac{1+\frac{4}{x}+\frac{4}{x^2}}{1+\frac{5}{x}+\frac{6}{x^2}}=\frac{\frac{x^2+4x+4}{x^2}\bullet x^2}{\frac{x^2+5x+6}{x^2}\bullet x^2}=$$

$$\frac{x^2+4x+4}{x^2+5x+6}=\frac{\cancel{(x+2)}(x+2)}{\cancel{(x+2)}(x+3)}=\frac{x+2}{x+3}$$

15.

$$\frac{3+\frac{1}{y+1}}{5-\frac{1}{y+1}}=\frac{\frac{3y+3+1}{y+1}\bullet(y+1)}{\frac{5y+5-1}{y+1}\bullet(y+1)}=\frac{3y+4}{5y+4}$$

17.

$$\frac{\frac{x}{4}-\frac{x}{8}}{\frac{2}{y^2}+\frac{2}{y}}=\frac{\frac{x}{8}\bullet 8y^2}{\frac{2+2y}{y^2}\bullet 8y^2}=\frac{xy^2}{8(2+2y)}=\frac{xy^2}{16(y+1)}$$

19.

$$\frac{\frac{x}{x+1}-\frac{2}{x}}{\frac{x}{3}}=\frac{\frac{x^2-2x-2}{x(x+1)}}{\frac{x}{3}}=\frac{x^2-2x-2}{x(x+1)}\div\frac{x}{3}=$$

$$\frac{x^2-2x-2}{x(x+1)}\bullet\frac{3}{x}=\frac{3(x^2-2x-2)}{x^2(x+1)}$$

Applications

21.

$$\frac{V}{\frac{1}{2R}+\frac{1}{2R+2}}=\frac{V(2R)(R+1)}{\left(\frac{1}{2R}+\frac{1}{2R+2}\right)(2R)(R+1)}=$$

$$\frac{2VR(R+1)}{R+1+R}=\frac{2VR(R+1)}{2R+1}$$

23. $\frac{E}{\frac{I}{9}}=\frac{E\bullet 9}{\frac{I}{9}\bullet 9}=\frac{9E}{I}$

25.

$$\frac{2}{\frac{1}{a}+\frac{1}{b}}=\frac{2}{\frac{b+a}{ab}}=\frac{2\bullet ab}{\frac{b+a}{ab}\bullet ab}=\frac{2ab}{a+b}$$

27.

$$\frac{w}{\left(1+\frac{h}{6400}\right)^2}=\frac{w}{\left(\frac{6400+h}{6400}\right)^2}=\frac{w}{\frac{(6400+h)^2}{6400^2}}=$$

$$\frac{w\bullet 6400^2}{\frac{(6400+h)^2}{6400^2}\bullet 6400^2}=\frac{6400^2 w}{(6400+h)^2}$$

7.5 Solving Rational Equations

Practice 7.5

1.

$$\frac{y}{2}-\frac{y}{3}=\frac{1}{12}$$

$$12\left(\frac{y}{2}-\frac{y}{3}\right)=\left(\frac{1}{12}\right)12$$

$$6y-4y=1$$

$$2y=1$$

$$\frac{2y}{2}=\frac{1}{2}$$

$$y=\frac{1}{2}$$

check :

$$\frac{\frac{1}{2}}{2}-\frac{\frac{1}{2}}{3}\stackrel{?}{=}\frac{1}{12}$$

$$\frac{1}{4}-\frac{1}{6}\stackrel{?}{=}\frac{1}{12}$$

$$\frac{1}{12}=\frac{1}{12}$$

2.

$$\frac{x-2}{5}-1=-\frac{2}{x}$$

$$5x\left(\frac{x-2}{5}-1\right)=\left(-\frac{2}{x}\right)5x$$

$$x^2-2x-5x=-10$$

$$x^2-7x+10=0$$

$$(x-2)(x-5)=0$$

$$x-2=0 \quad x-5=0$$

$$x=2 \qquad x=5$$

check :

$$\frac{2-2}{5}-1\stackrel{?}{=}-\frac{2}{2} \qquad \frac{5-2}{5}-1\stackrel{?}{=}-\frac{2}{5}$$

$$-1=-1 \qquad -\frac{2}{5}=-\frac{2}{5}$$

3.

$$\frac{4}{y+2}+\frac{2}{y-1}=\frac{12}{y^2+y-2}$$

$$(y+2)(y-1)\left(\frac{4}{y+2}+\frac{2}{y-1}\right)=\left(\frac{12}{y^2+y-2}\right)(y^2+y-2)$$

$$4(y-1)+2(y+2)=12$$

$$4y-4+2y+4=12$$

$$6y=12$$

$$y=2$$

check :

$$\frac{4}{2+2}+\frac{2}{2-1}\stackrel{?}{=}\frac{12}{2^2+2-2}$$

$$1+2\stackrel{?}{=}3$$

$$3=3$$

4.

$$x=\frac{9}{x+3}+\frac{3x}{x+3}$$

$$(x+3)x=\left(\frac{9}{x+3}+\frac{3x}{x+3}\right)(x+3)$$

$$x^2+3x=9+3x$$

$$x^2-9=0$$

$$(x+3)(x-3)=0$$

$$x+3=0 \quad x-3=0$$

$$x\neq-3 \qquad x=3$$

check :

$$-3\stackrel{?}{=}\frac{9}{-3+3}+\frac{3(-3)}{-3+3} \qquad 3\stackrel{?}{=}\frac{9}{3+3}+\frac{3(3)}{3+3}$$

$$-3\neq\text{undefined} \qquad 3=3$$

5.

	r	t	Part of task
Less power	$\frac{1}{10}$	t	$\frac{t}{10}$
More power	$\frac{1}{6}$	t	$\frac{t}{6}$

$$\frac{t}{10}+\frac{t}{6}=1$$

$$30\left(\frac{t}{10}+\frac{t}{6}\right)=1(30)$$

$$3t+5t=30$$

$$8t=30$$

$$t=3\frac{3}{4}$$

Working together, it will take both pumps $3\frac{3}{4}$ hours or 3 hours and 45 minutes to fill the tank.

6.

	d	r	t
Jet	1800	3r	$\frac{600}{r}$
Prop plane	300	r	$\frac{300}{r}$

$$\frac{600}{r}+\frac{300}{r}=6$$

$$600+300=6r$$

$$900=6r$$

$$150=r$$

The speed of the propeller plane was 150mph.

7.a.

$$D=\frac{500}{x}+50$$

$$D-50=\frac{500}{x}$$

$$x(D-50)=500$$

$$x=\frac{500}{D-50}$$

b.

$$x=\frac{500}{450-50}=\frac{500}{400}=\frac{5}{4}=1.25$$

The price is $1.25 per unit.

Exercises 7.5

1.

$$\frac{y}{2}+\frac{7}{10}=-\frac{4}{5}$$

$$10\left(\frac{y}{2}+\frac{7}{10}\right)=\left(-\frac{4}{5}\right)10$$

$$5y+7=-8$$

$$5y+7-7=-8-7$$

$$5y=-15$$

$$\frac{5y}{5}=\frac{-15}{5}$$

$$y=-3$$

check:

$$\frac{-3}{2}+\frac{7}{10}\stackrel{?}{=}-\frac{4}{5}$$

$$-\frac{4}{5}=-\frac{4}{5}$$

3.

$$\frac{1}{t}-\frac{7}{3}=-\frac{1}{3}$$

$$3t\left(\frac{1}{t}-\frac{7}{3}\right)=\left(-\frac{1}{3}\right)3t$$

$$3-7t=-t$$

$$3=6t$$

$$t=\frac{1}{2}$$

check:

$$\frac{1}{\frac{1}{2}}-\frac{7}{3}\stackrel{?}{=}-\frac{1}{3}$$

$$2-\frac{7}{3}\stackrel{?}{=}-\frac{1}{3}$$

$$-\frac{1}{3}=-\frac{1}{3}$$

5.

$$x+\frac{1}{x}=2$$

$$x\left(x+\frac{1}{x}\right)=(2)x$$

$$x^2+1=2x$$

$$x^2-2x+1=0$$

$$(x-1)^2=0$$

$$x-1=0$$

$$x=1$$

check:

$$1+\frac{1}{1}\stackrel{?}{=}2$$

$$2=2$$

7.

$$\frac{t+1}{2t-1}-\frac{5}{7}=0$$

$$7(2t-1)\left(\frac{t+1}{2t-1}-\frac{5}{7}\right)=0(7)(2t-1)$$

$$7(t+1)-(2t-1)(5)=0$$

$$7t+7-10t+5=0$$

$$-3t+12=0$$

$$-3t=-12$$

$$t=4$$

check:

$$\frac{4+1}{2(4)-1}-\frac{5}{7}\stackrel{?}{=}0$$

$$\frac{5}{7}-\frac{5}{7}\stackrel{?}{=}0$$

$$0=0$$

9.

$$\frac{t-2}{3}=4$$

$$3\left(\frac{t-2}{3}\right)=4(3)$$

$$t-2=12$$

$$t=14$$

check:

$$\frac{14-2}{3}\stackrel{?}{=}4$$

$$4=4$$

11.

$$\frac{4}{s-3}-\frac{3s}{s-3}=2$$

$$(s-3)\left(\frac{4-3s}{s-3}\right)-2(s-3)$$

$$4-3s=2s-6$$

$$-5s=-10$$

$$s=2$$

check:

$$\frac{4}{2-3}-\frac{3(2)}{2-3}\stackrel{?}{=}2$$

$$-4-(-6)\stackrel{?}{=}2$$

$$2=2$$

13

$$\frac{5x}{x+1}=\frac{x^2}{x+1}+2$$

$$(x+1)\left(\frac{5x}{x+1}\right)=\left(\frac{x^2}{x+1}+2\right)(x+1)$$

$$5x=x^2+2x+2$$

$$x^2-3x+2=0$$

$$(x-1)(x-2)=0$$

$$x-1=0 \quad x-2=0$$

$$x=1 \qquad x=2$$

check :

$$\frac{5(1)}{1+1}\stackrel{?}{=}\frac{1^2}{1+1}+2 \qquad \frac{5(2)}{2+1}\stackrel{?}{=}\frac{2^2}{2+1}+2$$

$$\frac{5}{2}\stackrel{?}{=}\frac{1}{2}+2 \qquad \frac{10}{3}\stackrel{?}{=}\frac{4}{3}+2$$

$$\frac{5}{2}=\frac{5}{2} \qquad \frac{10}{3}=\frac{10}{3}$$

15.

$$\frac{x}{x-3}-\frac{6}{x}=1$$

$$\left(x^2-3x\right)\left(\frac{x}{x-3}-\frac{6}{x}\right)=1\left(x^2-3x\right)$$

$$x(x)-(x-3)6=x^2-3x$$

$$x^2-6x+18=x^2-3x$$

$$-3x=-18$$

$$x=6$$

check :

$$\frac{6}{6-3}-\frac{6}{6}\stackrel{?}{=}1$$

$$2-1\stackrel{?}{=}1$$

$$1=1$$

17.

$$1+\frac{4}{x^2}=\frac{4}{x}$$

$$x^2\left(1+\frac{4}{x^2}\right)=\left(\frac{4}{x}\right)x^2$$

$$x^2+4=4x$$

$$x^2-4x+4=0$$

$$(x-2)^2=0$$

$$x-2=0$$

$$x=2$$

check :

$$1+\frac{4}{2^2}+\stackrel{?}{=}\frac{4}{2}$$

$$2=2$$

19.

$$\frac{2}{p+1}-\frac{1}{p-1}=\frac{2p}{p^2-1}$$

$$\left(p^2-1\right)\left(\frac{2}{p+1}-\frac{1}{p-1}\right)=\left(\frac{2p}{p^2-1}\right)\left(p^2-1\right)$$

$$2(p-1)-1(p+1)=2p$$

$$p-3=2p$$

$$-3=p$$

$$p=-3$$

check :

$$\frac{2}{-3+1}-\frac{1}{-3-1}\stackrel{?}{=}\frac{2(-3)}{(-3)^2-1}$$

$$-1+\frac{1}{4}\stackrel{?}{=}\frac{-6}{8}$$

$$-\frac{3}{4}=-\frac{3}{4}$$

21.

$$\frac{3}{x}-\frac{1}{x+4}=\frac{5}{x^2+4x}$$

$$\left(x^2+4x\right)\left(\frac{3}{x}-\frac{1}{x+4}\right)=\frac{5}{x^2+4x}\left(x^2+4x\right)$$

$$3(x+4)-1(x)=5$$

$$2x+12=5$$

$$2x=-7$$

$$x=-\frac{7}{2}$$

check :

$$\frac{3}{-\frac{7}{2}}-\frac{1}{-\frac{7}{2}+4}\stackrel{?}{=}\frac{5}{\left(-\frac{7}{2}\right)^2+4\left(-\frac{7}{2}\right)}$$

$$-\frac{6}{7}-\frac{2}{1}\stackrel{?}{=}\frac{5}{-\frac{7}{4}}$$

$$-\frac{20}{7}=-\frac{20}{7}$$

23.

$$1-\frac{6x}{(x-4)^2}=\frac{2x}{x-4}$$

$$(x-4)^2\left(1-\frac{6x}{(x-4)^2}\right)=\left(\frac{2x}{x-4}\right)(x-4)^2$$

$$(x-4)^2-6x=2x(x-4)$$

$$x^2-14x+16=2x^2-8x$$

$$x^2+6x-16=0$$

$$(x-2)(x+8)=0$$

$$x-2=0 \qquad x+8=0$$

$$x=2 \qquad x=-8$$

check :

$$1-\frac{6(2)}{(2-4)^2}\stackrel{?}{=}\frac{2(2)}{2-4} \qquad 1-\frac{6(-8)}{(-8-4)^2}\stackrel{?}{=}\frac{2(-8)}{-8-4}$$

$$1-3\stackrel{?}{=}-2 \qquad 1+\frac{1}{3}\stackrel{?}{=}\frac{-16}{-12}$$

$$-2=-2 \qquad \frac{4}{3}=\frac{4}{3}$$

25.

$$\frac{n+1}{n^2+2n-3}=\frac{n}{n+3}-\frac{1}{n-1}$$

$$\left(n^2+2n-3\right)\left(\frac{n+1}{n^2+2n-3}\right)=\left(\frac{n}{n+3}-\frac{1}{n-1}\right)\left(n^2+2n-3\right)$$

$$n+1=n(n-1)-1(n+3)$$

$$n+1=n^2-n-n-3$$

$$n^2-3n-4=0$$

$$(n+1)(n-4)=0$$

$$n+1=0 \qquad n-4=0$$

$$n=-1 \qquad n=4$$

check :

$$\frac{(-1)+1}{(-1)^2+2(-1)-3}\stackrel{?}{=}\frac{-1}{-1+3}-\frac{1}{-1-1} \qquad \frac{4+1}{4^2+2(4)-3}\stackrel{?}{=}\frac{4}{4+3}-\frac{1}{4-1}$$

$$\frac{0}{-4}\stackrel{?}{=}-\frac{1}{2}+\frac{1}{2} \qquad \frac{5}{21}\stackrel{?}{=}\frac{4}{7}-\frac{1}{3}$$

$$0=0 \qquad \frac{5}{21}=\frac{5}{21}$$

Applications

27.

$$\frac{x}{45}+\frac{x}{30}=1$$

$$90\left(\frac{x}{45}+\frac{x}{30}\right)=(1)90$$

$$2x+3x=90$$

$$5x=90$$

$$x=18$$

It will take them 18 minutes to clean the attic together.

29.

	R	t	Part of task
Secretary	$\frac{1}{x}$	3	$\frac{3}{x}$
Clerical	$\frac{1}{4x}$	3	$\frac{3}{4x}$

$$\frac{3}{x}+\frac{3}{4x}=1$$
$$4x\left(\frac{3}{x}+\frac{3}{4x}\right)=(1)4x$$
$$12+3=4x$$
$$4x=15$$
$$x=\frac{15}{4}$$
$$4x=15$$

It will take the clerical worker 15 hours to finish the job working alone.

31.

	r	t	d
Clear	2r	$\frac{60}{2r}$	60
Slippery	r	$\frac{60}{r}$	60

$$\frac{60}{r}+\frac{60}{2r}=3$$
$$2r\left(\frac{60}{r}+\frac{60}{2r}\right)=(3)2r$$
$$120+60=6r$$
$$6r=180$$
$$r=30$$
$$2r=60$$

The speed on the clear road was 60 mph.

33.

$$p=\frac{P}{LD}$$
$$Dp=\frac{P}{L\cancel{D}_1}\bullet\cancel{D}^1$$
$$Dp=\frac{P}{L}$$
$$(Dp)\left(\frac{1}{p}\right)=\left(\frac{P}{L}\right)\left(\frac{1}{p}\right)$$
$$D=\frac{P}{Lp}$$

7.6 Ratio and Proportion

Practice 7.6

1.

$$\frac{4}{5}=\frac{p}{10} \qquad \textit{check}:$$
$$5p=40 \qquad \frac{4}{5}\stackrel{?}{=}\frac{8}{10}$$
$$x=8 \qquad \frac{4}{5}=\frac{4}{5}$$

2.

$$\frac{80}{98}=\frac{x}{49} \qquad \textit{check}:$$
$$98x=3920 \qquad \frac{80}{98}\stackrel{?}{=}\frac{40}{49}$$
$$x=40 \qquad \frac{40}{49}=\frac{40}{49}$$

3.

$$\frac{900}{100{,}000}=\frac{900-x}{75{,}000}$$
$$90{,}000{,}000-100{,}000x=67{,}500{,}000$$
$$22{,}500{,}000=100{,}000x$$
$$225=x$$

check:

$$\frac{900}{100{,}000}\stackrel{?}{=}\frac{900-225}{75{,}000}$$
$$\frac{900}{100{,}000}\stackrel{?}{=}\frac{675}{75{,}000}$$
$$\frac{9}{1000}=\frac{9}{1000}$$

4.

$$\frac{8}{DE}=\frac{10}{5}$$
$$10DE=40$$
$$DE=4$$

5.

	d	**r**	**t**
With jet stream	1000	r+300	$\frac{1000}{r+300}$
Against jet stream	250	r-300	$\frac{250}{r-300}$

$\frac{1000}{r+300}=\frac{250}{r-300}$

$1000r-300,000=250r+75,000$

$750r=375,000$

$r=500$

The speed of the plane in still air is 500 mph.

Exercises 7.6

1.

$\frac{x}{10}=\frac{4}{5}$ *check*:

$5x=40$ $\frac{8}{10}\stackrel{?}{=}\frac{4}{5}$

$x=8$ $\frac{4}{5}=\frac{4}{5}$

3.

$\frac{n}{100}=\frac{4}{5}$ *check*:

$5n=400$ $\frac{80}{100}\stackrel{?}{=}\frac{4}{5}$

$n=80$ $\frac{4}{5}=\frac{4}{5}$

5.

$\frac{8}{7}=\frac{s}{21}$ *check*:

$7s=168$ $\frac{8}{7}\stackrel{?}{=}\frac{24}{21}$

$s=24$ $\frac{8}{7}=\frac{8}{7}$

7.

$\frac{8+x}{12}=\frac{22}{36}$ *check*:

$288+36x=264$ $\frac{8+\left(-\frac{2}{3}\right)}{12}\stackrel{?}{=}\frac{22}{36}$

$36x=-24$ $\frac{\frac{22}{3}}{12}\stackrel{?}{=}\frac{22}{36}$

$n=-\frac{24}{36}$ $\frac{22}{36}=\frac{22}{36}$

$n=-\frac{2}{3}$

9.

$\frac{y+3}{14}=\frac{y}{7}$

$7(y+3)=14y$

$7y+21=14y$

$7y=21$

$y=3$

check:

$\frac{3+3}{14}\stackrel{?}{=}\frac{3}{7}$

$\frac{6}{14}\stackrel{?}{=}\frac{3}{7}$

$\frac{3}{7}=\frac{3}{7}$

11.

$\frac{x-1}{8}=\frac{x+1}{12}$ *check*:

$12x-12=8x+8$ $\frac{5-1}{8}\stackrel{?}{=}\frac{5+1}{12}$

$4x=20$ $\frac{4}{8}\stackrel{?}{=}\frac{6}{12}$

$x=5$ $\frac{1}{2}=\frac{1}{2}$

13.

$\frac{x}{8}=\frac{2}{x}$

$x^2=16$

$x=\pm 4$

check:

$\frac{4}{8}\overset{?}{=}\frac{2}{4}$ $\quad \frac{-4}{8}\overset{?}{=}\frac{2}{-4}$

$\frac{1}{2}=\frac{1}{2}$ $\quad -\frac{1}{2}=-\frac{1}{2}$

15.

$\frac{2}{y}=\frac{y-4}{16}$

$y^2-4y=32$

$y^2-4y-32=0$ $\quad$ *check*:

$(y-8)(y+4)=0$ $\quad \frac{2}{8}\overset{?}{=}\frac{8-4}{16}$ $\quad \frac{2}{-4}\overset{?}{=}\frac{-4-4}{16}$

$y-8=0 \quad y+4=0$ $\quad \frac{2}{8}\overset{?}{=}\frac{4}{16}$ $\quad \frac{2}{-4}\overset{?}{=}\frac{-8}{16}$

$y=8 \quad y=-4$ $\quad \frac{1}{4}=\frac{1}{4}$ $\quad -\frac{1}{2}=-\frac{1}{2}$

17.

$\frac{a}{a+3}=\frac{4}{5a}$

$5a^2=4a+12$

$5a^2-4a-12=0$

$(5a+6)(a-2)=0$

$5a+6=0 \quad a-2=0$

$a=-\frac{6}{5} \quad a=2$

check:

$\frac{-\frac{6}{5}}{-\frac{6}{5}+3}\overset{?}{=}\frac{4}{5\left(-\frac{6}{5}\right)}$ $\quad \frac{2}{2+3}\overset{?}{=}\frac{4}{5(2)}$

$-\frac{2}{3}=-\frac{2}{3}$ $\quad \frac{2}{5}=\frac{2}{5}$

19.

$\frac{y+1}{y+6}=\frac{y}{y+6}$

$y^2+7y+6=y^2+6y$

$y\neq -6$

No solution. If -6 was substituted for y in the equation, a zero in the denominator would result. Since division by zero is undefined, there is no solution.

Applications

21.

$\frac{6}{2}=\frac{25}{x}$

$6x=50$

$x=\frac{50}{6}=\frac{25}{3}$

$x=8\frac{1}{3}$

It will take $8\frac{1}{3}$ minutes (or 8 minutes and 20 seconds) to print a 25-page report.

23.

$\frac{60}{8}=\frac{120}{x}$ $\quad$ *check*:

$60x=960$ $\quad \frac{60}{8}\overset{?}{=}\frac{120}{16}$

$x=16$ $\quad \frac{15}{2}=\frac{15}{2}$

It will take 16 gallons of gas to go 120 miles.

25.

r = rate of runner; r+14= rate of cyclist

$\frac{10}{r+14}=\frac{3}{r}$

$10r=3r+42$

$7r=42$

$r=6\ mph$

$r+14=20\ mph$

The speed of the cyclist was 20 mph.

27. speed of train = r + 30; speed of bus = r.

$\frac{400}{r+30}=\frac{250}{r}$

$400r=250r+7500$

$150r=7500$

$r=50$

$r+30=80$

The speed of the bus is 50 mph and the speed of the train is 80 mph.

29.

$\frac{y}{20}=\frac{10}{25}$

$25y=200$

$y=8$

y = AB = 8 feet

31.

$\frac{h}{6}=\frac{12}{4}$

$4h=72$

$h=18\ ft$

The height of the tree is 18 feet.

33.

$\frac{1}{2}=\frac{w+4}{24}$

$24=2w+8$

$16=2w$

$w=8$

There are 8 women at the party.

Chapter 7 Review Exercises

1.

a.

$x+1=0$

$x=-1$

b.

$x^2-x-6=0$

$(x-3)(x+2)=0$

$x-3=0 \quad x+2=0$

$x=3 \qquad x=-2$

2.

a. $\frac{10x^2y}{5xy^2}=\frac{2\cdot 5\cdot x^2\cdot y}{5\cdot x\cdot y^2}=\frac{2x}{y}$

Equivalent

b. $\frac{x^2-9}{x^2+6x+9}=\frac{(x-3)(x+3)}{(x+3)(x+3)}=\frac{x-3}{x+3}$

Equivalent

3. $\frac{12m}{20m^2}=\frac{3\cdot 4\cdot m}{5\cdot 4\cdot m^2}=\frac{3}{5m}$

4. $\frac{15n-18}{9n+6}=\frac{3(5n-6)}{3(3n+2)}=\frac{5n-6}{3n+2}$

5. $\frac{x^2+2x-8}{4-x^2}=\frac{(x+4)(x-2)}{(-1)(x+2)(x-2)}=-\frac{x+4}{x+2}$

6. $\frac{2x^2-3x-20}{3x^2-13x+4}=\frac{(2x+5)(x-4)}{(3x-1)(x-4)}=\frac{2x+5}{3x-1}$

7. $\frac{10mn}{3p^2}\bullet\frac{9np}{5m^2}=\frac{10\,m\,n}{3\,p^2}\bullet\frac{9\,n\,p}{5\,m^2}=\frac{6n^2}{pm}$

8.

$\frac{y-5}{4y+6}\bullet\frac{6y+9}{3y-15}=\frac{y-5}{2(2y+3)}\bullet\frac{3(2y+3)}{3(y-5)}=\frac{1}{2}$

9.

$\frac{x+6}{x^2+x-30}\bullet\frac{x^2-10x+25}{2x+5}=$

$\frac{x+6}{(x-5)(x+6)}\bullet\frac{(x-5)(x-5)}{2x+5}=\frac{x-5}{2x+5}$

10.

$$\frac{2a^2-2a-4}{4-a^2}\bullet\frac{2a^2+a-6}{4a^2-2a-6}=$$

$$\frac{\cancel{(2)}\cancel{(a-2)}\cancel{(a+1)}}{(-1)\cancel{(a-2)}\cancel{(a+2)}}\bullet\frac{\cancel{(a+2)}\cancel{(2a-3)}}{\cancel{(2)}\cancel{(2a-3)}\cancel{(a+1)}}=-1$$

11. $\frac{x^2y}{2x}\div xy^2=\frac{\cancel{x^2}\cancel{y}}{2\cancel{x}}\bullet\frac{1}{\cancel{x}\cancel{y^2}}=\frac{1}{2y}$

12.

$$\frac{5m+10}{2m-20}\div\frac{7m+14}{14m-20}=\frac{5m+10}{2m-20}\bullet\frac{14m-20}{7m+14}=$$

$$\frac{5\cancel{(m+2)}}{\cancel{2}(m-10)}\bullet\frac{\cancel{2}(7m-10)}{7\cancel{(m+2)}}=\frac{5(7m-10)}{7(m-10)}$$

13.

$$\frac{5y^2}{x^2-36}\div\frac{25xy-25y}{x^2-7x+6}=\frac{5y^2}{x^2-36}\bullet\frac{x^2-7x+6}{25xy-25y}=$$

$$\frac{\cancel{5}\cancel{y^2}}{\cancel{(x-6)}(x+6)}\bullet\frac{\cancel{(x-6)}\cancel{(x-1)}}{\cancel{25}\cancel{y}\cancel{(x-1)}}=\frac{y}{5(x+6)}$$

14.

$$\frac{2x^2+x-1}{x^2+8x+7}\div\frac{6x^2+x-2}{x^2+14x+49}=$$

$$\frac{\cancel{(2x-1)}\cancel{(x+1)}}{\cancel{(x+7)}\cancel{(x+1)}}\bullet\frac{\cancel{(x+7)}(x+7)}{(3x+2)\cancel{(2x-1)}}=\frac{(x+7)}{(3x+2)}$$

15.

$LCD=20x^2$

$$\frac{1}{5x}=\frac{1\bullet4x}{5x\bullet4x}=\frac{4x}{20x^2}$$

$$\frac{3}{20x^2}=\frac{3}{20x^2}$$

16.

$LCD=(n-1)(n+4)$

$$\frac{4}{n-1}=\frac{4(n+4)}{(n-1)(n+4)}$$

$$\frac{n}{n+4}=\frac{n(n-1)}{(n-1)(n+4)}$$

17.

$LCD=3(x+3)(x+1)$

$$\frac{1}{3x+9}=\frac{1(x+1)}{3(x+3)(x+1)}=\frac{x+1}{3(x+3)(x+1)}$$

$$\frac{x}{x^2+4x+3}=\frac{x(3)}{3(x+3)(x+1)}=\frac{3x}{3(x+3)(x+1)}$$

18.

$LCD=(3x+1)(x-2)(x+2)$

$$\frac{2}{3x^2-5x-2}=\frac{2}{(3x+1)(x-2)}=\frac{2(x+2)}{(3x+1)(x-2)(x+2)}$$

$$\frac{1}{4-x^2}=-\frac{1}{(x+2)(x-2)}=\frac{3x+1}{(3x+1)(x-2)(x+2)}$$

19.

$$\frac{3t+1}{2t}+\frac{t-1}{2t}=\frac{3t+1+t-1}{2t}=\frac{4t}{2t}=2$$

20.

$$\frac{5y}{y+7}-\frac{y-28}{y+7}=\frac{5y-y+28}{y+7}=$$

$$\frac{4y+28}{y+7}=\frac{4(y+7)}{y+7}=4$$

21.

$$\frac{5y+4}{4y^2-2y}-\frac{2}{2y-1}=\frac{5y+4}{2y(2y-1)}-\frac{2(2y)}{2y(2y-1)}=$$

$$\frac{5y+4-4y}{2y(2y-1)}=\frac{y+4}{2y(2y-1)}$$

22.

$$\frac{n}{3n+15}+\frac{n-2}{n^2+5n}=\frac{n}{3(n+5)}+\frac{n-2}{n(n+5)}=$$

$$\frac{n(n)}{3n(n+5)}+\frac{3(n-2)}{3n(n+5)}=\frac{n^2+3n-6}{3n(n+5)}$$

23.

$$\frac{4}{x-3}-\frac{4x+1}{9-x^2}=\frac{4}{x-3}-\frac{4x+1}{(-1)(x-3)(x+3)}=$$

$$\frac{4(x+3)}{x-3}+\frac{4x+1}{(x-3)(x+3)}=\frac{8x+13}{(x-3)(x+3)}$$

24.

$$\frac{y+3}{4-y^2}+\frac{1}{2-y}=\frac{y+3}{(2-y)(2+y)}+\frac{1(2+y)}{(2-y)(2+y)}=$$

$$\frac{y+3+2+y}{(2-y)(2+y)}=\frac{2y+5}{(2-y)(2+y)}$$

25.

$$\frac{2}{m+1}+\frac{6m-2}{m^2-2m-3}=$$

$$\frac{2(m-3)}{(m-3)(m+1)}+\frac{6m-2}{(m-3)(m+1)}=$$

$$\frac{2m-6+6m-2}{(m-3)(m+1)}=\frac{8m-8}{(m-3)(m+1)}=\frac{8(m-1)}{(m-3)(m+1)}$$

26.

$$\frac{3x-2}{x^2-x-12}-\frac{x+3}{x-4}=$$

$$\frac{3x-2}{(x+3)(x-4)}-\frac{(x+3)(x+3)}{(x-4)(x+3)}=$$

$$\frac{3x-2-x^2-6x-9}{(x+3)(x-4)}=\frac{-x^2-3x-11}{(x+3)(x-4)}$$

27.

$$\frac{2x}{x^2+4x+4}-\frac{x-1}{x^2-2x-8}=$$

$$\frac{2x(x-4)}{(x+2)^2(x-4)}-\frac{(x-1)(x+2)}{(x-4)(x+2)^2}=$$

$$\frac{2x^2-8x-x^2-x+2}{(x+2)^2(x-4)}=\frac{x^2-9x+2}{(x+2)^2(x-4)}$$

28.

$$\frac{n+4}{2n^2-3n+1}+\frac{n+1}{2n^2+5n-3}=$$

$$\frac{(n+4)(n+3)}{(2n-1)(n-1)(n+3)}+\frac{(n+1)(n-1)}{(2n-1)(n-1)(n+3)}=$$

$$\frac{n^2+7n+12+n^2-1}{(2n-1)(n-1)(n+3)}=\frac{2n^2+7n+11}{(2n-1)(n-1)(n+3)}$$

29.

$$\frac{\frac{x}{2}}{\frac{3x^2}{7}}=\frac{\frac{x}{2}\bullet 14}{\frac{3x^2}{7}\bullet 14}=\frac{7x}{6x^2}=\frac{7}{6x}$$

30.

$$\frac{1-\frac{9}{y}}{1-\frac{81}{y^2}}=\frac{\frac{y-9}{y}}{\frac{y^2-81}{y^2}}=\frac{\frac{y-9}{y}\bullet y^2}{\frac{y^2-81}{y^2}\bullet y^2}=$$

$$\frac{y^2-9y}{y^2-81}=\frac{y\,\overset{1}{\cancel{(y-9)}}}{(y+9)\,\underset{1}{\cancel{(y-9)}}}=\frac{y}{y+9}$$

31.

$$\frac{\frac{1}{x}+\frac{1}{y}}{\frac{1}{2x}+\frac{1}{2y}}=\frac{\left(\frac{1}{x}+\frac{1}{y}\right)2xy}{\left(\frac{1}{2x}+\frac{1}{2y}\right)2xy}=\frac{2(y+x)}{y+x}=2$$

32.

$$\frac{4-\frac{3}{x}-\frac{1}{x^2}}{2-\frac{5}{x}+\frac{3}{x^2}}=\frac{\frac{4x^2-3x-1}{x^2}\bullet x^2}{\frac{2x^2-5x+3}{x^2}\bullet x^2}=$$

$$\frac{4x^2-3x-1}{2x^2-5x+3}=\frac{\overset{1}{\cancel{(x-1)}}(4x+1)}{\underset{1}{\cancel{(x-1)}}(2x-3)}=\frac{4x+1}{2x-3}$$

33.

$$\frac{2x}{x-4}=5-\frac{1}{x-4}$$

$$(x-4)\left(\frac{2x}{x-4}\right)=\left(5-\frac{1}{x-4}\right)(x-4)$$

$$2x=5x-20-1$$

$$-3x=-21$$

$$x=7$$

check:

$$\frac{2(7)}{7-4}\overset{?}{=}5-\frac{1}{7-4}$$

$$\frac{14}{3}=\frac{14}{3}$$

34.

$$\frac{y+1}{y}+\frac{1}{2y}=4$$

$$2y\left(\frac{y+1}{y}+\frac{1}{2y}\right)=4(2y)$$

$$2(y+1)+1=8y$$

$$2y+3=8y$$

$$3=6y$$

$$\frac{1}{2}=y$$

check :

$$\frac{\frac{1}{2}+1}{\frac{1}{2}}+\frac{1}{2\left(\frac{1}{2}\right)}\stackrel{?}{=}4$$

$$3+1\stackrel{?}{=}4$$

$$4=4$$

35.

$$\frac{5}{2x}+\frac{3}{x+1}=\frac{7}{x}$$

$$(2x)(x+1)\left(\frac{5}{2x}+\frac{3}{x+1}\right)=\frac{7}{x}(2x)(x+1)$$

$$(x+1)5+3(2x)=7(2)(x+1)$$

$$11x+5=14x+14$$

$$-3x=9$$

$$x=-3$$

check :

$$\frac{5}{2(-3)}+\frac{3}{(-3)+1}\stackrel{?}{=}\frac{7}{(-3)}$$

$$-\frac{14}{6}\stackrel{?}{=}-\frac{7}{3}$$

$$-\frac{7}{3}=-\frac{7}{3}$$

36.

$$\frac{y-2}{y-4}=\frac{1}{y+2}+\frac{y+3}{y^2-2y-8}$$

$$(y+2)(y-4)\left(\frac{y-2}{y-4}\right)=\left(\frac{1}{y+2}+\frac{y+3}{y^2-2y-8}\right)(y+2)(y-4)$$

$$y^2-4=y-4+y+3$$

$$y^2-2y-3=0$$

$$(y-3)(y+1)0$$

$$y-3=0 \quad y+1=0$$

$$y=3 \qquad y=-1$$

check :

$$\frac{3-2}{3-4}\stackrel{?}{=}\frac{1}{3+2}+\frac{3+3}{3^2-2(3)-8} \qquad \frac{-1-2}{-1-4}\stackrel{?}{=}\frac{1}{-1+2}+\frac{-1+3}{(-1)^2-2(-1)-8}$$

$$-1\stackrel{?}{=}\frac{1}{5}+\left(-\frac{6}{5}\right) \qquad \frac{3}{5}\stackrel{?}{=}1+\left(-\frac{2}{5}\right)$$

$$-1=-1 \qquad \frac{3}{5}=\frac{3}{5}$$

37.

$$\frac{x}{x+2}-\frac{2}{2-x}=\frac{x+6}{x^2-4}$$

$$(x^2-4)\left(\frac{x}{x+2}-\left(-\frac{2}{x-2}\right)\right)=\frac{x+6}{x^2-4}(x^2-4)$$

$$(x-2)x+2(x+2)=x+6$$

$$x^2+4=x+6$$

$$x^2-x-2=0$$

$$(x-2)(x+1)=0$$

$$x-2=0 \quad x+1=0$$

$$x\neq 2 \qquad x=-1$$

check :

$$\frac{-1}{-1+2}-\frac{2}{2-(-1)}\stackrel{?}{=}\frac{-1+6}{(-1)^2-4}$$

$$-1-\frac{2}{3}\stackrel{?}{=}-\frac{5}{3}$$

$$-\frac{5}{3}=-\frac{5}{3}$$

2 cannot be a solution for this equation because it results in a zero in the denominator. Division by zero is undefined under the set of real numbers.

38.

$$\frac{3}{n^2-5n+4}-\frac{1}{n^2-4n+3}=\frac{n-3}{n^2-7n+12}$$
$$\frac{3}{(n-4)(n-1)}-\frac{1}{(n-3)(n-1)}=\frac{n-3}{(n-3)(n-4)}$$
$$3(n-3)-1(n-4)=(n-3)(n-1)$$
$$3n-9-n+4=n^2-4n+3$$
$$n^2-6n+8=0$$
$$(n-4)(n-2)=0$$
$$n-4=0 \quad n-2=0$$
$$n\neq 4 \quad n=2$$
check :
$$\frac{3}{2^2-5(2)+4}-\frac{1}{2^2-4(2)+3}\stackrel{?}{=}\frac{2-3}{2^2-7(2)+12}$$
$$-\frac{3}{2}+1\stackrel{?}{=}-\frac{1}{2}$$
$$-\frac{1}{2}=-\frac{1}{2}$$

39.

$$\frac{8}{5}=\frac{72}{x}$$
$$8x=360$$
$$x=45$$
check :
$$\frac{8}{5}\stackrel{?}{=}\frac{72}{45}$$
$$8(45)\stackrel{?}{=}360$$
$$360=360$$

40.

$$\frac{28}{x+3}=\frac{7}{9}$$
$$7x+21=252$$
$$7x=231$$
$$x=33$$
check :
$$\frac{28}{33+3}\stackrel{?}{=}\frac{7}{9}$$
$$36(7)\stackrel{?}{=}252$$
$$252=252$$

41.

$$\frac{5}{3+y}=\frac{3}{7y+1}$$
$$35y+5=9+3y$$
$$32y=4$$
$$y=\frac{4}{32}=\frac{1}{8}$$
check :
$$\frac{5}{3+\frac{1}{8}}\stackrel{?}{=}\frac{3}{7\left(\frac{1}{8}\right)+1}$$
$$\frac{8}{5}=\frac{8}{5}$$

42.

$$\frac{11}{x-2}=\frac{x+7}{2}$$
$$x^2+5x-14=22$$
$$x^2+5x-36=0$$
$$(x-4)(x+9)=0$$
$$x-4=0 \quad x+9=0$$
$$x=4 \quad x=-9$$
check :
$$\frac{11}{4-2}\stackrel{?}{=}\frac{4+7}{2} \qquad \frac{11}{-9-2}=\frac{-9+7}{2}$$
$$\frac{11}{2}=\frac{11}{2} \qquad -1=-1$$

Mixed Applications

43.

$$0.72+\frac{200}{x}=\frac{0.72x+200}{x}\text{ dollars}$$

44.

$$\frac{c}{4}-10=\frac{c}{5}$$
$$5c-200=4c$$
$$c=\$200$$
The total cost of the car rental is $200.

45.

$$\frac{2d}{\frac{d}{r}+\frac{d}{s}}=\frac{(2d)rs}{\left(\frac{d}{r}+\frac{d}{s}\right)rs}=\frac{2drs}{ds+dr}=\frac{2rs}{s+r}$$

46.

$$\frac{x}{10}-\frac{x}{15}=1$$

$$3x-2x=30$$

$$x=30$$

It will take 30 minutes to fill the tub.

47.

$$\frac{d}{50}+\frac{400-d}{60}=7$$

$$6d+2000-5d=2100$$

$$d=100$$

The family drove 100 miles at 50 mph.

48.

$$\frac{1}{x}+\frac{1}{x+1}=\frac{1(x+1)+1(x)}{x(x+1)}=\frac{2x+1}{x(x+1)}$$ of the job will be done in one hour

49.

$$\frac{31,000}{15,000}=\frac{x}{20,000}$$

$$15,000x=620,000,000$$

$$x\approx 41333$$

She should expect to spend about $41,333.

50.

$$\frac{1}{n}+\frac{1}{n+1}+\frac{1}{n+2}$$

Chapter 7 Posttest

1. The expression is undefined when x = 8.

$$x-8=0$$

$$x=8$$

2.

$$-\frac{3y-y^2}{y^2}=-\frac{y(3-y)}{y^2}=-\frac{\overset{1}{\cancel{y}}(3-y)}{\underset{y}{\cancel{y^2}}}=$$

$$\frac{-(3-y)}{y}=\frac{y-3}{y}$$

3. $$\frac{15a^3b}{12ab^2}=\frac{5\cdot\overset{1}{\cancel{3}}\cdot\overset{a^2}{\cancel{a^3}}\cdot\overset{1}{\cancel{b}}}{4\cdot\underset{1}{\cancel{3}}\cdot\underset{1}{\cancel{a}}\cdot\underset{b}{\cancel{b^2}}}=\frac{5a^2}{4b}$$

4. $$\frac{x^2-4x}{xy-4y}=\frac{x\overset{1}{\cancel{(x-4)}}}{y\underset{1}{\cancel{(x-4)}}}=\frac{x}{y}$$

5. $$\frac{3b^2-27}{b^2-4b-21}=\frac{3\overset{1}{\cancel{(b+3)}}(b-3)}{\underset{1}{\cancel{(b+3)}}(b-7)}=\frac{3(b-3)}{(b-7)}$$

6.

$$\frac{\frac{3}{x^2}-\frac{1}{x}}{\frac{9}{x^2}-1}=\frac{\frac{3-x}{x^2}}{\frac{9-x^2}{x^2}}=\frac{3-x}{x^2}\div\frac{9-x^2}{x^2}=$$

$$\frac{\overset{1}{\cancel{(3-x)}}}{\underset{1}{\cancel{x^2}}}\cdot\frac{\overset{1}{\cancel{x^2}}}{\underset{1}{\cancel{(3-x)}}(3+x)}=\frac{1}{3+x}$$

7.

$$LCD=4(n+8)(n-2)$$

$$\frac{4n-1}{n^2+6n-16}=\frac{4n-1}{(n+8)(n-2)}=\frac{4(4n-1)}{4(n+8)(n-2)}$$

$$\frac{2}{n+8}=\frac{(2)(4)(n-2)}{4(n+8)(n-2)}=\frac{8(n-2)}{4(n+8)(n-2)}$$

$$\frac{n}{4n-8}=\frac{n}{4(n-2)}=\frac{n(n+8)}{4(n+8)(n-2)}$$

8.

$$\frac{7x-10}{x+6}-\frac{5x-22}{x+6}=$$

$$\frac{7x-10-5x+22}{x+6}=\frac{2x+12}{x+6}=$$

$$\frac{2\cancel{(x+6)}^{1}}{\cancel{(x+6)}_{1}}=2$$

9.

$$\frac{3}{2y-8}+\frac{2}{4y^2-16y}=\frac{3}{2(y-4)}+\frac{2}{4y(y-4)}=$$

$$\frac{3(2y)+2}{4y(y-4)}=\frac{6y+2}{4y(y-4)}=\frac{\overset{1}{\cancel{2}}(3y+1)}{\underset{2}{\cancel{4}}\,y(y-4)}=\frac{3y+1}{2y(y-4)}$$

10.

$$\frac{5}{d-3}-\frac{d-4}{d^2-d-6}=\frac{5}{d-3}-\frac{d-4}{(d+2)(d-3)}=$$

$$\frac{5(d+2)-d+4}{(d+2)(d-3)}=\frac{4d+14}{(d+2)(d-3)}=\frac{2(2d+7)}{(d+2)(d-3)}$$

11.

$$\frac{5}{2x^2-3x-2}-\frac{x}{4-x^2}=$$

$$\frac{5}{(2x+1)(x-2)}-\left(-\frac{x}{(x-2)(x+2)}\right)=$$

$$\frac{5(x+2)+x(2x+1)}{(2x+1)(x-2)(x+2)}=\frac{2x^2+6x+10}{(2x+1)(x-2)(x+2)}$$

12.

$$\frac{n+1}{3n-18}\bullet\frac{n-6}{6n^3-6n}=\frac{\overset{1}{\cancel{(n+1)}}}{3\underset{1}{\cancel{(n-6)}}}\bullet\frac{\overset{1}{\cancel{(n-6)}}}{6n(n-1)\underset{1}{\cancel{(n+1)}}}=$$

$$\frac{1}{18n(n-1)}$$

13.

$$\frac{a^2-25}{a^2-2a-24}\div\frac{a^2+a-30}{a^2-36}=$$

$$\frac{\overset{1}{\cancel{(a-5)}}(a+5)}{\underset{1}{\cancel{(a-6)}}(a+4)}\bullet\frac{\overset{1}{\cancel{(a-6)}}\overset{1}{\cancel{(a+6)}}}{\underset{1}{\cancel{(a+6)}}\underset{1}{\cancel{(a-5)}}}=$$

$$\frac{a+5}{a+4}$$

14.

$$\frac{x^2+6x+8}{x^2+x-2}\div\frac{x+4}{2x^2+12x+16}=$$

$$\frac{\overset{1}{\cancel{(x+2)}}\overset{1}{\cancel{(x+4)}}}{\underset{1}{\cancel{(x+2)}}(x-1)}\bullet\frac{2(x+2)(x+4)}{\underset{1}{\cancel{(x+4)}}}=$$

$$\frac{2(x+2)(x+4)}{x-1}$$

15.

$$\frac{1}{y-5}+\frac{y+4}{25-y^2}=\frac{1}{y+5}$$

$$\left(y^2-25\right)\left(\frac{1}{y-5}-\frac{y+4}{y^2-25}\right)=\frac{1}{y+5}\left(y^2-25\right)$$

$$(y+5)-(y+4)=y-5$$

$$1=y-5$$

$$6=y$$

$$y=6$$

check:

$$\frac{1}{6-5}+\frac{6+4}{25-6^2}\overset{?}{=}\frac{1}{6+5}$$

$$1-\frac{10}{11}\overset{?}{=}\frac{1}{11}$$

$$\frac{1}{11}=\frac{1}{11}$$

16.

$\frac{2y}{y-4}-2=\frac{4}{y+5}$

$(y-4)(y+5)\left(\frac{2y}{y-4}-2\right)=\frac{4}{y+5}(y-4)(y+5)$

$2y^2+10y-2y^2-2y+40=4y-16$

$8y+40=4y-16$

$4y=-56$

$y=-14$

check :

$\frac{2(-14)}{-14-4}-2\overset{?}{=}\frac{4}{-14+5}$

$\frac{14}{9}-2\overset{?}{=}-\frac{4}{9}$

$-\frac{4}{9}=-\frac{4}{9}$

17.

$\frac{x}{x+6}=\frac{1}{x+2}$

$x^2+2x=x+6$

$x^2+x-6=0$

$(x+3)(x-2)=0$

$x+3=0 \quad x-2=0$

$x=-3 \quad x=2$

check :

$\frac{-3}{-3+6}\overset{?}{=}\frac{1}{-3+2} \quad \frac{2}{2+6}\overset{?}{=}\frac{1}{2+2}$

$-1=-1 \quad \frac{1}{4}=\frac{1}{4}$

18.

$\frac{1}{R}=\frac{1}{R_1}+\frac{1}{R_2}$

$R_1R_2=RR_2+RR_1$

$R_1R_2-RR_1=RR_2$

$R_1(R_2-R)=RR_2$

$R_1=\frac{RR_2}{R_2-R}$

19.

$\frac{20}{x}+\frac{20}{2x}=1$

$2x\left(\frac{20}{x}+\frac{20}{2x}\right)=(1)2x$

$40+20=2x$

$2x=60$

$x=30$

Working alone, the newer machine can process 1000 pieces of mail in 30 minutes and the older machine can process 1000 pieces of mail in 60 minutes.

20.

$\frac{h}{1.5}=\frac{120}{3}$

$3h=180$

$h=60$

The height of the tree is 60 meters.

Cumulative Review Exercises

1.

$y-[2y-3(y-1)]=y-[2y-3y+3]=$

$y-[-y+3]=2y-3$

2.

$4n-5(n+2)<-7$

$4n-5n-10<-7$

$-n<3$

$n>-3$

3. $y \ge 2$

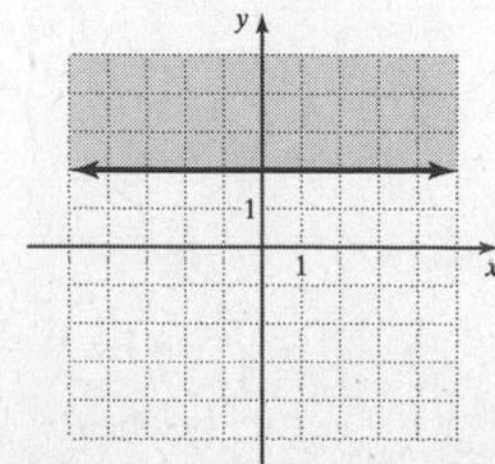

4.

$x+y=8$

$\underline{\quad y=2x-1}$

$x+(2x-1)=8$

$3x-1=8$

$3x=9$

$x=3$

$y=2(3)-1=5 \qquad (3,5)$

check :

$3+5\overset{?}{=}8 \qquad 5\overset{?}{=}2(3)-1$

$8=8 \qquad 5=5$

5.

$(n+5)(n-2)=0$

$n+5=0 \quad n-2=0$

$n=-5 \qquad n=2$

check :

$(-5+5)(-5-2)\overset{?}{=}0 \quad (2+5)(2-2)\overset{?}{=}0$

$(0)(-7)\overset{?}{=}0 \qquad (7)(0)\overset{?}{=}0$

$0=0 \qquad 0=0$

6.

$100y^2-81=(10y)^2-9^2=(10y+9)(10y+9)$

7.

$\frac{c}{c+6}=\frac{1}{c+2}$

$c^2+2c=c+6$

$c^2+c-6=0$

$(c+3)(c-2)=0$

$c+3=0 \quad c-2=0$

$c=-3 \qquad c=2$

check :

$\frac{-3}{-3+6}\overset{?}{=}\frac{1}{-3+2} \qquad \frac{2}{2+6}\overset{?}{=}\frac{1}{2+2}$

$-1=-1 \qquad \frac{1}{4}=\frac{1}{4}$

8.

P	R	I
2x	0.06	0.12x
x	0.04	0.04x

$0.12x+0.04x\geq 1200$

$0.16x\geq 1200$

$x\geq \$7500$

$2x\geq \$15,000$

She should invest $7500 in the fund at 4% and $15,000 in the fund at 6%.

9.

	r	t	d
Arlington bus	40	t+2	40(t+2)
Reston bus	60	t	60t

$40(t+2)=60t$

$40t+80=60t$

$20t=80$

$t=4$

It will take the Reston bus 4 hours to overtake the Arlington bus.

10.

$\frac{7}{3}=\frac{20}{x}$

$7x=60$

$x\approx 9$

It will take approximately 9 minutes to print a 20-page report.

Chapter 8 Radical Expressions and Equations

Chapter 8 Pretest

1. $\sqrt{81}=9$

2. $-\sqrt{27}=-\sqrt{9}\bullet\sqrt{3}=-3\sqrt{3}$

3. $\sqrt{45a^2}=\sqrt{9}\bullet\sqrt{5}\bullet\sqrt{a^2}=3a\sqrt{5}$

4. $\sqrt{\dfrac{x}{64}}=\dfrac{\sqrt{x}}{\sqrt{64}}=\dfrac{\sqrt{x}}{8}$

5. $6\sqrt{2}+\sqrt{2}-3\sqrt{2}=4\sqrt{2}$

6.

$\sqrt{12}+2\sqrt{75}=\sqrt{4}\bullet\sqrt{3}+2\sqrt{25}\bullet\sqrt{3}=$

$2\sqrt{3}+10\sqrt{3}=12\sqrt{3}$

7.

$\sqrt{9x^3}-4x\sqrt{x}+x\sqrt{36x}=$

$\sqrt{9}\bullet\sqrt{x^2}\bullet\sqrt{x}-4x\sqrt{x}+x\sqrt{36}\sqrt{x}=$

$3x\sqrt{x}-4x\sqrt{x}+6x\sqrt{x}=5x\sqrt{x}$

8. $\sqrt{6}\bullet\sqrt{3}=\sqrt{18}=\sqrt{9}\bullet\sqrt{2}=3\sqrt{2}$

9.

$\sqrt{2xy}\bullet\sqrt{10xy^3}=\sqrt{20x^2y^4}=$

$\sqrt{4}\bullet\sqrt{5}\bullet\sqrt{x^2}\bullet\sqrt{y^4}=2xy^2\sqrt{5}$

10. $\dfrac{\sqrt{30}}{\sqrt{5}}=\sqrt{\dfrac{30}{5}}=\sqrt{6}$

11. $\sqrt{n}\left(\sqrt{n}+2\right)=\sqrt{n^2}+2\sqrt{n}=n+2\sqrt{n}$

12.

$\left(\sqrt{3}-1\right)\left(\sqrt{3}+4\right)=\sqrt{3}\left(\sqrt{3}+4\right)-1\left(\sqrt{3}+4\right)=$

$3+4\sqrt{3}-\sqrt{3}-4=-1+3\sqrt{3}$

13.

$\sqrt{\dfrac{5x}{6}}=\dfrac{\sqrt{5x}}{\sqrt{6}}=\dfrac{\sqrt{5x}\cdot\sqrt{6}}{\sqrt{6}\cdot\sqrt{6}}=\dfrac{\sqrt{30x}}{6}$

14.

$\dfrac{\sqrt{40x^3}}{\sqrt{2x}}=\sqrt{\dfrac{40x^3}{2x}}=\sqrt{20x^2}=$

$\sqrt{4}\bullet\sqrt{5}\bullet\sqrt{x^2}=2x\sqrt{5}$

15.

$\dfrac{8+\sqrt{7}}{\sqrt{2}}=\dfrac{\left(8+\sqrt{7}\right)\sqrt{2}}{\sqrt{2}\cdot\sqrt{2}}=\dfrac{8\sqrt{2}+\sqrt{14}}{2}$

16.

$\sqrt{x}-1=5$ *check*:

$\sqrt{x}=6$ $\sqrt{36}-1\stackrel{?}{=}5$

$x=36$ $5=5$

17.

$y=\sqrt{4y-3}$

$y^2=4y-3$

$y^2-4y+3=0$

$(y-1)(y-3)=0$

$y-1=0$ $y-3=0$

$y=1$ $y=3$

check:

$1\stackrel{?}{=}\sqrt{4(1)-3}$ $3\stackrel{?}{=}\sqrt{4(3)-3}$

$1\stackrel{?}{=}\sqrt{1}$ $3\stackrel{?}{=}\sqrt{9}$

$1=1$ $3=3$

18.

$\sqrt{(2)\left(2\,m/\text{sec}^2\right)(100m)}=$

$\sqrt{400\,m^2/\text{sec}^2}=20\,m/\text{sec}$

The velocity of the car is 20 m/sec.

19.

$d^2=12^2+12^2$

$d^2=288$

$d=\sqrt{288}$

$d=12\sqrt{2}\approx 17.0$

The gymnast covers $12\sqrt{2}$ meters, or approximately 17.0 meters in the tumbling sequence.

20.

$S=\sqrt{30fL}$

$S^2=30fL$

$\dfrac{S^2}{30f}=L$

$L=\dfrac{S^2}{30f}$

8.1 Introduction to Radical Expressions

Practice 8.1

1

a. $\sqrt{4}=\sqrt{2^2}=2$

b. $-5\sqrt{49}=-5\sqrt{7^2}=-5(7)=-35$

2. $\sqrt{10}=3.162$

3.

a. $\left(\sqrt{6}\right)^2=6$

b. $\left(\sqrt{5}\right)^2=5$

c. $\sqrt{7^2}=7$

d. $\sqrt{1^2}=1$

4.

a. $\sqrt{x^4}=\sqrt{\left(x^2\right)^2}=x^2$

b. $\sqrt{64t^{10}}=\sqrt{\left(8t^5\right)^2}=8t^5$

c. $-\sqrt{121x^2y^2}=-\sqrt{11^2x^2y^2}=-11xy$

5.

a. $\sqrt{72}=\sqrt{36}\bullet\sqrt{2}=6\sqrt{2}$

b. $2\sqrt{40}=2\sqrt{4}\bullet\sqrt{10}=2\bullet2\sqrt{10}=4\sqrt{10}$

c. $\dfrac{\sqrt{75}}{15}=\dfrac{\sqrt{25}\bullet\sqrt{3}}{15}=\dfrac{\overset{1}{\cancel{5}}\sqrt{3}}{\underset{3}{\cancel{15}}}=\dfrac{\sqrt{3}}{3}$

6.

a. $\sqrt{x^3}=\sqrt{x^2}\bullet\sqrt{x}=x\sqrt{x}$

b. $\sqrt{18n^4}=\sqrt{9}\bullet\sqrt{2}\bullet\sqrt{\left(n^2\right)^2}=3n^2\sqrt{2}$

c. $-\sqrt{50ab^2}=-\sqrt{25}\bullet\sqrt{2}\bullet\sqrt{a}\bullet\sqrt{b^2}=-5b\sqrt{2a}$

7.

a. $\sqrt{\dfrac{1}{16}}=\sqrt{\left(\dfrac{1}{4}\right)^2}=\dfrac{1}{4}$

b. $\sqrt{\dfrac{y^2}{4}}=\sqrt{\left(\dfrac{y}{2}\right)^2}=\dfrac{y}{2}$

c. $\sqrt{\dfrac{x^4}{y^6}}=\sqrt{\left(\dfrac{x^2}{y^3}\right)^2}=\dfrac{x^2}{y^3}$

8.

a. $\sqrt{\dfrac{3}{16}}=\dfrac{\sqrt{3}}{\sqrt{16}}=\dfrac{\sqrt{3}}{4}$

b. $\sqrt{\dfrac{2y}{49}}=\dfrac{\sqrt{2y}}{\sqrt{49}}=\dfrac{\sqrt{2y}}{7}$

c. $\sqrt{\dfrac{5x^7y^2}{4}}=\dfrac{\sqrt{x^6y^2}\bullet\sqrt{5x}}{\sqrt{4}}=\dfrac{x^3y\sqrt{5x}}{2}$

9. The force will be 1000 lb.

$\sqrt{600^2+800^2}=\sqrt{1{,}000{,}000}=\sqrt{10^6}=$

$\sqrt{\left(10^3\right)^2}=10^3=1000$

Exercises 8.1

1. $\sqrt{36}=6$

3. $\sqrt{1}=1$

5. $-\sqrt{100}=-10$

7. $3\sqrt{49}=(3)(7)=21$

9. $\sqrt{5}=2.236$

11. $3\sqrt{2}=3\bullet1.414=4.243$

13. $\left(\sqrt{16}\right)^2=16$

15. $\left(\sqrt{11}\right)^2=11$

17. $\left(\sqrt{5x}\right)^2=5x$

19. $\sqrt{2^2}=2$

21. $\sqrt{9^2}=9$

23. $\sqrt{n^8}=\sqrt{\left(n^4\right)^2}=n^4$

25. $\sqrt{49y^2}=\sqrt{49}\bullet\sqrt{y^2}=7y$

27. $\sqrt{9x^4}=\sqrt{9}\bullet\sqrt{\left(x^2\right)^2}=3x^2$

29. $\sqrt{25x^2y^{10}}=\sqrt{25}\bullet\sqrt{x^2}\bullet\sqrt{\left(y^5\right)^2}=5xy^5$

31. $\sqrt{32}=\sqrt{16}\bullet\sqrt{2}=4\sqrt{2}$

33. $-\sqrt{108}=-\sqrt{36}\bullet\sqrt{3}=-6\sqrt{3}$

35. $6\sqrt{27}=6\bullet\sqrt{9}\bullet\sqrt{3}=6\bullet3\bullet\sqrt{3}=18\sqrt{3}$

37. $\frac{\sqrt{48}}{12}=\frac{\sqrt{16}\bullet\sqrt{3}}{12}=\frac{\not{4}^{1}\sqrt{3}}{\not{12}_{3}}=\frac{\sqrt{3}}{3}$

39. $\sqrt{11x^2}=\sqrt{11}\bullet\sqrt{x^2}=x\sqrt{11}$

41. $\sqrt{n^5}=\sqrt{(n^2)^2n}=n^2\sqrt{n}$

43. $\sqrt{20x^3}=\sqrt{4}\bullet\sqrt{5}\bullet\sqrt{x^2}\bullet\sqrt{x}=2x\sqrt{5x}$

45. $\sqrt{12p^2q}=\sqrt{4}\bullet\sqrt{3}\sqrt{p^2}\bullet\sqrt{q}=2p\sqrt{3q}$

47.

$9\sqrt{10x^3y^4}=9\sqrt{10}\bullet\sqrt{x^2x}\bullet\sqrt{(y^2)^2}=9xy^2\sqrt{10x}$

51. $-\sqrt{\frac{1}{4}}=-\frac{\sqrt{1}}{\sqrt{4}}=-\frac{1}{2}$

53. $\sqrt{\frac{81}{n^6}}=\frac{\sqrt{81}}{\sqrt{(n^3)^2}}=\frac{9}{n^3}$

55. $\sqrt{\frac{x^4}{y^2}}=\frac{\sqrt{x^4}}{\sqrt{y^2}}=\frac{\sqrt{(x^2)^2}}{\sqrt{y^2}}=\frac{x^2}{y}$

57. $\sqrt{\frac{3}{4}}=\frac{\sqrt{3}}{\sqrt{4}}=\frac{\sqrt{3}}{2}$

59. $\sqrt{\frac{5n}{16}}=\frac{\sqrt{5n}}{\sqrt{16}}=\frac{\sqrt{5n}}{4}$

61. $\sqrt{\frac{3x^2y^6}{4}}=\frac{\sqrt{3x^2(y^3)^2}}{\sqrt{4}}=\frac{xy^3\sqrt{3}}{2}$

63. $\sqrt{\frac{27x^6y}{16}}=\frac{\sqrt{9\bullet3(x^3)^2y}}{\sqrt{16}}=\frac{3x^3\sqrt{3y}}{4}$

Applications

65.

a. $m=\sqrt{a\cdot b}$

b. $m=\sqrt{2\cdot8}=\sqrt{16}=4$

67.

a. $\sqrt{\frac{20}{5}}=\sqrt{4}=2$ seconds

It takes 2 seconds for the object to reach the ground.

b. $\sqrt{\frac{2\bullet20}{5}}=\sqrt{8}=2\sqrt{2}\approx2.83$

$2.83\neq2\bullet2$

It takes $2\sqrt{2}$ seconds to reach the ground. No, it is not two times the answer to part a.

69. $S=2\sqrt{5(180)}=2\sqrt{900}=60$

The speed of the car was approximately 60 miles per hour at the time of the accident.

71.

$11^2+5^2=d^2$

$121+25=d^2$

$146=d^2$

$d=\sqrt{146}\approx12$

The distance between the towns is $\sqrt{146}$ miles or about 12 miles.

8.2 Addition and Subtraction of Radical Expressions

Practice 8.2

1.

a. $8\sqrt{5}-2\sqrt{5}=(8-2)\sqrt{5}=6\sqrt{5}$

b. $3\sqrt{n}+\sqrt{n}+7\sqrt{n}=(3+1+7)\sqrt{n}=11\sqrt{n}$

c.

$10\sqrt{t^2-3}+\sqrt{t^2-3}=(10+1)\sqrt{t^2-3}=11\sqrt{t^2-3}$

d. $4\sqrt{6}-2\sqrt{2}$; Cannot be combined because they are not like radicals.

2.a.

$\sqrt{50}+\sqrt{98}=\sqrt{25}\sqrt{2}+\sqrt{49}\sqrt{2}=$

$5\sqrt{2}+7\sqrt{2}=(5+7)\sqrt{2}=12\sqrt{2}$

b.

$\sqrt{12}+2\sqrt{75}-6\sqrt{27}=\sqrt{4}\sqrt{3}+2\sqrt{25}\sqrt{3}-6\sqrt{9}\sqrt{3}=$

$2\sqrt{3}+2\bullet5\sqrt{3}-6\bullet3\sqrt{3}=2\sqrt{3}+10\sqrt{3}-18\sqrt{3}=$

$(2+10-18)\sqrt{3}=-6\sqrt{3}$

c.

$-3\sqrt{16t}+\sqrt{9t}=-3\sqrt{16}\sqrt{t}+\sqrt{9}\sqrt{t}=$

$-3\cdot4\sqrt{t}+3\sqrt{t}=-12\sqrt{t}+3\sqrt{t}=$

$(-12+3)\sqrt{t}=-9\sqrt{t}$

d.

$\sqrt{25ab^4}+7b^2\sqrt{a}=\sqrt{25}\sqrt{a}\sqrt{(b^2)^2}+7b^2\sqrt{a}=$

$5b^2\sqrt{a}+7b^2\sqrt{a}=(5b^2+7b^2)\sqrt{a}=12b^2\sqrt{a}$

3.

$x=\sqrt{1800}-\sqrt{200}=30\sqrt{2}-10\sqrt{2}=$

$20\sqrt{2}\approx 28$

The length of the front yard is $20\sqrt{2}$ meters or approximately 28 meters.

Exercises 8.2

1. $5\sqrt{7}+3\sqrt{7}=(5+3)\sqrt{7}=8\sqrt{7}$

3. $3\sqrt{2}-8\sqrt{2}=(3-8)\sqrt{2}=-5\sqrt{2}$

5. Cannot be combined

7.

$-5\sqrt{11}-10\sqrt{11}+2\sqrt{11}=$

$(-5-10+2)\sqrt{11}=-13\sqrt{11}$

9. $7t\sqrt{3}+2t\sqrt{3}=(7t+2t)\sqrt{3}=9t\sqrt{3}$

11. $13\sqrt{x}+10\sqrt{x}=(13+10)\sqrt{x}=23\sqrt{x}$

13. $6\sqrt{x+1}-\sqrt{x+1}=(6-1)\sqrt{x+1}=5\sqrt{x+1}$

15.

$\sqrt{8}-\sqrt{32}=\sqrt{4}\sqrt{2}-\sqrt{16}\sqrt{2}=$

$2\sqrt{2}-4\sqrt{2}=(2-4)\sqrt{2}=-2\sqrt{2}$

17.

$\sqrt{50}+\sqrt{72}=\sqrt{25}\sqrt{2}+\sqrt{36}\sqrt{2}=$

$5\sqrt{2}+6\sqrt{2}=(5+6)\sqrt{2}=11\sqrt{2}$

19.

$-\sqrt{12}+5\sqrt{3}=-\sqrt{4}\sqrt{3}+5\sqrt{3}=$

$-2\sqrt{3}+5\sqrt{3}=(-2+5)\sqrt{3}=3\sqrt{3}$

21.

$6\sqrt{75}-2\sqrt{12}=6\sqrt{25}\sqrt{3}-2\sqrt{4}\sqrt{3}=$

$6\cdot5\sqrt{3}-2\cdot2\sqrt{3}=(30-4)\sqrt{3}=26\sqrt{3}$

23.

$5\sqrt{8}-3\sqrt{12}+\sqrt{2}=5\sqrt{4}\sqrt{2}-3\sqrt{4}\sqrt{3}+\sqrt{2}=$

$5\cdot2\sqrt{2}-3\cdot2\sqrt{3}+\sqrt{2}=-6\sqrt{3}+(10+1)\sqrt{2}=$

$11\sqrt{2}-6\sqrt{3}$

25.

$2\sqrt{16y}+3\sqrt{4y}=2\sqrt{16}\sqrt{y}+3\sqrt{4}\sqrt{y}=$

$2\cdot4\sqrt{y}+3\cdot2\sqrt{y}=(8+6)\sqrt{y}=14\sqrt{y}$

27.

$\sqrt{9x}-\sqrt{16x^3}=\sqrt{9}\sqrt{x}-\sqrt{16}\sqrt{x^2}\sqrt{x}=$

$3\sqrt{x}-4x\sqrt{x}=(3-4x)\sqrt{x}$

29.

$\sqrt{25p}+\sqrt{64p}+\sqrt{p}=\sqrt{25}\sqrt{p}+\sqrt{64}\sqrt{p}+\sqrt{p}=$

$(5+8+1)\sqrt{p}=14\sqrt{p}$

31.

$-5x\sqrt{2x^3y^4}+x\sqrt{2x^5y^2}=-5x\sqrt{2}\sqrt{x^2}\sqrt{x}\sqrt{(y^2)^2}+x\sqrt{2}\sqrt{(x^2)^2}$

$-5x^2y^2\sqrt{2x}+x^3y\sqrt{2x}=(-5x^2y^2+x^3y)\sqrt{2x}$

$or\ (x^3y-5x^2y^2)\sqrt{2x}$

Applications

33.

a.

$s^2=6^2+5^2=61$

$s=\sqrt{61}$

Both sides are $\sqrt{61}$ units.

b. The perimeter of the triangle is $10+2\sqrt{61}$ units.

35.

a. The side of the larger square is $3\sqrt{10}$ inches long. The side of the smaller square is $2\sqrt{10}$ inches long.

$s_1{}^2=90$ $\qquad$ $s_2{}^2=40$

$s_1=\sqrt{90}=3\sqrt{10}$ $\qquad$ $s_2=\sqrt{40}=2\sqrt{10}$

b. $3\sqrt{10}-2\sqrt{10}=\sqrt{10}$

The side of the larger tile is $\sqrt{10}$ inches longer than the side of the smaller tile.

37.

$P_{1000} = 9\sqrt{1000} = 9 \cdot 10\sqrt{10} = 90\sqrt{10}$

$P_{4000} = 9\sqrt{4000} = 9 \cdot 20\sqrt{10} = 180\sqrt{10}$

difference:

$180\sqrt{10} - 90\sqrt{10} = 90\sqrt{10}$

The manufacturer will charge $90\sqrt{10}$ more dollars for 4000 machine parts than for 1000 machine parts.

8.3 Multiplication and Division of Radical Expressions

Practice 8.3

1.a. $\sqrt{5} \cdot \sqrt{5} = \left(\sqrt{5}\right)^2 = 5$

b. $\left(\sqrt{2y^3}\right)^2 = 2y^3$

c. $\left(\sqrt{t+1}\right)^2 = t+1$

2. a. $\sqrt{7} \cdot \sqrt{10} = \sqrt{70}$

b.

$\left(9\sqrt{6}\right)\left(-4\sqrt{3}\right) = (9)(-4)\sqrt{6 \cdot 3} = -36\sqrt{18} =$

$-36\sqrt{9 \cdot 2} = -36 \cdot 3\sqrt{2} = -108\sqrt{2}$

c. $\sqrt{8y} \cdot \sqrt{2y^5} = \sqrt{8y \cdot 2y^5}\sqrt{16y^6} = 4y^3$

3. a.

$\sqrt{6}\left(3\sqrt{3} - \sqrt{8}\right) = \sqrt{6} \cdot 3\sqrt{3} - \sqrt{6}\sqrt{8} =$

$3\sqrt{18} - \sqrt{48} = 9\sqrt{2} - 4\sqrt{3}$

b.

$\sqrt{a}\left(\sqrt{b} + 3\right) = \sqrt{a} \cdot \sqrt{b} + \sqrt{a} \cdot 3 = \sqrt{ab} + 3\sqrt{a}$

4.a.

$\left(2\sqrt{3} + 4\right)\left(\sqrt{3} - 1\right) = 2\sqrt{3}\left(\sqrt{3} - 1\right) + 4\left(\sqrt{3} - 1\right) =$

$2\left(\sqrt{3}\right)^2 - 2\sqrt{3} + 4\sqrt{3} - 4(1) = 6 + 2\sqrt{3} - 4 =$

$2 + 2\sqrt{3}$

b.

$\left(\sqrt{x} + 2\right)\left(3\sqrt{x} - 2\right) = \sqrt{x}\left(3\sqrt{x} - 2\right) + 2\left(3\sqrt{x} - 2\right) =$

$3x - 2\sqrt{x} + 6\sqrt{x} - 4 = 3x + 4\sqrt{x} - 4$

5.

a.

$\left(\sqrt{7} - 3\right)\left(\sqrt{7} + 3\right) = \sqrt{7}\left(\sqrt{7} + 3\right) + (-3)\left(\sqrt{7} + 3\right) =$

$7 + 3\sqrt{7} - 3\sqrt{7} - 9 = 7 - 9 = -2$

b.

$\left(\sqrt{p} + \sqrt{q}\right)\left(\sqrt{p} - \sqrt{q}\right) = \left(\sqrt{p}\right)^2 - \left(\sqrt{q}\right)^2 = p - q$

6.

a.

$\left(\sqrt{2} + b\right)^2 = \left(\sqrt{2}\right)^2 + 2\sqrt{2} \cdot b + b^2 =$

$2 + 2b\sqrt{2} + b^2 \quad or \quad b^2 + 2b\sqrt{2} + 2$

b.

$\left(\sqrt{x} - 6\right)^2 = \left(\sqrt{x}\right)^2 + 2\sqrt{x}(-6) + (-6)^2 =$

$x - 12\sqrt{x} + 36$

7. $0.2\left(\sqrt{60}\right)^3 = 0.2(60)\sqrt{60} \approx 93$

It takes Mercury about 93 twenty-four hour days to revolve around the sun.

8.

a. $\dfrac{\sqrt{21}}{\sqrt{3}} = \sqrt{\dfrac{21}{3}} = \sqrt{7}$

b. $\dfrac{\sqrt{4x^5}}{\sqrt{x}} = \sqrt{\dfrac{4x^5}{x}} = \sqrt{4x^4} = 2x^2$

c.

$\dfrac{\sqrt{2y}}{10\sqrt{8y^5}} = \dfrac{1}{10} \cdot \sqrt{\dfrac{2y}{8y^5}} = \dfrac{1}{10}\sqrt{\dfrac{1}{4y^4}} =$

$\dfrac{1}{10} \cdot \dfrac{1}{2y^2} = \dfrac{1}{20y^2}$

9.

a. $\sqrt{\dfrac{m^5}{n^4}} = \dfrac{\sqrt{m^4 \cdot m}}{\sqrt{n^4}} = \dfrac{m^2\sqrt{m}}{n^2}$

b. $\sqrt{\dfrac{y^6}{25x^2}} = \dfrac{\sqrt{y^6}}{\sqrt{25}\sqrt{x^2}} = \dfrac{y^3}{5x}$

c. $\sqrt{\dfrac{5a^2}{9}} = \dfrac{\sqrt{5 \cdot a^2}}{\sqrt{9}} = \dfrac{a\sqrt{5}}{3}$

10.

a. $\frac{1}{\sqrt{2}}=\frac{1\bullet\sqrt{2}}{\sqrt{2}\sqrt{2}}=\frac{\sqrt{2}}{2}$

b. $\frac{\sqrt{5}}{\sqrt{s}}=\frac{\sqrt{5}\bullet\sqrt{s}}{\sqrt{s}\bullet\sqrt{s}}=\frac{\sqrt{5s}}{s}$

c.

$\frac{\sqrt{49r^4}}{\sqrt{12}}=\frac{\sqrt{49}\bullet\sqrt{r^4}}{\sqrt{4}\bullet\sqrt{3}}=\frac{7r^2}{2\sqrt{3}}=$

$\frac{7r^2\bullet\sqrt{3}}{2\sqrt{3}\sqrt{3}}=\frac{7r^2\sqrt{3}}{6}$

11.

a. $\sqrt{\frac{1}{6}}=\frac{\sqrt{1}}{\sqrt{6}}=\frac{1}{\sqrt{6}}=\frac{1\cdot\sqrt{6}}{\sqrt{6}\cdot\sqrt{6}}=\frac{\sqrt{6}}{6}$

b. $\sqrt{\frac{n}{20}}=\frac{\sqrt{n}}{\sqrt{4\cdot 5}}=\frac{\sqrt{n}\cdot\sqrt{5}}{2\sqrt{5}\cdot\sqrt{5}}=\frac{\sqrt{5n}}{10}$

12.

a.

$\frac{\sqrt{5}-1}{\sqrt{3}}=\frac{(\sqrt{5}-1)(\sqrt{3})}{\sqrt{3}(\sqrt{3})}=$

$\frac{(\sqrt{5})(\sqrt{3})-1(\sqrt{3})}{3}=\frac{\sqrt{15}-\sqrt{3}}{3}$

b.

$\frac{\sqrt{c}+2}{\sqrt{b}}=\frac{(\sqrt{c}+2)(\sqrt{b})}{\sqrt{b}(\sqrt{b})}=$

$\frac{(\sqrt{c})(\sqrt{b})+2(\sqrt{b})}{b}=\frac{\sqrt{bc}+2\sqrt{b}}{b}$

13.

a.

$\frac{8}{3-\sqrt{2}}=\frac{8(3+\sqrt{2})}{(3-\sqrt{2})(3+\sqrt{2})}=$

$\frac{8(3)+8\sqrt{2}}{9-2}=\frac{24+8\sqrt{2}}{7}$

b.

$\frac{a}{\sqrt{b}+\sqrt{5}}=\frac{a(\sqrt{b}-\sqrt{5})}{(\sqrt{b}+\sqrt{5})(\sqrt{b}-\sqrt{5})}=$

$\frac{a(\sqrt{b})-a\sqrt{5}}{b-5}$

14.

$P=\frac{590}{\sqrt{72}}=\frac{590}{\sqrt{36\cdot 2}}=\frac{\overset{295}{\cancel{590}}}{\underset{3}{\cancel{6}}\sqrt{2}}=$

$\frac{295\sqrt{2}}{3\sqrt{2}\cdot\sqrt{2}}=\frac{295\sqrt{2}}{6}\approx 70$

The pulse rate is $\frac{295\sqrt{2}}{6}$ beats per minute, or approximately 70 beats per minute.

Exercises 8.3

1. $\sqrt{21}\bullet\sqrt{21}=(\sqrt{21})^2=21$

3. $(\sqrt{3n})^2=3n$

5. $(4\sqrt{x-1})^2=16(x-1)=16x-16$

7. $\sqrt{18}\bullet\sqrt{3}=\sqrt{54}=\sqrt{9}\sqrt{6}=3\sqrt{6}$

9.

$(-2\sqrt{5})(7\sqrt{10})=-14\sqrt{50}=$

$(-14)\sqrt{25}\sqrt{2}=(-14)(5)\sqrt{2}=$

$-70\sqrt{2}$

11. $\sqrt{8x^3}\bullet\sqrt{2x}=\sqrt{16x^4}=\sqrt{16}\sqrt{x^4}=4x^2$

13. $\sqrt{3r}\bullet\sqrt{5r}=\sqrt{15r^2}=\sqrt{15}\sqrt{r^2}=r\sqrt{15}$

15.

$\sqrt{2x}\bullet\sqrt{5}\bullet\sqrt{10y}=\sqrt{100xy}=$

$\sqrt{100}\sqrt{x}\sqrt{y}=10\sqrt{xy}$

17.

$\sqrt{3}(\sqrt{3}-1)=\sqrt{3}\sqrt{3}-\sqrt{3}(1)=3-\sqrt{3}$

19.

$\sqrt{x}(\sqrt{x}-7)=\sqrt{x}\sqrt{x}+\sqrt{x}(-7)=x-7\sqrt{x}$

21.
$\sqrt{a}\left(4\sqrt{b}+1\right)=\sqrt{a}\cdot 4\sqrt{b}+\sqrt{a}\,(1)=$
$4\sqrt{ab}+\sqrt{a}$

23.
$\left(\sqrt{5}+3\right)\left(\sqrt{5}+2\right)=\sqrt{5}\left(\sqrt{5}+2\right)+3\left(\sqrt{5}+2\right)=$
$5+2\sqrt{5}+3\sqrt{5}+6=11+5\sqrt{5}$

25.
$\left(8\sqrt{3}+1\right)\left(5\sqrt{3}-2\right)=8\sqrt{3}\left(5\sqrt{3}-2\right)+1\left(5\sqrt{3}-2\right)=$
$120-16\sqrt{3}+5\sqrt{3}-2=118-11\sqrt{3}$

27.
$\left(\sqrt{n}+5\right)\left(3\sqrt{n}-1\right)=\sqrt{n}\left(3\sqrt{n}-1\right)+5\left(3\sqrt{n}-1\right)=$
$3n-\sqrt{n}+15\sqrt{n}-5=3n+14\sqrt{n}-5$

29.
$\left(6-\sqrt{3}\right)\left(6+\sqrt{3}\right)=6\left(6+\sqrt{3}\right)+\left(-\sqrt{3}\right)\left(6+\sqrt{3}\right)=$
$36+6\sqrt{3}-6\sqrt{3}-3=33$

31.
$\left(5+2\sqrt{3}\right)\left(5-2\sqrt{3}\right)=5^2-\left(2\sqrt{3}\right)^2=$
$25-(4)(3)=25-12=13$

33. $\left(\sqrt{x}+2\right)\left(\sqrt{x}-2\right)=\left(\sqrt{x}\right)^2-2^2=x-4$

35. $\left(\sqrt{a}+\sqrt{b}\right)\left(\sqrt{a}-\sqrt{b}\right)=\left(\sqrt{a}\right)^2-\left(\sqrt{b}\right)^2=a-b$

37.
$\left(\sqrt{3x}-\sqrt{y}\right)\left(\sqrt{3x}+\sqrt{y}\right)=\left(\sqrt{3x}\right)^2-\left(\sqrt{y}\right)^2=3x-y$

39.
$\left(\sqrt{2}-x\right)^2=\left(\sqrt{2}\right)^2+2\left(\sqrt{2}\right)(-x)+(-x)^2=$
$2-2x\sqrt{2}+x^2$ or $x^2-2x\sqrt{2}+2$

41.
$\left(\sqrt{x}-1\right)^2=\left(\sqrt{x}\right)^2+2\left(\sqrt{x}\right)(-1)+(-1)^2=$
$x-2\sqrt{x}+1$

43. $\dfrac{\sqrt{15}}{\sqrt{3}}=\sqrt{\dfrac{15}{3}}=\sqrt{5}$

45. $\dfrac{\sqrt{5}}{\sqrt{125}}=\sqrt{\dfrac{5}{125}}=\sqrt{\dfrac{1}{25}}=\dfrac{1}{5}$

47. $\dfrac{\sqrt{4a^3}}{\sqrt{a}}=\sqrt{\dfrac{4a^3}{a}}=\sqrt{4a^2}=2a$

49.
$\dfrac{4\sqrt{5y}}{\sqrt{45y^5}}=\sqrt{\dfrac{5y}{45y^5}}=\sqrt{\dfrac{1}{9y^4}}=\dfrac{1}{3y^2}$

51.
$\dfrac{\sqrt{a^4}}{\sqrt{b^6}}=\dfrac{a^2}{b^3}$

53. $\sqrt{\dfrac{16x^{12}}{y^8}}=\dfrac{\sqrt{6x^{12}}}{\sqrt{y^8}}=\dfrac{4x^6}{y^4}$

55.
$\sqrt{\dfrac{5x^{10}}{36}}=\dfrac{\sqrt{5x^{10}}}{\sqrt{36}}=\dfrac{x^5\sqrt{5}}{6}$

57.
$\dfrac{2}{\sqrt{3}}=\dfrac{2\cdot\sqrt{3}}{\sqrt{3}\cdot\sqrt{3}}=\dfrac{2\sqrt{3}}{3}$

59.
$\dfrac{\sqrt{5}}{\sqrt{y}}=\dfrac{\sqrt{5}\cdot\sqrt{y}}{\sqrt{y}\cdot\sqrt{y}}=\dfrac{\sqrt{5y}}{y}$

61.
$\sqrt{\dfrac{2}{11}}=\dfrac{\sqrt{2}}{\sqrt{11}}=\dfrac{\sqrt{2}\cdot\sqrt{11}}{\sqrt{11}\cdot\sqrt{11}}=\dfrac{\sqrt{22}}{11}$

63.
$\sqrt{\dfrac{x^2}{5}}=\dfrac{\sqrt{x^2}}{\sqrt{5}}=\dfrac{x\cdot\sqrt{5}}{\sqrt{5}\cdot\sqrt{5}}=\dfrac{x\sqrt{5}}{5}$

65.
$\sqrt{\dfrac{t}{50}}=\dfrac{\sqrt{t}}{\sqrt{25\cdot 2}}=\dfrac{\sqrt{t}\cdot\sqrt{2}}{5\sqrt{2}\cdot\sqrt{2}}=\dfrac{\sqrt{2t}}{10}$

67.
$\sqrt{\dfrac{a}{2}}=\dfrac{\sqrt{a}}{\sqrt{2}}=\dfrac{\sqrt{a}\cdot\sqrt{2}}{\sqrt{2}\cdot\sqrt{2}}=\dfrac{\sqrt{2a}}{2}$

69.
$\dfrac{\sqrt{5}+2}{\sqrt{3}}=\dfrac{\left(\sqrt{5}+2\right)\sqrt{3}}{\sqrt{3}\cdot\sqrt{3}}=$
$\dfrac{\sqrt{5}\sqrt{3}+2\sqrt{3}}{3}=\dfrac{\sqrt{15}+2\sqrt{3}}{3}$

71.

$$\frac{\sqrt{n}-1}{\sqrt{m}}=\frac{\left(\sqrt{n}-1\right)\sqrt{m}}{\sqrt{m}\bullet\sqrt{m}}=\frac{\sqrt{mn}-\sqrt{m}}{m}$$

73.

$$\frac{15}{4+\sqrt{6}}=\frac{15\left(4-\sqrt{6}\right)}{\left(4+\sqrt{6}\right)\left(4-\sqrt{6}\right)}=\frac{60-15\sqrt{6}}{16-6}=$$

$$\frac{60-15\sqrt{6}}{10}=\frac{\overset{1}{\cancel{5}}\left(12-3\sqrt{6}\right)}{\underset{2}{\cancel{10}}}=\frac{12-3\sqrt{6}}{2}$$

75.

$$\frac{11}{4-\sqrt{5}}=\frac{11\left(4+\sqrt{5}\right)}{\left(4-\sqrt{5}\right)\left(4+\sqrt{5}\right)}=$$

$$\frac{44+11\sqrt{5}}{16-5}=\frac{\overset{1}{\cancel{11}}\left(4+\sqrt{5}\right)}{\underset{1}{\cancel{11}}}=4+\sqrt{5}$$

77.

$$\frac{4}{\sqrt{5}-\sqrt{3}}=\frac{4\left(\sqrt{5}+\sqrt{3}\right)}{\left(\sqrt{5}-\sqrt{3}\right)\left(\sqrt{5}+\sqrt{3}\right)}=$$

$$\frac{4\left(\sqrt{5}+\sqrt{3}\right)}{5-3}=\frac{\overset{2}{\cancel{4}}\left(\sqrt{5}+\sqrt{3}\right)}{\underset{1}{\cancel{2}}}=2\sqrt{5}+2\sqrt{3}$$

79.

$$\frac{a}{\sqrt{b}-\sqrt{3}}=\frac{a\left(\sqrt{b}+\sqrt{3}\right)}{\left(\sqrt{b}-\sqrt{3}\right)\left(\sqrt{b}+\sqrt{3}\right)}=\frac{a\sqrt{b}+a\sqrt{3}}{b-3}$$

Applications

81.

$A=\left(6\sqrt{2}\right)^2=36(2)=72$

Area = 72 square centimeters

83. The distance from (0,0) to (3,5) is $\sqrt{34}$, and the distance from (0,0) to (6,10) is $2\sqrt{34}$.

$d_1=\sqrt{(0-6)^2+(0-10)^2}=\sqrt{36+100}=$

$\sqrt{136}=\sqrt{4\cdot34}=2\sqrt{34}$

$d_2=\sqrt{(0-3)^2+(0-5)^2}=\sqrt{9+25}=\sqrt{34}$

85. $t=\sqrt{\frac{500}{16}}=\frac{\sqrt{500}}{\sqrt{16}}=\frac{\overset{5}{\cancel{10}}\sqrt{5}}{\underset{2}{\cancel{4}}}=\frac{5\sqrt{5}}{2}$

It will take the hailstone $\frac{5\sqrt{5}}{2}$ seconds or about 5.6 seconds to drop 500 feet.

87.

$$\sqrt{\frac{V}{\pi h}}=\frac{\sqrt{V}\bullet\sqrt{\pi h}}{\sqrt{\pi h}\bullet\sqrt{\pi h}}=\frac{\sqrt{V\pi h}}{\pi h}$$

8.4 Solving Radical Equations

Practice 8.4

1.

$\sqrt{y}+3=7$	*check* :
$\sqrt{y}=4$	$\sqrt{16}+3\overset{?}{=}7$
$\left(\sqrt{y}\right)^2=(4)^2$	$4+3\overset{?}{=}7$
$y=16$	$7=7$

The solution is 16.

2.

$\sqrt{2t-5}+7=0$	
$\sqrt{2t-5}=-7$	*check* :
$\left(\sqrt{2t-5}\right)^2=(-7)^2$	$\sqrt{2(27)-5}+7\overset{?}{=}0$
$2t-5=49$	$\sqrt{49}+7\overset{?}{=}0$
$2t=54$	$7+7\overset{?}{=}0$
$t=27$	$14\neq0$

There is no solution.

3.

$\sqrt{4x+7}=\sqrt{6x-11}$	
$\left(\sqrt{4x+7}\right)^2=\left(\sqrt{6x-11}\right)^2$	*check* :
$4x+7=6x-11$	$\sqrt{4(9)+7}\overset{?}{=}\sqrt{6(9)-11}$
$-2x=-18$	$\sqrt{36+7}\overset{?}{=}\sqrt{54-11}$
$x=9$	$\sqrt{43}=\sqrt{43}$

4.

$y-\sqrt{y-2}=2$

$y-2=\sqrt{y-2}$

$(y-2)^2=\left(\sqrt{y-2}\right)^2$

$y^2-4y+4=y-2$

$y^2-5y+6=0$

$(y-2)(y-3)=0$

$y-2=0 \quad y-3=0$

$y=2 \quad\quad y=3$

check :

$2-\sqrt{2-2}\stackrel{?}{=}2 \quad 3-\sqrt{3-2}\stackrel{?}{=}2$

$2-0\stackrel{?}{=}2 \quad\quad 3-1\stackrel{?}{=}2$

$2=2 \quad\quad 2=2$

5.

$\sqrt{n^2+11}=3\sqrt{n-1}$ *check* :

$\left(\sqrt{n^2+11}\right)^2=\left(3\sqrt{n-1}\right)^2 \quad \sqrt{4^2+11}\stackrel{?}{=}3\sqrt{4-1}$

$n^2+11=9(n-1) \quad \sqrt{27}\stackrel{?}{=}3\sqrt{3}$

$n^2+11=9n-9 \quad 3\sqrt{3}=3\sqrt{3}$

$n^2-9n+20=0$

$(n-4)(n-5)=0 \quad \sqrt{5^2+11}\stackrel{?}{=}3\sqrt{5-1}$

$n-4=0 \quad n-5=0 \quad \sqrt{36}\stackrel{?}{=}3\sqrt{4}$

$n=4 \quad\quad n=5 \quad\quad 6=6$

6.

$50=\sqrt{2.5r}$

$50^2=\left(\sqrt{2.5r}\right)^2$

$2500=2.5r$

$1000=r$

A radius of 1000 feet will permit a maximum safe speed of 50 mph.

7.

$d=\sqrt{\dfrac{3h}{2}}$

$d^2=\left(\sqrt{\dfrac{3h}{2}}\right)^2$

$d^2=\dfrac{3h}{2}$

$2d^2=3h$

$\dfrac{2d^2}{3}=h \quad or \quad h=\dfrac{2d^2}{3}$

Exercises 8.4

1.

$\sqrt{x}=3$ *check* :

$\left(\sqrt{x}\right)^2=3^2 \quad \sqrt{9}\stackrel{?}{=}3$

$x=9 \quad\quad 3=3$

9 is the solution

3.

$\sqrt{2x}=8$ *check* :

$\left(\sqrt{2x}\right)^2=(8)^2 \quad \sqrt{2(32)}\stackrel{?}{=}8$

$2x=64 \quad\quad \sqrt{64}\stackrel{?}{=}8$

$x=32 \quad\quad 8\neq 8$

32 is the solution

5. No solution

7.

$\sqrt{a}-4=4$ *check* :

$\sqrt{a}=8 \quad\quad \sqrt{64}-4\stackrel{?}{=}4$

$\left(\sqrt{a}\right)^2=(8)^2 \quad 8-4\stackrel{?}{=}4$

$a=64 \quad\quad 4=4$

64 is the solution

9.

$$\sqrt{x+3}=3 \qquad \textit{check}:$$
$$\left(\sqrt{x+3}\right)^2=3^2 \qquad \sqrt{6+3}\stackrel{?}{=}3$$
$$x+3=9 \qquad \sqrt{9}\stackrel{?}{=}3$$
$$x=6 \qquad 3=3$$

6 is the solution

11.

$$\sqrt{3t+7}=4 \qquad \textit{check}:$$
$$\left(\sqrt{3t+7}\right)^2=4^2 \qquad \sqrt{3(3)+7}\stackrel{?}{=}4$$
$$3t+7=16 \qquad \sqrt{16}\stackrel{?}{=}4$$
$$3t=9 \qquad 4=4$$
$$t=3$$

3 is the solution

13.

$$\sqrt{9t-14}=\sqrt{2t} \qquad \textit{check}:$$
$$\left(\sqrt{9t-14}\right)^2=\left(\sqrt{2t}\right)^2 \qquad \sqrt{9(2)-14}\stackrel{?}{=}\sqrt{2(2)}$$
$$9t-14=2t \qquad \sqrt{4}\stackrel{?}{=}\sqrt{4}$$
$$7t=14 \qquad 2=2$$
$$t=2$$

2 is the solution

15.

$$\sqrt{4x+1}=\sqrt{2x+7}$$
$$\left(\sqrt{4x+1}\right)^2=\left(\sqrt{2x+7}\right)^2 \qquad \textit{check}:$$
$$4x+1=2x+7 \qquad \sqrt{4(3)+1}\stackrel{?}{=}\sqrt{2(3)+7}$$
$$2x=6 \qquad \sqrt{12+1}\stackrel{?}{=}\sqrt{6+7}$$
$$x=3 \qquad \sqrt{13}=\sqrt{13}$$

3 is the solution

17.

$$-2\sqrt{n-2}=\sqrt{3n+4}$$
$$\left(-2\sqrt{n-2}\right)^2=\left(\sqrt{3n+4}\right)^2$$
$$4(n-2)=3n+4$$
$$4n-8=3n+4$$
$$n-8=4$$
$$n=-4$$

check:

$$-2\sqrt{-4-2}\stackrel{?}{=}\sqrt{3(-4)+4}$$
$$-2\sqrt{-6}\stackrel{?}{=}\sqrt{-8}$$

No solution

19.

$$\sqrt{y-1}+4=6$$
$$\sqrt{y-1}=2 \qquad \textit{check}:$$
$$\left(\sqrt{y-1}\right)^2=2^2 \qquad \sqrt{5-1}+4\stackrel{?}{=}6$$
$$y-1=4 \qquad \sqrt{4}+4\stackrel{?}{=}6$$
$$y=5 \qquad 6=6$$

5 is the solution

21.

$$3-\sqrt{3x+1}=2 \qquad \textit{check}:$$
$$-\sqrt{3x+1}=-1 \qquad 3-\sqrt{3(0)+1}\stackrel{?}{=}2$$
$$\left(-\sqrt{3x+1}\right)^2=(-1)^2 \qquad 3-\sqrt{1}\stackrel{?}{=}2$$
$$3x+1=1 \qquad 3-1\stackrel{?}{=}2$$
$$3x=0 \qquad 2=2$$
$$x=0$$

0 is the solution

23.

$\sqrt{x+6}-x=4$
$\sqrt{x+6}=x+4$
$\left(\sqrt{x+6}\right)^2=(x+4)^2$
$x+6=x^2+8x+16$
$x^2+7x+10=0$
$(x+5)(x+2)=0$
$x+5=0 \quad x+2=0$
$x\neq -5 \quad x=-2$

check:

$\sqrt{-5+6}-(-5)\overset{?}{=}4 \qquad \sqrt{-2+6}-(-2)\overset{?}{=}4$

$\sqrt{1}+(-5)\overset{?}{=}4 \qquad \sqrt{4}+2\overset{?}{=}4$

$1+(-5)\overset{?}{=}4 \qquad 2+2\overset{?}{=}4$

$-4\neq 4 \qquad 4=4$

-2 is the only solution to the original equation

25.

$7+\sqrt{2x+9}=x+4$

$\sqrt{2x+9}=x-3$

$\left(\sqrt{2x+9}\right)^2=(x-3)^2$

$2x+9=x^2-6x+9$

$x^2-8x=0$

$x(x-8)=0$

$x\neq 0 \quad x-8=0$

$x=8$

check:

$7+\sqrt{2(0)+9}\overset{?}{=}0+4 \qquad 7+\sqrt{2(8)+9}\overset{?}{=}8+4$

$7+\sqrt{9}\overset{?}{=}0 \qquad 7+\sqrt{25}\overset{?}{=}12$

$7+3\overset{?}{=}0 \qquad 7+5\overset{?}{=}12$

$10\neq 0 \qquad 12=12$

8 is the only solution

27.

$7\sqrt{v}+v=-10$

$v+10=-7\sqrt{v}$

$(v+10)^2=\left(-7\sqrt{v}\right)^2$

$v^2+20v+100=49v$

$v^2-29v+100=0$

$(v-25)(v-4)=0$

$v-25=0 \quad v-4=0$

$v\neq 25 \quad v\neq 4$

check:

$7\sqrt{25}+25\overset{?}{=}-10 \qquad 7\sqrt{4}+4\overset{?}{=}-10$

$7(5)+25\overset{?}{=}-10 \qquad 7(2)+4\overset{?}{=}-10$

$60\neq -10 \qquad 18\neq -10$

No solution

29.

$n-3\sqrt{n+2}=-4$

$n+4=3\sqrt{n+2}$

$(n+4)^2=\left(3\sqrt{n+2}\right)^2$

$n^2+8n+16=9(n+2)$

$n^2+8n+16=9n+18$

$n^2-n-2=0$

$(n-2)(n+1)=0$

$n-2=0 \quad n+1=0$

$n=2 \quad n=-1$

check:

$2-3\sqrt{2+2}\overset{?}{=}-4 \qquad -1-3\sqrt{-1+2}\overset{?}{=}-4$

$2-3(2)\overset{?}{=}-4 \qquad -1-3(1)\overset{?}{=}-4$

$-4=-4 \qquad -4=-4$

-1 and 2 are the solutions to the original equation

31.

$5\sqrt{y-6}=\sqrt{y^2-14}$

$\left(5\sqrt{y-6}\right)^2=\left(\sqrt{y^2-14}\right)^2$

$25(y-6)=y^2-14$

$25y-150=y^2-14$

$y^2-25y+136=0$

$(x-8)(x-17)=0$

$x-8=0 \quad x-17=0$

$x=8 \qquad x=17$

$check:$

$5\sqrt{8-6}\stackrel{?}{=}\sqrt{8^2-14} \qquad 5\sqrt{17-6}\stackrel{?}{=}\sqrt{17^2-14}$

$5\sqrt{2}\stackrel{?}{=}\sqrt{50} \qquad 5\sqrt{11}\stackrel{?}{=}\sqrt{275}$

$5\sqrt{2}=5\sqrt{2} \qquad 5\sqrt{11}=5\sqrt{11}$

8 and 17 are the solutions

35.

$\sqrt{4x+13}-2x=-1$

$\sqrt{4x+13}=2x-1$

$\left(\sqrt{4x+13}\right)^2=(2x-1)^2$

$4x+13=4x^2-4x+1$

$4x^2-8x-12=0$

$x^2-2x-3=0$

$(x+1)(x-3)=0$

$x+1=0 \quad x-3=0$

$x\neq-1 \qquad x=3$

$check:$

$\sqrt{4(-1)+13}-2(-1)\stackrel{?}{=}-1 \qquad \sqrt{4(3)+13}-2(3)\stackrel{?}{=}-1$

$\sqrt{9}+2\stackrel{?}{=}-1 \qquad \sqrt{25}-6\stackrel{?}{=}-1$

$5\neq-1 \qquad -1=-1$

3 is the only solution to the original equation

Applications

37.

$10=\sqrt{\dfrac{P}{25}}$

$100=\dfrac{P}{25}$

$(25)(100)=P$

$P=2500$

The power of the appliance is 2500 watts.

39.

$2=2\pi\sqrt{\dfrac{m}{8}}$

$4=4\pi^2\dfrac{m}{8}$

$32=4\pi^2 m$

$\dfrac{32}{4\pi^2}=m$

$\dfrac{8}{\pi^2}=m$

A mass of $\dfrac{8}{\pi^2}$ grams will produce a period of 2 seconds.

41.

$f=120\sqrt{p}$

$\dfrac{f}{120}=\sqrt{p}$

$\left(\dfrac{f}{120}\right)^2=\left(\sqrt{p}\right)^2$

$\dfrac{f^2}{14,400}=p$

$p=\dfrac{f^2}{14,400}$

Chapter 8 Review Exercises

1. $-\sqrt{49}=-7$

2. $\sqrt{6^2}=6$

3. $\left(\sqrt{7x}\right)^2 = 7x$

4. $\sqrt{28} = \sqrt{4} \bullet \sqrt{7} = 2\sqrt{7}$

5. $-3\sqrt{18} = (-3)\sqrt{9}\sqrt{2} = (-3)(3)\sqrt{2} = -9\sqrt{2}$

6. $\sqrt{32x^3} = \sqrt{16} \bullet \sqrt{2} \bullet \sqrt{x^2} \bullet \sqrt{x} = 4x\sqrt{2x}$

7. $\sqrt{\frac{9}{25}} = \frac{\sqrt{9}}{\sqrt{25}} = \frac{3}{5}$

8. $-\sqrt{\frac{3t}{16}} = -\frac{\sqrt{3t}}{\sqrt{16}} = -\frac{\sqrt{3t}}{4}$

9. $\sqrt{\frac{144}{x^{100}}} == \frac{\sqrt{144}}{\sqrt{x^{100}}} = \frac{12}{x^{50}}$

10.

$2\sqrt{25a^5b^3} = 2\sqrt{25} \bullet \sqrt{a^4} \bullet \sqrt{a} \bullet \sqrt{b^2} \bullet \sqrt{b} =$

$2(5)a^2b\sqrt{a} \bullet \sqrt{b} = 10a^2b\sqrt{ab}$

11. $2\sqrt{5} + \sqrt{5} = (2+1)\sqrt{5} = 3\sqrt{5}$

12. $\sqrt{n} + 7\sqrt{n} = (1+7)\sqrt{n} = 8\sqrt{n}$

13. $4x\sqrt{3} - 3x\sqrt{3} = (4x - 3x)\sqrt{3} = x\sqrt{3}$

14.

$\sqrt{27} - 2\sqrt{75} = 3\sqrt{3} - 2(5)\sqrt{3} = 3\sqrt{3} - 10\sqrt{3}$

$(3-10)\sqrt{3} = -7\sqrt{3}$

15.

$x\sqrt{4x} + \sqrt{9x^3} = x\sqrt{4} \bullet \sqrt{x} + \sqrt{9} \bullet \sqrt{x^2} \bullet \sqrt{x} =$

$2x\sqrt{x} + 3x\sqrt{x} = (2x + 3x)\sqrt{x} = 5x\sqrt{x}$

16.

$\sqrt{50a} - 3\sqrt{8a} + 8\sqrt{2a} =$

$5\sqrt{2a} - (3)2\sqrt{2a} + 8\sqrt{2a} =$

$5\sqrt{2a} - 6\sqrt{2a} + 8\sqrt{2a} = 7\sqrt{2a}$

17. $\sqrt{5} \bullet \sqrt{3} = \sqrt{5 \cdot 3} = \sqrt{15}$

18. $\sqrt{8n} \bullet \sqrt{2n} = \sqrt{8n \cdot 2n} = \sqrt{16n^2} = 4n$

19.

$\sqrt{5a^2b^3} \bullet \sqrt{10ab^3} = \sqrt{5a^2b^3 \cdot 10ab^3} =$

$\sqrt{50a^3b^6} = 5ab^3\sqrt{2a}$

20. $\sqrt{x}\left(\sqrt{x} - 4\right) = \sqrt{x} \bullet \sqrt{x} - 4 \bullet \sqrt{x} = x - 4\sqrt{x}$

21. $\left(\sqrt{7} - 1\right)\left(\sqrt{7} + 1\right) = \left(\sqrt{7}\right)^2 - 1^2 = 7 - 1 = 6$

22.

$\left(\sqrt{y} + 1\right)\left(2\sqrt{y} - 3\right) =$

$\sqrt{y}\left(2\sqrt{y} - 3\right) + 1\left(2\sqrt{y} - 3\right) =$

$2y - 3\sqrt{y} + 2\sqrt{y} - 3 = 2y - \sqrt{y} - 3$

23.

$\left(\sqrt{y} + 5\right)^2 = \left(\sqrt{y}\right)^2 + 2 \bullet 5\sqrt{y} + 5^2 = y + 10\sqrt{y} + 25$

24. $\frac{\sqrt{54}}{\sqrt{6}} = \sqrt{\frac{54}{6}} = \sqrt{9} = 3$

25. $\frac{\sqrt{3}}{\sqrt{12}} = \sqrt{\frac{3}{12}} = \sqrt{\frac{1}{4}} = \frac{1}{2}$

26. $\frac{\sqrt{48x}}{\sqrt{3x}} = \sqrt{\frac{48x}{3x}} = \sqrt{16} = 4$

27.

$\frac{\sqrt{24a^6}}{\sqrt{2a^3}} = \sqrt{\frac{24a^6}{2a^3}} = \sqrt{12a^3} =$

$\sqrt{4} \bullet \sqrt{3} \bullet \sqrt{a^2} \bullet \sqrt{a} = 2a\sqrt{3a}$

28. $\frac{2}{\sqrt{11}} = \frac{2 \bullet \sqrt{11}}{\sqrt{11} \bullet \sqrt{11}} = \frac{2\sqrt{11}}{11}$

29.

$\frac{\sqrt{3x^2}}{\sqrt{6x}} = \sqrt{\frac{3x^2}{6x}} = \sqrt{\frac{x}{2}} = \frac{\sqrt{x}}{\sqrt{2}} = \frac{\sqrt{x} \cdot \sqrt{2}}{\sqrt{2} \cdot \sqrt{2}} = \frac{\sqrt{2x}}{2}$

30.

$\frac{4\sqrt{8} + \sqrt{2}}{\sqrt{2}} = \frac{\left(4\sqrt{8} + \sqrt{2}\right)\sqrt{2}}{\sqrt{2} \bullet \sqrt{2}} =$

$\frac{4\sqrt{16} + 2}{2} = \frac{16 + 2}{2} = \frac{18}{2} = 9$

31.

$\frac{10}{\sqrt{7} - 1} = \frac{10\left(\sqrt{7} + 1\right)}{\left(\sqrt{7} - 1\right)\left(\sqrt{7} + 1\right)} = \frac{10\left(\sqrt{7} + 1\right)}{7 - 1} =$

$\frac{10\left(\sqrt{7} + 1\right)}{6} = \frac{5\left(\sqrt{7} + 1\right)}{3} = \frac{5\sqrt{7} + 5}{3}$

32.

$\sqrt{n}-3=5$ *check*:

$\sqrt{n}=8$ $\sqrt{64}-3\stackrel{?}{=}5$

$(\sqrt{n})^2=8^2$ $8-3\stackrel{?}{=}5$

$n=64$ $5=5$

33.

$\sqrt{2x+1}=4$ *check*:

$(\sqrt{2x+1})^2=4^2$ $\sqrt{2\left(\frac{15}{2}\right)+1}\stackrel{?}{=}4$

$2x+1=16$ $\sqrt{15+1}\stackrel{?}{=}4$

$2x=15$ $\sqrt{16}-4$

$x=\frac{15}{2}$ $4=4$

34.

$\sqrt{4n-5}=\sqrt{n+10}$ *check*:

$(\sqrt{4n-5})^2=(\sqrt{n+10})^2$ $\sqrt{4(5)-5}\stackrel{?}{=}\sqrt{5+10}$

$4n-5=n+10$ $\sqrt{20-5}\stackrel{?}{=}\sqrt{5+10}$

$3n=15$ $\sqrt{15}\stackrel{?}{=}\sqrt{15}$

$n=5$

35.

$x-\sqrt{3x+1}=3$

$-\sqrt{3x+1}=3-x$

$(-\sqrt{3x+1})^2=(3-x)^2$

$3x+1=9-6x+x^2$

$x^2-9x+8=0$

$(x-8)(x-1)=0$

$x-8=0 \quad x-1=0$

$x=8 \quad x\neq 1$

check:

$8-\sqrt{3(8)+1}\stackrel{?}{=}3$ $1-\sqrt{3(1)+1}=3$

$8-\sqrt{25}\stackrel{?}{=}3$ $1-\sqrt{4}\stackrel{?}{=}3$

$8-5\stackrel{?}{=}3$ $1-2\stackrel{?}{=}3$

$3=3$ $-1\neq 3$

8 is the only solution to the original equation

Mixed Applications

36. $A=(50\sqrt{2})^2=2500(2)=5000$

The area of the city block is 5000 square feet.

37. $d=\sqrt{\frac{(2100)(6)}{2640}}=\sqrt{\frac{12600}{2640}}\approx 2$

You can see approximately 2 miles.

38.

$d=\sqrt{5^2+4^2+3^2}=\sqrt{25+16+9}=\sqrt{50}\approx 7.07$

No; an 8-inch screwdriver will not fit diagonally in the box.

39.

$A=s^2$

$s^2+s^2=(8\sqrt{2})^2$

$2s^2=(8\sqrt{2})^2$

$2s^2=64(2)$

$s^2=64$ sq m

40.

$r=\sqrt{\frac{S}{4\pi}}=\frac{\sqrt{S}}{\sqrt{4\pi}}=\frac{\sqrt{S}\cdot\sqrt{\pi}}{2\sqrt{\pi}\cdot\sqrt{\pi}}=\frac{\sqrt{S\pi}}{2\pi}$

41.

a. $D=\frac{2\pi r}{2\sqrt{\pi(\pi r^2)}}=\frac{2\pi r}{2\sqrt{\pi^2 r^2}}=\frac{2\pi r}{2\pi r}=1$

b. $D=\frac{L}{2\sqrt{\pi A}}=\frac{L\sqrt{\pi A}}{2\sqrt{\pi A}\bullet\sqrt{\pi A}}=\frac{L\sqrt{\pi A}}{2\pi A}$

c.

$D=\frac{L}{2\sqrt{\pi A}}$

$\sqrt{\pi A}=\frac{L}{2D}$

$\pi A=\frac{L^2}{4D^2}$

$A=\frac{L^2}{4\pi D^2}$

$A=\frac{L^2}{4\pi(1.58)^2}$

42. $\sqrt{\frac{V}{15}}=\frac{\sqrt{V}}{\sqrt{15}}=\frac{\sqrt{V}\cdot\sqrt{15}}{\sqrt{15}\cdot\sqrt{15}}=\frac{\sqrt{15V}}{15}$

Chapter 8 Posttest

1. $2\sqrt{36}=2\cdot6=12$

2. $-3\sqrt{18}=-3\cdot\sqrt{9\cdot2}=-3\cdot3\sqrt{2}=-9\sqrt{2}$

3. $\sqrt{32x^3y}=\sqrt{16\cdot2}\cdot\sqrt{x^2}\cdot\sqrt{x}\cdot\sqrt{y}=4x\sqrt{2xy}$

4. $\sqrt{\frac{5n}{16}}=\frac{\sqrt{5n}}{\sqrt{16}}=\frac{\sqrt{5n}}{4}$

5. $\sqrt{x}-3\sqrt{x}+5\sqrt{x}=(1-3+5)\sqrt{x}=3\sqrt{x}$

6.

$\sqrt{8}-4\sqrt{50}+\sqrt{18}=2\sqrt{2}-(4)5\sqrt{2}+3\sqrt{2}=$

$2\sqrt{2}-20\sqrt{2}+3\sqrt{2}=(2-20+3)\sqrt{2}=-15\sqrt{2}$

7.

$t\sqrt{4t}+2\sqrt{16t^3}=(t)2\sqrt{t}+(2)4t\sqrt{t}=2t\sqrt{t}+8t\sqrt{t}=$

$(2t+8t)\sqrt{t}=10t\sqrt{t}$

8. $(\sqrt{12})(\sqrt{2})=\sqrt{24}=\sqrt{4\cdot6}=2\sqrt{6}$

9. $(\sqrt{5x^3y})(\sqrt{5x^2y})=\sqrt{25x^5y^2}=5x^2y\sqrt{x}$

10. $\frac{\sqrt{75}}{\sqrt{3}}=\sqrt{\frac{75}{3}}=\sqrt{25}=5$

11.

$\sqrt{y}(5\sqrt{y}-1)=\sqrt{y}(5\sqrt{y})+\sqrt{y}(-1)=5y-\sqrt{y}$

12.

$(\sqrt{3}+4)^2=(\sqrt{3})^2+2\cdot4\cdot\sqrt{3}+4^2=$

$3+8\sqrt{3}+16=19+8\sqrt{3}$

13. $\sqrt{\frac{2p}{5}}=\frac{\sqrt{2p}}{\sqrt{5}}=\frac{\sqrt{2p}\cdot\sqrt{5}}{\sqrt{5}\cdot\sqrt{5}}=\frac{\sqrt{10p}}{5}$

14. $\frac{\sqrt{48x^4}}{\sqrt{2x}}=\sqrt{\frac{48x^4}{2x}}=\sqrt{24x^3}=2x\sqrt{6x}$

15.

$\frac{\sqrt{4}-3\sqrt{10}}{\sqrt{2}}=\frac{(2-3\sqrt{10})\sqrt{2}}{(\sqrt{2})\sqrt{2}}=\frac{2\sqrt{2}-3\sqrt{20}}{2}=$

$\frac{2\sqrt{2}-6\sqrt{5}}{2}=\frac{\overset{1}{\cancel{2}}(\sqrt{2}-3\sqrt{5})}{\underset{1}{\cancel{2}}}=\sqrt{2}-3\sqrt{5}$

16.

$\sqrt{5x+1}-2=4$

$\sqrt{5x+1}=6$ *check*:

$(\sqrt{5x+1})^2=(6)^2$ $\sqrt{5(7)+1}-2\overset{?}{=}4$

$5x+1=36$ $\sqrt{36}-2\overset{?}{=}4$

$5x=35$ $6-2\overset{?}{=}4$

$x=7$ $4=4$

17.

$\sqrt{x+11}=\sqrt{7x-1}$ *check*:

$(\sqrt{x+11})^2=(\sqrt{7x-1})^2$ $\sqrt{2+11}\overset{?}{=}\sqrt{7(2)-1}$

$x+11=7x-1$ $\sqrt{13}=\sqrt{13}$

$-6x=-12$

$x=2$

18.

$WCT=91+0.08(3.7\sqrt{16}+6-0.3(16))(42-91)=$

$91+0.08(14.8+6-4.8)(-49)=$

$91+0.08(16)(-49)\approx28$

The windchill temperature is 28° F.

19.

$l^2=24^2+6^2$

$l^2=612$

$l=\sqrt{612}$

$l=\sqrt{36\cdot17}$

$l=6\sqrt{17}$

The length of the ladder is $6\sqrt{17}$ feet.

20.

$20^2 = \sqrt{l^2 + 12^2}$

$400 = l^2 + 144$

$256 = l^2$

$l = \sqrt{256}$

$l = 16$

The length of the rectangle is 16 inches.

Cumulative Review Exercises

1.

$c = -2(c-1)$ *check* :

$c = -2c + 2$ $\frac{2}{3} \stackrel{?}{=} -2\left(\frac{2}{3} - 1\right)$

$3c = 2$ $\frac{2}{3} \stackrel{?}{=} -2\left(-\frac{1}{3}\right)$

$c = \frac{2}{3}$ $\frac{2}{3} = \frac{2}{3}$

2. $m = \frac{5-0}{3-4} = \frac{5}{-1} = -5$

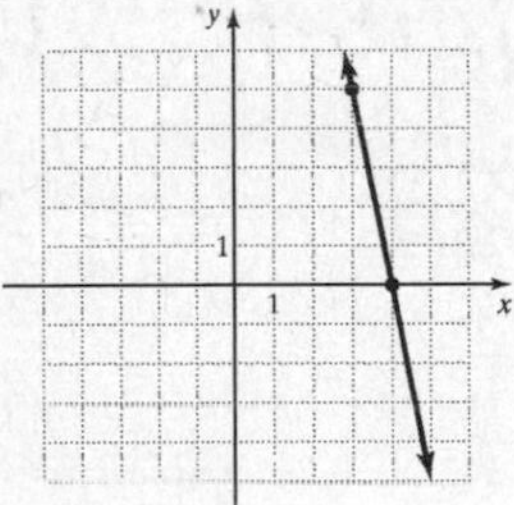

3. The lines intersect in one point. There is one solution to this system of equations.

4. $(4x^2 - 3)(4x - 1) = 16x^3 - 4x^2 - 12x + 3$

5. $y^2 - 12y + 32 = (y-4)(y-8)$

6.

$$\frac{9}{c-2} + \frac{c-3}{c^2 - c - 2} = \frac{9(c+1)}{c-2} + \frac{c-3}{c^2 - c - 2} =$$

$$\frac{9c+9}{c-2} + \frac{c-3}{c^2 - c - 2} = \frac{9c+9+c-3}{c^2 - c - 2} =$$

$$\frac{10c+6}{c^2 - c - 2} = \frac{2(5c+3)}{c^2 - c - 2}$$

7.

$\sqrt{n+6} + 7 = 9$ *check* :

$\sqrt{n+6} = 2$ $\sqrt{-2+6} + 7 \stackrel{?}{=} 9$

$\left(\sqrt{n+6}\right)^2 = 2^2$ $\sqrt{4} + 7 \stackrel{?}{=} 9$

$n + 6 = 4$ $9 = 9$

$n = -2$

8.

$343 = 0.6t + 331$

$12 = 0.6t$

$20 = t$

$t = 20$

The speed is 343 meters per second when the temperature is 20° Celsius.

9.

$(-x^2 + 47x + 396) - (x^2 + 58x + 483) =$

$-x^2 + 47x + 396 - x^2 - 58x - 483 =$

$-2x^2 - 11x - 87$

The balance of payments (in billions of dollars) can be modeled by the polynomial $-2x^2 - 11x - 87$.

10.

$d^2 = 1700^2 + 1500^2$

$d = \sqrt{1700^2 + 1500^2}$

$d = \sqrt{5{,}140{,}000}$

$d = \sqrt{514 \cdot 10^4} = 100\sqrt{514}$

$d \approx 2267$

The distance from the camp to the waterfall is $100\sqrt{514}$ meters or approximately 2267 meters.

Chapter 9 Quadratic Equations

Chapter 9 Pretest

1.

$3x^2 = 54$

$x^2 = 18$

$x = \pm\sqrt{18}$

$x = 3\sqrt{2} \quad x = -3\sqrt{2}$

check :

$3\left(3\sqrt{2}\right)^2 \stackrel{?}{=} 54 \qquad 3\left(-3\sqrt{2}\right)^2 \stackrel{?}{=} 54$

$3(18) \stackrel{?}{=} 54 \qquad 3(18) \stackrel{?}{=} 54$

$54 = 54 \qquad 54 = 54$

2.

$A = 4\pi r^2$

$\frac{A}{4\pi} = r^2$

$\sqrt{\frac{A}{4\pi}} = r$

$r = \frac{\sqrt{A}}{\sqrt{4\pi}}$

$r = \frac{\sqrt{A}\cdot\sqrt{\pi}}{2\sqrt{\pi}\cdot\sqrt{\pi}}$

$r = \frac{\sqrt{A\pi}}{2\pi}$

3. $x^2 + 12x + [\ ]$

$[\ ] = \left[\frac{1}{2}(12)\right]^2 = 36$

$x^2 + 12x + 36$

4.

$3x^2 - 12x + 6 = 0$

$x^2 - 4x = -2$

$x^2 - 4x + 4 = 2$

$(x-2)^2 = 2$

$x - 2 = \pm\sqrt{2}$

$x = 2 + \sqrt{2} \quad x = 2 - \sqrt{2}$

check :

$3\left(2+\sqrt{2}\right)^2 - 12\left(2+\sqrt{2}\right) + 6 \stackrel{?}{=} 0$

$3\left(4+4\sqrt{2}+2\right) - 12\left(2+\sqrt{2}\right) + 6 \stackrel{?}{=} 0$

$12 + 12\sqrt{2} + 6 - 24 - 12\sqrt{2} + 6 \stackrel{?}{=} 0$

$0 = 0$

$3\left(2-\sqrt{2}\right)^2 - 12\left(2-\sqrt{2}\right) + 6 \stackrel{?}{=} 0$

$3\left(4-4\sqrt{2}+2\right) - 12\left(2-\sqrt{2}\right) + 6 \stackrel{?}{=} 0$

$12 - 12\sqrt{2} + 6 - 24 + 12\sqrt{2} + 6 \stackrel{?}{=} 0$

$0 = 0$

5.

$x = \frac{-2 \pm \sqrt{2^2 - 4(-2)(5)}}{2(-2)}$

$x = \frac{-2 \pm \sqrt{4+40}}{-4}$

$x = \frac{-2 \pm \sqrt{44}}{-4}$

$x = \frac{-2 + 2\sqrt{11}}{-4} \qquad x = \frac{-2 - 2\sqrt{11}}{-4}$

$x = \frac{1-\sqrt{11}}{2} \qquad x = \frac{1+\sqrt{11}}{2}$

check :

$-2\left(\frac{1-\sqrt{11}}{2}\right)^2 + 2\left(\frac{1-\sqrt{11}}{2}\right) + 5 \stackrel{?}{=} 0$

$-6 + \sqrt{11} + 1 - \sqrt{11} + 5 \stackrel{?}{=} 0$

$0 = 0$

$-2\left(\frac{1+\sqrt{11}}{2}\right)^2 + 2\left(\frac{1+\sqrt{11}}{2}\right) + 5 \stackrel{?}{=} 0$

$-6 - \sqrt{11} + 1 + \sqrt{11} + 5 \stackrel{?}{=} 0$

$0 = 0$

6.

$(x-6)^2 = 25$

$x-6 = \pm\sqrt{25}$

$x-6=5 \quad x-6=-5$

$x=11 \quad x=1$

check :

$(11-6)^2 \stackrel{?}{=} 25 \quad (1-6)^2 \stackrel{?}{=} 25$

$5^2 \stackrel{?}{=} 25 \quad (-5)^2 \stackrel{?}{=} 25$

$25 = 25 \quad 25 = 25$

7.

$n^2 + 7n + 12 = 0$

$(n+3)(n+4) = 0$

$n+3=0 \quad n+4=0$

$n=-3 \quad n=-4$

check :

$(-3)^2 + 7(-3) + 12 \stackrel{?}{=} 0 \quad (-4)^2 + 7(-4) + 12 \stackrel{?}{=} 0$

$9 - 21 + 12 \stackrel{?}{=} 0 \quad 16 - 28 + 12 \stackrel{?}{=} 0$

$0=0 \quad 0=0$

8.

$y^2 - 6y = -4$

$y^2 - 6y + 4 = 0$

$$y = \frac{6 \pm \sqrt{(-6)^2 - 4(1)(4)}}{2(1)}$$

$$y = \frac{6 \pm \sqrt{36-16}}{2}$$

$$y = \frac{6 \pm \sqrt{20}}{2}$$

$$y = \frac{6 \pm 2\sqrt{5}}{2}$$

$y = 3+\sqrt{5} \quad y = 3-\sqrt{5}$

check :

$(3+\sqrt{5})^2 - 6(3+\sqrt{5}) \stackrel{?}{=} -4$

$9 + 6\sqrt{5} + 5 - 18 - 6\sqrt{5} \stackrel{?}{=} -4$

$-4 = -4$

$(3-\sqrt{5})^2 - 6(3-\sqrt{5}) \stackrel{?}{=} -4$

$9 - 6\sqrt{5} + 5 - 18 + 6\sqrt{5} \stackrel{?}{=} -4$

$-4 = -4$

9.

$4x^2 + 9 = 21$

$4x^2 = 12$

$x^2 = 3$

$x = \sqrt{3} \quad x = -\sqrt{3}$

check :

$4(\sqrt{3})^2 + 9 \stackrel{?}{=} 21 \quad 4(-\sqrt{3})^2 + 9 \stackrel{?}{=} 21$

$12 + 9 \stackrel{?}{=} 21 \quad 12 + 9 \stackrel{?}{=} 21$

$21 = 21 \quad 21 = 21$

10.

$5n^2 - 10n - 4 = 0$

$$n = \frac{10 \pm \sqrt{(-10)^2 - 4(5)(-4)}}{2(5)}$$

$$n = \frac{10 \pm \sqrt{100+80}}{10}$$

$$n = \frac{10 \pm \sqrt{180}}{10}$$

$$y = \frac{10 \pm 6\sqrt{5}}{10}$$

$$y = \frac{5+3\sqrt{5}}{5} \quad y = \frac{5-3\sqrt{5}}{5}$$

check :

$$5\left(\frac{5+3\sqrt{5}}{5}\right)^2-10\left(\frac{5+3\sqrt{5}}{5}\right)-4\overset{?}{=}0$$

$$14+6\sqrt{5}-10-6\sqrt{5}-4\overset{?}{=}0$$

$$0=0$$

$$5\left(\frac{5-3\sqrt{5}}{5}\right)^2-10\left(\frac{5-3\sqrt{5}}{5}\right)-4\overset{?}{=}0$$

$$14-6\sqrt{5}-10+6\sqrt{5}-4\overset{?}{=}0$$

$$0=0$$

11.

$$(x+2)(x+2)=18$$

$$(x+2)^2=18$$

$$x+2=\pm\sqrt{18}$$

$$x+2=3\sqrt{2} \qquad x+2=-3\sqrt{2}$$

$$x=-2+3\sqrt{2} \qquad x=-2-3\sqrt{2}$$

check :

$$\left(-2+3\sqrt{2}+2\right)\left(-2+3\sqrt{2}+2\right)\overset{?}{=}18$$

$$\left(3\sqrt{2}\right)\left(3\sqrt{2}\right)\overset{?}{=}18$$

$$18=18$$

$$\left(-2-3\sqrt{2}+2\right)\left(-2-3\sqrt{2}+2\right)\overset{?}{=}18$$

$$\left(-3\sqrt{2}\right)\left(-3\sqrt{2}\right)\overset{?}{=}18$$

$$18=18$$

12.

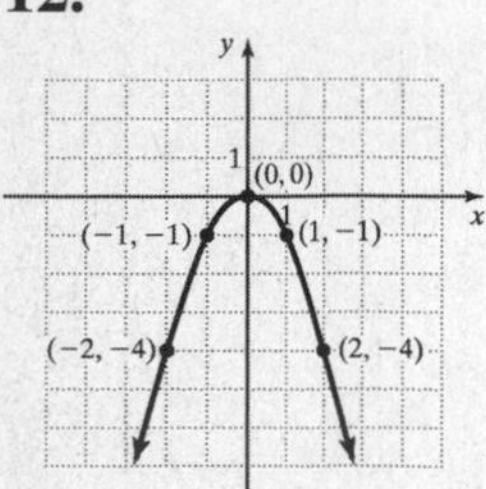

13.

$$x=\frac{-b}{2a}=\frac{6}{2}=3$$

$$y=3^2-6(3)+7=-2$$

vertex = (x, y)=(3, -2)
axis of symmetry: x = 3

14.

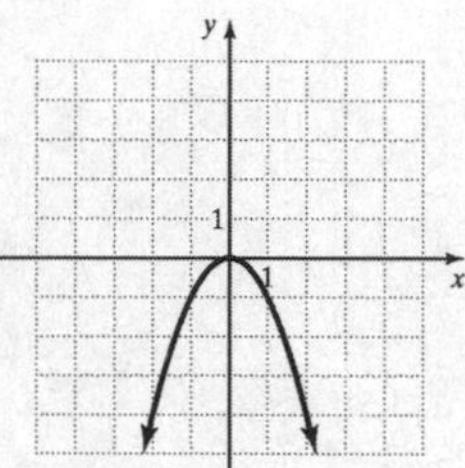

15.

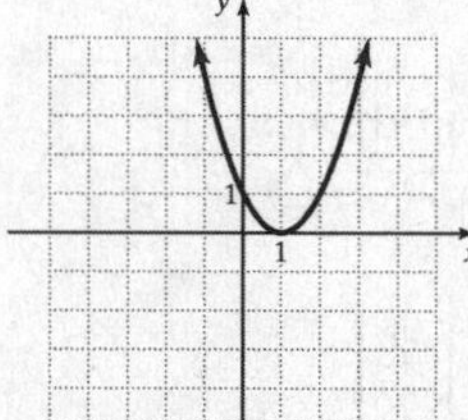

16.

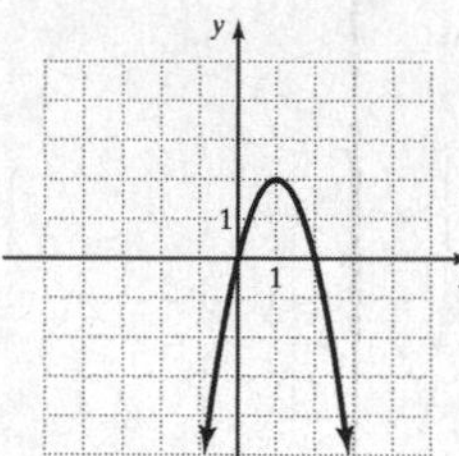

17.

a.

$$v^2=2as$$

$$v=\sqrt{2as}$$

b.

$$v=\sqrt{2as}$$

$$v=\sqrt{2(1.5)(100)}$$

$$v=\sqrt{300}$$

$$v\approx 17$$

The velocity of the car was approximately 17 meters per second.

18.

$(36-2x)(30-2x)=720$

$1080-132x+4x^2=720$

$4x^2-132x+360=0$

$x^2-33x+90=0$

$(x-3)(x-30)=0$

$x-3=0 \quad x-30=0$

$x=3 \qquad x=30$

3-inch by 3-inch squares should be cut from the corners to meet the specifications.

19.

	r	t	d
With wind	x+25	$\frac{420}{x+25}$	420
Against wind	x-15	$\frac{216}{x-15}$	216

$\frac{420}{x+25}+\frac{216}{x-15}=4$

$420(x-15)+216(x+25)=4(x+25)(x-15)$

$636x-900=4x^2+40x-1500$

$4x^2-596x-600=0$

$x^2-149x-150=0$

$(x-150)(x+1)=0$

$x-150=0 \quad x+1=0$

$x=150 \qquad x=-1$

The speed of the plane in still air is 150 miles per hour.

20.

a.

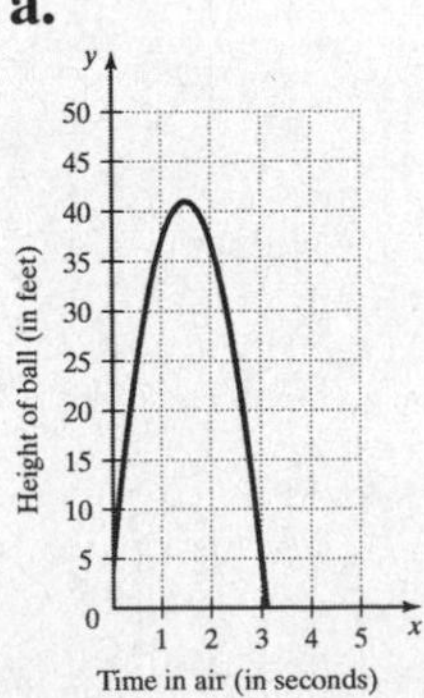

b.

$t=\frac{-b}{2a}=\frac{-48}{2(-16)}=\frac{-48}{-32}=\frac{3}{2}=1.5$

$h=-16(1.5)^2+48(1.5)+5=41$

The ball reaches a maximum height of 41 feet.

9.1 Solving Quadratic Equations by Using the Square Root Property

Practice 9.1

1.

a.

$x^2=100$

$x=\pm\sqrt{100}$

$x=10 \quad x=-10$

check:

$10^2 \stackrel{?}{=} 100 \quad (-10)^2 \stackrel{?}{=} 100$

$100=100 \qquad 100=100$

b.

$2x^2=36$

$x^2=18$

$x=\pm\sqrt{18}$

$x=3\sqrt{2} \quad x=-3\sqrt{2}$

check:

$(3\sqrt{2})^2 \stackrel{?}{=} 36 \quad (-3\sqrt{2})^2 \stackrel{?}{=} 36$

$36=36 \qquad 36=36$

c.

$$25-81r^2=0$$
$$-81r^2=-25$$
$$r^2=\frac{-25}{-81}$$
$$r=\pm\sqrt{\frac{25}{81}}$$
$$r=\frac{5}{9} \quad r=-\frac{5}{9}$$

check:

$$25-81\left(\frac{5}{9}\right)^2\overset{?}{=}0 \quad 25-81\left(-\frac{5}{9}\right)^2\overset{?}{=}0$$
$$25-25\overset{?}{=}0 \qquad 25-25\overset{?}{=}0$$
$$0=0 \qquad 0=0$$

2.

a.

$$(n+2)^2=9$$
$$\sqrt{(n+2)^2}=\sqrt{9}$$
$$n+2=\pm 3$$
$$n+2=3 \quad n+2=-3$$
$$n=1 \qquad n=-5$$

check:

$$(1+2)^2\overset{?}{=}9 \quad (-5+2)^2\overset{?}{=}9$$
$$(3)^2\overset{?}{=}9 \quad (-3)^2\overset{?}{=}9$$
$$9=9 \qquad 9=9$$

b.

$$3(x-4)^2=15$$
$$(x-4)^2=5$$
$$\sqrt{(x-4)^2}=\pm\sqrt{5}$$
$$x-4=\pm\sqrt{5}$$
$$x-4=\sqrt{5} \quad x-4=-\sqrt{5}$$
$$x=4+\sqrt{5} \qquad x=4-\sqrt{5}$$

check:

$$3\left(4+\sqrt{5}-4\right)^2\overset{?}{=}15 \quad 3\left(4-\sqrt{5}-4\right)^2\overset{?}{=}15$$
$$3\left(\sqrt{5}\right)^2\overset{?}{=}15 \quad 3\left(-\sqrt{5}\right)^2\overset{?}{=}15$$
$$15=15 \qquad 15=15$$

c.

$$(4m-1)^2=54$$
$$\sqrt{(4m-1)^2}=\sqrt{54}$$
$$4m-1=\pm\sqrt{54}$$
$$4m-1=\sqrt{54} \quad 4m-1=-\sqrt{54}$$
$$4m=1+3\sqrt{6} \qquad 4m=1-3\sqrt{6}$$
$$m=\frac{1+3\sqrt{6}}{4} \qquad m=\frac{1-3\sqrt{6}}{4}$$

check:

$$\left(4\left(\frac{1+3\sqrt{6}}{4}\right)-1\right)^2\overset{?}{=}54$$
$$\left(3\sqrt{6}\right)^2\overset{?}{=}54$$
$$54=54$$

$$\left(4\left(\frac{1-3\sqrt{6}}{4}\right)-1\right)^2\overset{?}{=}54$$
$$\left(-3\sqrt{6}\right)^2\overset{?}{=}54$$
$$54=54$$

3.

$$x+4=\frac{2}{x+4}$$
$$(x+4)^2=\left(\frac{2}{x+4}\right)(x+4)$$
$$(x+4)^2=2$$
$$x+4=\pm\sqrt{2}$$
$$x=-4+\sqrt{2} \quad x=-4-\sqrt{2}$$
check :
$$-4+\sqrt{2}+4\overset{?}{=}\frac{2}{-4+\sqrt{2}+4}$$
$$\sqrt{2}\overset{?}{=}\frac{2}{\sqrt{2}}$$
$$\sqrt{2}=\sqrt{2}$$

$$-4-\sqrt{2}+4\overset{?}{=}\frac{2}{-4-\sqrt{2}+4}$$
$$-\sqrt{2}\overset{?}{=}\frac{2}{-\sqrt{2}}$$
$$-\sqrt{2}=-\sqrt{2}$$

4.

a.
$$E=mc^2$$
$$\frac{E}{m}=c^2$$
$$c^2=\frac{E}{m}$$
$$c=\sqrt{\frac{E}{m}}=\frac{\sqrt{Em}}{m}$$

b.
$$c^2=a^2+b^2$$
$$c^2-b^2=a^2$$
$$a^2=c^2-b^2$$
$$a=\sqrt{c^2-b^2}$$

5.

a.
$$t^2=\frac{d}{16}$$
$$t=\sqrt{\frac{d}{16}}=\frac{\sqrt{d}}{\sqrt{16}}$$
$$t=\frac{\sqrt{d}}{4}$$

b.
$$t=\frac{\sqrt{400}}{4}=\frac{20}{4}=5$$
It takes the object 5 seconds to fall 400 feet.

6.

a.
$$B=(d-4)^2$$
$$(d-4)^2=B$$
$$d-4=\pm\sqrt{B}$$
$$d=4\pm\sqrt{B}$$

b.
$$d=4\pm\sqrt{16}$$
$$d=4+4$$
$$d=8$$
The diameter of the log is 8 inches.

Exercises 9.1

1.
$$y^2=9$$
$$y=\pm\sqrt{9}$$
$$y=3 \qquad y=-3$$
check :
$$(3)^2\overset{?}{=}9 \quad (-3)^2\overset{?}{=}9$$
$$9=9 \qquad 9=9$$

3.

$p^2 = 2$

$p = \pm\sqrt{2}$

$p = \sqrt{2} \qquad p = -\sqrt{2}$

check :

$(\sqrt{2})^2 \stackrel{?}{=} 2 \quad (-\sqrt{2})^2 \stackrel{?}{=} 2$

$2 = 2 \qquad 2 = 2$

5.

$n^2 = \frac{1}{4}$

$n = \pm\sqrt{\frac{1}{4}}$

$n = \frac{1}{2} \qquad n = -\frac{1}{2}$

check :

$\left(\frac{1}{2}\right)^2 \stackrel{?}{=} \frac{1}{4} \quad \left(-\frac{1}{2}\right)^2 \stackrel{?}{=} \frac{1}{4}$

$\frac{1}{4} = \frac{1}{4} \qquad \frac{1}{4} = \frac{1}{4}$

7.

$5t^2 = 20$

$t^2 = 4$

$t = \pm\sqrt{4}$

$t = 2 \qquad t = -2$

check :

$5(2)^2 \stackrel{?}{=} 20 \quad 5(-2)^2 \stackrel{?}{=} 20$

$5(4) \stackrel{?}{=} 20 \quad 5(4) \stackrel{?}{=} 20$

$20 = 20 \qquad 20 = 20$

9.

$4x^2 = 28$

$x^2 = 7$

$x = \pm\sqrt{7}$

$x = \sqrt{7} \qquad x = -\sqrt{7}$

check :

$4(\sqrt{7})^2 \stackrel{?}{=} 28 \quad 4(-\sqrt{7})^2 \stackrel{?}{=} 28$

$4(7) \stackrel{?}{=} 28 \qquad 4(7) \stackrel{?}{=} 28$

$28 = 28 \qquad 28 = 28$

11.

$3x^2 = 36$

$x^2 = 12$

$x = \pm\sqrt{12}$

$x = 2\sqrt{3} \qquad x = -2\sqrt{3}$

check :

$3(2\sqrt{3})^2 \stackrel{?}{=} 36 \quad 3(-2\sqrt{3})^2 \stackrel{?}{=} 36$

$3(12) \stackrel{?}{=} 36 \qquad 3(12) \stackrel{?}{=} 36$

$36 = 36 \qquad 36 = 36$

13.

$\frac{1}{4}p^2 = 20$

$p^2 = 80$

$p = \pm\sqrt{80}$

$p = 4\sqrt{5} \qquad s = -4\sqrt{5}$

check :

$\frac{1}{4}(4\sqrt{5})^2 \stackrel{?}{=} 20 \quad \frac{1}{4}(-4\sqrt{5})^2 \stackrel{?}{=} 20$

$\frac{1}{4}(80) \stackrel{?}{=} 20 \qquad \frac{1}{4}(80) \stackrel{?}{=} 20$

$20 = 20 \qquad 20 = 20$

15.

$5 + t^2 = 11$

$t^2 = 6$

$t = \pm\sqrt{6}$

$t = \sqrt{6} \quad t = -\sqrt{6}$

check :

$5 + \left(\sqrt{6}\right)^2 \stackrel{?}{=} 11 \quad 5 + \left(-\sqrt{6}\right)^2 \stackrel{?}{=} 11$

$5 + 6 \stackrel{?}{=} 11 \qquad 5 + 6 \stackrel{?}{=} 11$

$11 = 11 \qquad 11 = 11$

17.

$x^2 - 8 = 9$

$x^2 = 17$

$x = \pm\sqrt{17}$

$x = \sqrt{17} \quad x = -\sqrt{17}$

check :

$\left(\sqrt{17}\right)^2 - 8 \stackrel{?}{=} 9 \quad \left(-\sqrt{17}\right)^2 - 8 \stackrel{?}{=} 9$

$17 - 8 \stackrel{?}{=} 9 \qquad 17 - 8 \stackrel{?}{=} 9$

$9 = 9 \qquad 9 = 9$

19.

$4n^2 - 9 = 0$

$4n^2 = 9$

$n^2 = \frac{9}{4}$

$n = \pm\sqrt{\frac{9}{4}}$

$n = \frac{3}{2} \quad n = -\frac{3}{2}$

check :

$4\left(\frac{3}{2}\right)^2 - 9 \stackrel{?}{=} 0 \quad 4\left(-\frac{3}{2}\right)^2 - 9 \stackrel{?}{=} 0$

$9 - 9 \stackrel{?}{=} 0 \qquad 9 - 9 \stackrel{?}{=} 0$

$0 = 0 \qquad 0 = 0$

21.

$15 - 16x^2 = 3$

$-16x^2 = -12$

$x^2 = \frac{-12}{-16}$

$x = \pm\sqrt{\frac{12}{16}}$

$x = \frac{2\sqrt{3}}{4} \quad x = -\frac{2\sqrt{3}}{4}$

$x = \frac{\sqrt{3}}{2} \quad x = -\frac{\sqrt{3}}{2}$

check :

$15 - 16\left(\frac{\sqrt{3}}{2}\right)^2 \stackrel{?}{=} 3 \quad 15 - 16\left(-\frac{\sqrt{3}}{2}\right)^2 \stackrel{?}{=} 3$

$15 - 12 \stackrel{?}{=} 3 \qquad 15 - 12 \stackrel{?}{=} 3$

$3 = 3 \qquad 3 = 3$

23.

$(n+2)^2 = 9$

$n + 2 = \pm\sqrt{9}$

$n + 2 = 3 \quad n + 2 = -3$

$n = 1 \qquad n = -5$

check :

$(1+2)^2 \stackrel{?}{=} 9 \quad (-5+2)^2 \stackrel{?}{=} 9$

$3^2 \stackrel{?}{=} 9 \qquad (-3)^2 \stackrel{?}{=} 9$

$9 = 9 \qquad 9 = 9$

25.

$(x-7)^2 = 49$

$x - 7 = \pm\sqrt{49}$

$x - 7 = 7 \quad x - 7 = -7$

$x = 14 \qquad x = 0$

check :

$(14-7)^2 \stackrel{?}{=} 49 \quad (0-7)^2 \stackrel{?}{=} 49$

$7^2 \stackrel{?}{=} 49 \qquad (-7)^2 \stackrel{?}{=} 49$

$49 = 49 \qquad 49 = 49$

27.

$(x+6)^2=5$

$x+6=\pm\sqrt{5}$

$x+6=\sqrt{5} \quad x+6=-\sqrt{5}$

$x=-6+\sqrt{5} \qquad x=-6-\sqrt{5}$

check :

$\left(-6+\sqrt{5}+6\right)^2\overset{?}{=}5 \quad \left(-6-\sqrt{5}+6\right)^2\overset{?}{=}5$

$\left(\sqrt{5}\right)^2\overset{?}{=}5 \qquad \left(-\sqrt{5}\right)^2\overset{?}{=}5$

$5=5 \qquad 5=5$

29.

$(5-s)^2=\frac{9}{16}$

$5-s=\pm\sqrt{\frac{9}{16}}$

$5-s=\frac{3}{4} \quad 5-s=-\frac{3}{4}$

$-s=-\frac{17}{4} \qquad -s=-\frac{23}{4}$

$s=\frac{17}{4} \qquad s=\frac{23}{4}$

check :

$\left(5-\frac{17}{4}\right)^2\overset{?}{=}\frac{9}{16} \quad \left(5-\frac{23}{4}\right)^2\overset{?}{=}\frac{9}{16}$

$\left(\frac{3}{4}\right)^2\overset{?}{=}\frac{9}{16} \qquad \left(-\frac{3}{4}\right)^2\overset{?}{=}\frac{9}{16}$

$\frac{9}{16}=\frac{9}{16} \qquad \frac{9}{16}=\frac{9}{16}$

31.

$(y-4)^2=9$

$y-4=\pm\sqrt{9}$

$y-4=3 \quad y-4=-3$

$y=7 \qquad y=1$

check :

$(7-4)^2\overset{?}{=}9 \quad (1-4)^2\overset{?}{=}9$

$3^2\overset{?}{=}9 \qquad (-3)^2\overset{?}{=}9$

$9=9 \qquad 9=9$

33.

$2(p-5)^2=6$

$(p-5)^2=3$

$p-5=\pm\sqrt{3}$

$p=5+\sqrt{3} \quad p=5-\sqrt{3}$

check :

$2\left(5+\sqrt{3}-5\right)^2\overset{?}{=}6 \quad 2\left(5-\sqrt{3}-5\right)^2\overset{?}{=}6$

$2\left(\sqrt{3}\right)^2\overset{?}{=}6 \qquad 2\left(-\sqrt{3}\right)^2\overset{?}{=}6$

$6=6 \qquad 6=6$

35.

$5(x+1)^2=40$

$(x+1)^2=8$

$x+1=\pm\sqrt{8}$

$x=-1+2\sqrt{2} \quad x=-1-2\sqrt{2}$

check :

$5\left(-1+2\sqrt{2}+1\right)^2\overset{?}{=}40$

$5\left(2\sqrt{2}\right)^2\overset{?}{=}40$

$5(8)\overset{?}{=}40$

$40=40$

$5\left(-1-2\sqrt{2}+1\right)^2\overset{?}{=}40$

$5\left(-2\sqrt{2}\right)^2\overset{?}{=}40$

$5(8)\overset{?}{=}40$

$40=40$

37.

$(3x+1)^2=4$

$3x+1=\pm\sqrt{4}$

$3x+1=2 \quad 3x+1=-2$

$3x=1 \qquad 3x=-3$

$x=\frac{1}{3} \qquad x=-1$

check :

$\left(3\left(\frac{1}{3}\right)+1\right)^2 \stackrel{?}{=} 4 \quad (3(-1)+1)^2 \stackrel{?}{=} 4$

$(2)^2 \stackrel{?}{=} 4 \qquad (-2)^2 \stackrel{?}{=} 4$

$4=4 \qquad 4=4$

39.

$(4y+5)^2=3$

$4y+5=\pm\sqrt{3}$

$4y+5=\sqrt{3} \quad 4y+5=-\sqrt{3}$

$4y=-5+\sqrt{3} \qquad 4y=-5-\sqrt{3}$

$y=\frac{-5+\sqrt{3}}{4} \qquad y=\frac{-5-\sqrt{3}}{4}$

check :

$\left(4\bullet\frac{-5+\sqrt{3}}{4}+5\right)^2 \stackrel{?}{=} 3 \quad \left(4\bullet\frac{-5-\sqrt{3}}{4}+5\right)^2 \stackrel{?}{=} 3$

$\left(\sqrt{3}\right)^2 \stackrel{?}{=} 3 \qquad \left(-\sqrt{3}\right)^2 \stackrel{?}{=} 3$

$3=3 \qquad 3=3$

41.

$(2x-7)^2=20$

$2x-7=\pm\sqrt{20}$

$2x-7=2\sqrt{5} \quad 2x-7=-2\sqrt{5}$

$2x=7+2\sqrt{5} \qquad 2x=7-2\sqrt{5}$

$x=\frac{7+2\sqrt{5}}{2} \qquad x=\frac{7-2\sqrt{5}}{2}$

check :

$\left(2\bullet\frac{7+2\sqrt{5}}{2}-7\right)^2 \stackrel{?}{=} 20 \quad \left(2\bullet\frac{7-2\sqrt{5}}{2}-7\right)^2 \stackrel{?}{=} 20$

$\left(2\sqrt{5}\right)^2 \stackrel{?}{=} 20 \qquad \left(-2\sqrt{5}\right)^2 \stackrel{?}{=} 20$

$20=20 \qquad 20=20$

43.

$\left(\frac{1}{2}x-5\right)^2=10$

$\frac{1}{2}x-5=\pm\sqrt{10}$

$\frac{1}{2}x=5+\sqrt{10} \qquad \frac{1}{2}x=5-\sqrt{10}$

$x=10+2\sqrt{10} \qquad x=10-2\sqrt{10}$

check :

$\left(\frac{1}{2}\bullet\left(10+2\sqrt{10}\right)-5\right)^2 \stackrel{?}{=} 10$

$\left(\sqrt{10}\right)^2 \stackrel{?}{=} 10$

$10=10$

$\left(\frac{1}{2}\bullet\left(10-2\sqrt{10}\right)-5\right)^2 \stackrel{?}{=} 10$

$\left(-\sqrt{10}\right)^2 \stackrel{?}{=} 10$

$10=10$

45.

$(x+1)^2+49=0$

$(x+1)^2=-49$

$x+1=\pm\sqrt{-49}$

No real solution. The square root of a negative number is undefined under the set of real numbers.

47.

$$a+8=\frac{5}{a+8}$$

$$(a+8)(a+8)=\frac{5}{a+8}(a+8)$$

$$(a+8)^2=5$$

$$a+8=\pm\sqrt{5}$$

$$a=-8+\sqrt{5} \quad a=-8-\sqrt{5}$$

check:

$$-8+\sqrt{5}+8\overset{?}{=}\frac{5}{-8+\sqrt{5}+8}$$

$$\sqrt{5}\overset{?}{=}\frac{5}{\sqrt{5}}$$

$$\sqrt{5}=\sqrt{5}$$

$$-8-\sqrt{5}+8\overset{?}{=}\frac{5}{-8-\sqrt{5}+8}$$

$$-\sqrt{5}\overset{?}{=}\frac{5}{-\sqrt{5}}$$

$$-\sqrt{5}=-\sqrt{5}$$

49.

$$x-3=\frac{24}{x-3}$$

$$(x-3)^2=24$$

$$x-3=\pm\sqrt{24}$$

$$x-3=2\sqrt{6} \quad x-3=-2\sqrt{6}$$

$$x=3+2\sqrt{6} \quad x=3-2\sqrt{6}$$

check:

$$3\pm2\sqrt{6}-3\overset{?}{=}\frac{24}{3\pm2\sqrt{6}-3}$$

$$\pm2\sqrt{6}\overset{?}{=}\pm\frac{24}{2\sqrt{6}}$$

$$\pm2\sqrt{6}\overset{?}{=}\pm\frac{12}{\sqrt{6}}$$

$$\pm2\sqrt{6}=\pm2\sqrt{6}$$

51.

$$(y-1)(y+1)=4$$

$$y^2-1=4$$

$$y^2=5$$

$$y=\pm\sqrt{5}$$

$$y=\sqrt{5} \quad y=-\sqrt{5}$$

check:

$$\left(\sqrt{5}-1\right)\left(\sqrt{5}+1\right)\overset{?}{=}4 \quad \left(-\sqrt{5}-1\right)\left(-\sqrt{5}+1\right)\overset{?}{=}4$$

$$\left(\sqrt{5}\right)^2-1^2\overset{?}{=}4 \quad \left(-\sqrt{5}\right)^2-1^2\overset{?}{=}4$$

$$4=4 \quad 4=4$$

53.

$$ax^2-b=0$$

$$ax^2=b$$

$$x^2=\frac{b}{a}$$

$$x=\pm\sqrt{\frac{b}{a}}$$

$$x=\pm\frac{\sqrt{ab}}{a}$$

55.

$$K=\frac{4\pi^2}{v^2r}$$

$$v^2K=\frac{4\pi^2}{r}$$

$$v^2=\frac{4\pi^2}{Kr}$$

$$v=\pm\sqrt{\frac{4\pi^2}{Kr}}$$

$$v=\pm\frac{\sqrt{4\pi^2}}{\sqrt{Kr}}$$

$$v=\pm\frac{2\pi\sqrt{Kr}}{Kr}$$

57.

$$\frac{x^2}{16}-\frac{y^2}{25}=1$$
$$25x^2-16y^2=400$$
$$16y^2=25x^2-400$$
$$y^2=\frac{25x^2-400}{16}$$
$$y^2=\frac{25(x^2-16)}{16}$$
$$y=\sqrt{\frac{25(x^2-16)}{16}}$$
$$y=\pm\frac{5\sqrt{x^2-16}}{4}$$

59.

$$(45t)^2+(60t)^2=300^2$$
$$2025t^2+3600t^2=90,000$$
$$5625t^2=90,000$$
$$t^2=16$$
$$t=4 \text{ minutes}$$

They can hear each other for 4 minutes.

Applications

61.

a.

$$s=\frac{1}{2}at^2$$
$$2s=at^2$$
$$t^2=\frac{2s}{a}$$
$$t=\sqrt{\frac{2s}{a}}=\frac{\sqrt{2as}}{a}$$

b.

$$t=\frac{\sqrt{2(32)(64)}}{32}$$
$$t=\frac{\sqrt{4096}}{32}$$
$$t=\frac{64}{32}=2$$ The time is 2 seconds.

63.

$$x^2+y^2=r^2$$
$$y^2=r^2-x^2$$
$$y=\pm\sqrt{r^2-x^2}$$

65.

$$I=\frac{4050}{d^2}$$
$$d^2I=4050$$
$$d^2=\frac{4050}{I}$$
$$d=\sqrt{\frac{4050}{I}}$$
$$d=\frac{\sqrt{4050}}{\sqrt{I}}$$
$$d=\frac{45\sqrt{2}}{\sqrt{I}}$$
$$d=45\frac{\sqrt{2I}}{I}$$

9.2 Solving Quadratic Equations by Completing the Square

Practice 9.2

1.

a.

$$\frac{1}{2}(-12)=-6$$
$$(-6)^2=36$$
$$x^2-12x+36=(x-6)^2$$

b.

$$\frac{1}{2}(5)=\frac{5}{2}$$
$$\left(\frac{5}{2}\right)^2=\frac{25}{4}$$
$$n^2+5n+\frac{25}{4}=\left(n+\frac{5}{2}\right)^2$$

2.

a.

$$y^2+4y=21$$
$$y^2+4y+4=21+4$$
$$(y+2)^2=25$$
$$y+2=\pm\sqrt{25}$$
$$y+2=5 \quad y+2=-5$$
$$y=3 \quad y=-7$$

check :

$$3^2+4(3)\overset{?}{=}21 \quad (-7)^2+4(-7)\overset{?}{=}21$$
$$9+12\overset{?}{=}21 \quad 49-28\overset{?}{=}21$$
$$21=21 \quad 21=21$$

b.

$$n^2+7n+5=0$$
$$n^2+7n=-5$$
$$n^2+7n+\frac{49}{4}=-5+\frac{49}{4}$$
$$\left(n+\frac{7}{2}\right)^2=\frac{29}{4}$$
$$n+\frac{7}{2}=\pm\sqrt{\frac{29}{4}}$$
$$n+\frac{7}{2}=\frac{\sqrt{29}}{2} \quad n+\frac{7}{2}=-\frac{\sqrt{29}}{2}$$
$$n=\frac{-7+\sqrt{29}}{2} \quad n=\frac{-7-\sqrt{29}}{2}$$

check :

$$\left(\frac{-7\pm\sqrt{29}}{2}\right)^2+7\left(\frac{-7\pm\sqrt{29}}{2}\right)+5\overset{?}{=}0$$
$$\frac{39\pm7\sqrt{29}}{2}+\frac{-49\pm7\sqrt{29}}{2}+5\overset{?}{=}0$$
$$-\frac{10}{2}+5\overset{?}{=}0$$
$$0=0$$

3.

a.

$$2y^2-16y=8$$
$$y^2-8y=4$$
$$y^2-8y+16=4+16$$
$$(y-4)^2=20$$
$$y-4=\pm\sqrt{20}$$
$$y-4=2\sqrt{5} \quad y-4=-2\sqrt{5}$$
$$y=4+2\sqrt{5} \quad y=4-2\sqrt{5}$$

check :

$$2\left(4+2\sqrt{5}\right)^2-16\left(4+2\sqrt{5}\right)\overset{?}{=}8$$
$$72+32\sqrt{5}-64-32\sqrt{5}\overset{?}{=}8$$
$$8=8$$
$$2\left(4-2\sqrt{5}\right)^2-16\left(4-2\sqrt{5}\right)\overset{?}{=}8$$
$$72-32\sqrt{5}-64+32\sqrt{5}\overset{?}{=}8$$
$$8=8$$

b.

$$4x^2-4x-1=0$$
$$x^2-x=\frac{1}{4}$$
$$x^2-x+\frac{1}{4}=\frac{1}{4}+\frac{1}{4}$$
$$\left(x-\frac{1}{2}\right)^2=\frac{1}{2}$$
$$x-\frac{1}{2}=\pm\sqrt{\frac{1}{2}}$$
$$x-\frac{1}{2}=\frac{\sqrt{2}}{2} \quad x-\frac{1}{2}=-\frac{\sqrt{2}}{2}$$
$$x=\frac{1+\sqrt{2}}{2} \quad x=\frac{1-\sqrt{2}}{2}$$

check :

$$4\left(\frac{1+\sqrt{2}}{2}\right)^2-4\left(\frac{1+\sqrt{2}}{2}\right)-1\overset{?}{=}0$$
$$3+2\sqrt{2}-2-2\sqrt{2}-1\overset{?}{=}0$$
$$0=0$$
$$4\left(\frac{1-\sqrt{2}}{2}\right)^2-4\left(\frac{1-\sqrt{2}}{2}\right)-1\overset{?}{=}0$$
$$3-2\sqrt{2}-2+2\sqrt{2}-1\overset{?}{=}0$$
$$0=0$$

4.

$9x^2 - 9x - 4 = 0$

$x^2 - x = \frac{4}{9}$

$x^2 - x + \frac{1}{4} = \frac{4}{9} + \frac{1}{4}$

$\left(x - \frac{1}{2}\right)^2 = \frac{25}{36}$

$x - \frac{1}{2} = \pm\sqrt{\frac{25}{36}}$

$x - \frac{1}{2} = \frac{5}{6} \quad x - \frac{1}{2} = -\frac{5}{6}$

$x = \frac{4}{3} \quad x = -\frac{1}{3}$

check :

$9\left(\frac{4}{3}\right)^2 - 9\left(\frac{4}{3}\right) - 4 \stackrel{?}{=} 0$

$16 - 12 - 4 \stackrel{?}{=} 0$

$0 = 0$

$9\left(-\frac{1}{3}\right)^2 - 9\left(-\frac{1}{3}\right) - 4 \stackrel{?}{=} 0$

$1 + 3 - 4 \stackrel{?}{=} 0$

$0 = 0$

5. a. The length of the table is 2w – 1.

b.

$w(2w-1) = 36$

$2w^2 - w = 36$

$w^2 - \frac{1}{2}w = 18$

$w^2 - \frac{1}{2}w + \frac{1}{16} = 18 + \frac{1}{16}$

$\left(w - \frac{1}{4}\right)^2 = \frac{289}{16}$

$w - \frac{1}{4} = \frac{17}{4} \quad w - \frac{1}{4} = -\frac{17}{4}$

$w = 4.5 \quad w = -4$

$2w - 1 = 8$

The length is 8 feet and the width is 4.5 feet.

Exercises 9.2

1.

$x^2 + 6x + [\ ]$

$\frac{1}{2}(6) = 3$

$(3)^2 = 9$

$x^2 + 6x + 9 = (x-3)^2$

3.

$n^2 - 10n + [\ ]$

$\frac{1}{2}(-10) = -5$

$(-5)^2 = 25$

$n^2 - 10n + 25 = (n-5)^2$

5.

$x^2 + 5x + [\ ]$

$\frac{1}{2}(5) = \frac{5}{2}$

$\left(\frac{5}{2}\right)^2 = \frac{25}{4}$

$x^2 + 5x + \frac{25}{4} = \left(x + \frac{5}{2}\right)^2$

7.

$y^2 - y + [\ ]$

$\frac{1}{2}(-1) = -\frac{1}{2}$

$\left(-\frac{1}{2}\right)^2 = \frac{1}{4}$

$y^2 - y + \frac{1}{4} = \left(y - \frac{1}{2}\right)^2$

9.

$x^2+4x=0$

$x^2+4x+4=0+4$

$(x+2)^2=4$

$x+2=\pm\sqrt{4}$

$x+2=2 \quad x+2=-2$

$x=0 \quad x=-4$

check :

$(0)^2+4(0)\stackrel{?}{=}0 \quad (-4)^2+4(-4)\stackrel{?}{=}0$

$0+0\stackrel{?}{=}0 \quad 16+(-16)\stackrel{?}{=}0$

$0=0 \quad 0=0$

11.

$b^2-10b=-1$

$b^2-10b+25=-1+25$

$(x-5)^2=24$

$x-5=\pm\sqrt{24}$

$x-5=2\sqrt{6} \quad x-5=-2\sqrt{6}$

$x=5+2\sqrt{6} \quad x=5-2\sqrt{6}$

check :

$\left(5+2\sqrt{6}\right)^2-10\left(5+2\sqrt{6}\right)\stackrel{?}{=}-1$

$49+20\sqrt{6}-50-20\sqrt{6}\stackrel{?}{=}-1$

$-1=-1$

$\left(5-2\sqrt{6}\right)^2-10\left(5-2\sqrt{6}\right)\stackrel{?}{=}-1$

$49-20\sqrt{6}-50+20\sqrt{6}\stackrel{?}{=}-1$

$-1=-1$

13.

$y^2+14y-15=0$

$y^2+14y=15$

$y^2+14t+49=15+49$

$(y+7)^2=64$

$y+7=\pm\sqrt{64}$

$y+7=8 \quad y+7=-8$

$y=1 \quad y=-15$

check :

$(1)^2+14(1)-15\stackrel{?}{=}0$

$1+14-15\stackrel{?}{=}0$

$0=0$

$(-15)^2+14(-15)-15\stackrel{?}{=}0$

$225-210-15\stackrel{?}{=}0$

$0=0$

15.

$x^2-6x-4=0$

$x^2-6x+9=4+9$

$(x-3)^2=13$

$x-3=\pm\sqrt{13}$

$x-3=\sqrt{13} \quad x-3=-\sqrt{13}$

$x=3+\sqrt{13} \quad x=3-\sqrt{13}$

check :

$\left(3+\sqrt{13}\right)^2-6\left(3+\sqrt{13}\right)-4\stackrel{?}{=}0$

$22+6\sqrt{13}-18-6\sqrt{13}-4\stackrel{?}{=}0$

$0=0$

$\left(3-\sqrt{13}\right)^2-6\left(3-\sqrt{13}\right)-4\stackrel{?}{=}0$

$22-6\sqrt{13}-18+6\sqrt{13}-4\stackrel{?}{=}0$

$0=0$

17.

$n^2+8n+20=0$

$n^2+8n+16=-20+16$

$(n+4)^2=-4$

$n+4=\pm\sqrt{-4}$

No real solution. The square root of a negative number is undefined under the set of real numbers.

19.

$x^2-x-3=0$

$x^2-x+\frac{1}{4}=3+\frac{1}{4}$

$\left(y-\frac{1}{2}\right)^2=\frac{13}{4}$

$y-\frac{1}{2}=\pm\frac{\sqrt{13}}{2}$

$y-\frac{1}{2}=\frac{\sqrt{13}}{2}$ $\quad y-\frac{1}{2}=-\frac{\sqrt{13}}{2}$

$y=\frac{1+\sqrt{13}}{2}$ $\quad y=\frac{1-\sqrt{13}}{2}$

check :

$\left(\frac{1+\sqrt{13}}{2}\right)^2-\left(\frac{1+\sqrt{13}}{2}\right)-3\overset{?}{=}0$

$\frac{14+2\sqrt{13}}{4}-\frac{2+2\sqrt{13}}{4}-3\overset{?}{=}0$

$0=0$

$\left(\frac{1-\sqrt{13}}{2}\right)^2-\left(\frac{1-\sqrt{13}}{2}\right)-3\overset{?}{=}0$

$\frac{14-2\sqrt{13}}{4}-\frac{2-2\sqrt{13}}{4}-3\overset{?}{=}0$

$0=0$

21.

$2y^2+8y=24$

$y^2+4y=12$

$y^2+4y+4=12+4$

$(y+2)^2=16$

$y+2=\pm\sqrt{16}$

$x+2=4 \quad x+2=-4$

$x=2 \quad\quad x=-6$

check :

$2(2)^2+8(2)\overset{?}{=}24 \quad 2(-6)^2+8(-6)\overset{?}{=}24$

$8+16\overset{?}{=}24 \quad\quad 72-48\overset{?}{=}24$

$24=24 \quad\quad 24=24$

23.

$6x^2+12x-5=0$

$x^2+2x=\frac{5}{6}$

$x^2+2x+1=\frac{5}{6}+1$

$(x+1)^2=\frac{11}{6}$

$x+1=\pm\sqrt{\frac{11}{6}}$

$x+1=\pm\frac{\sqrt{66}}{6}$

$x=-1\pm\frac{\sqrt{66}}{6}$

check :

$6\left(-1\pm\frac{\sqrt{66}}{6}\right)^2+12\left(-1\pm\frac{\sqrt{66}}{6}\right)-5\overset{?}{=}0$

$\frac{30}{6}-5\overset{?}{=}0$

$0=0$

25.

$4n^2 - 20n + 7 = 0$

$n^2 - 5n = -\frac{7}{4}$

$n^2 - 5n + \frac{25}{4} = -\frac{7}{4} + \frac{25}{4}$

$\left(n - \frac{5}{2}\right)^2 = \frac{18}{4}$

$n - \frac{5}{2} = \pm\sqrt{\frac{18}{4}}$

$n = \frac{5}{2} + \frac{3\sqrt{2}}{2} \quad n = \frac{5}{2} - \frac{3\sqrt{2}}{2}$

$n = \frac{5 + 3\sqrt{2}}{2} \quad n = \frac{5 - 3\sqrt{2}}{2}$

check :

$4\left(\frac{5+3\sqrt{2}}{2}\right)^2 - 20\left(\frac{5+3\sqrt{2}}{2}\right) + 7 \overset{?}{=} 0$

$43 + 30\sqrt{2} - 50 - 30\sqrt{2} + 7 \overset{?}{=} 0$

$0 = 0$

$4\left(\frac{5-3\sqrt{2}}{2}\right)^2 - 20\left(\frac{5-3\sqrt{2}}{2}\right) + 7 \overset{?}{=} 0$

$43 - 30\sqrt{2} - 50 + 30\sqrt{2} + 7 \overset{?}{=} 0$

$0 = 0$

27.

$4n^2 - 3n - 4 = 0$

$n^2 - \frac{3}{4}n = 1$

$n^2 - \frac{3}{4}n + \frac{9}{64} = 1 + \frac{9}{64}$

$\left(n - \frac{3}{8}\right)^2 = \frac{73}{64}$

$n - \frac{3}{8} = \pm\sqrt{\frac{73}{64}}$

$n = \frac{3}{8} + \frac{\sqrt{73}}{8} \quad n = \frac{3}{8} - \frac{\sqrt{73}}{8}$

$n = \frac{3 + \sqrt{73}}{8} \quad n = \frac{3 - \sqrt{73}}{8}$

check :

$4\left(\frac{3+\sqrt{73}}{8}\right)^2 - 3\left(\frac{3+\sqrt{73}}{8}\right) - 4 \overset{?}{=} 0$

$\frac{32}{8} - 4 \overset{?}{=} 0$

$0 = 0$

$4\left(\frac{3-\sqrt{73}}{8}\right)^2 - 3\left(\frac{3-\sqrt{73}}{8}\right) - 4 \overset{?}{=} 0$

$\frac{32}{8} - 4 \overset{?}{=} 0$

$0 = 0$

29.

$(x-3)(x+1) = 1$

$x^2 - 2x - 3 = 1$

$x^2 - 2x + 1 = 5$

$(x-1)^2 = 5$

$x - 1 = \pm\sqrt{5}$

$x = 1 + \sqrt{5} \quad x = 1 - \sqrt{5}$

check :

$\left(1 + \sqrt{5} - 3\right)\left(1 + \sqrt{5} + 1\right) \overset{?}{=} 1$

$\left(-2 + \sqrt{5}\right)\left(2 + \sqrt{5}\right) \overset{?}{=} 1$

$-4 + 5 \overset{?}{=} 1$

$1 = 1$

$\left(1 - \sqrt{5} - 3\right)\left(1 - \sqrt{5} + 1\right) \overset{?}{=} 1$

$\left(-2 - \sqrt{5}\right)\left(2 - \sqrt{5}\right) \overset{?}{=} 1$

$-4 + 5 \overset{?}{=} 1$

$1 = 1$

Applications

31.

$d^2+12d+6=51$

$d^2+12d+36=51-6+36$

$d^2+12d+36=81$

$(d+6)^2=81$

$d+6=9$

$d=3$

There are 51 insects after 3 days.

33. a.

$-16t^2+32t=16$

$t^2-2t=-1$

$t^2-2t+1=-1+1$

$t^2-2t+1=0$

$(t-1)^2=0$

$t-1=0$

$t=1$

The fireworks will be 16 feet above the ground one second after being shot into the air.

b.

$-16t^2+32t=8$

$t^2-2t=-\frac{1}{2}$

$t^2-2t+1=-\frac{1}{2}+1$

$t^2-2t+1=\frac{1}{2}$

$(t-1)^2=\frac{1}{2}$

$t-1=\pm\frac{\sqrt{2}}{2}$

$t=1+\frac{\sqrt{2}}{2}\quad t=1-\frac{\sqrt{2}}{2}$

They will be 8 feet above the ground in approximately 0.3 seconds and again in approximately 1.7 seconds.

35.

$x^2+(x+2)^2=10^2$

$x^2+x^2+4x+4=100$

$2x^2+4x=96$

$x^2+2x=48$

$x^2+2x+1=48+1$

$(x+1)^2=49$

$x+1=\pm\sqrt{49}$

$x+1=7\quad x+1=-7$

$x=6\quad x=-8$

$x+2=8$

One friend was 6 miles from the party and the other was 8 miles from the party.

37.

$\frac{4}{x}+\frac{4}{x+3}=1$

$4x+12+4x=x(x+3)$

$8x+12=x^2+3x$

$x^2-5x-12=0$

$x^2-5x=12$

$x^2-5x+\frac{25}{4}=12+\frac{25}{4}$

$\left(x-\frac{5}{2}\right)^2=\frac{73}{4}$

$x-\frac{5}{2}=\pm\sqrt{\frac{73}{4}}$

$x=\frac{5}{2}+\frac{\sqrt{73}}{2}\quad x=\frac{5}{2}-\frac{\sqrt{73}}{2}$

$x\approx 7\quad x\approx -2$

It would take the faster machine about 7 minutes to do the job alone.

9.3 Solving Quadratic Equations by Using the Quadratic Formula

Practice 9.3

1.a.

$$n=\frac{1\pm\sqrt{(-1)^2-4(1)(0)}}{2(1)}$$

$$n=\frac{1\pm\sqrt{1}}{2}$$

$$n=\frac{1+1}{2}\quad n=\frac{1-1}{2}$$

$$n=\frac{2}{2}\quad n=\frac{0}{2}$$

$$n=1\quad n=0$$

check :

$$(1)^2-(1)\stackrel{?}{=}0\quad 0^2-0\stackrel{?}{=}0$$

$$0=0\quad 0=0$$

b.

$$x=\frac{-1\pm\sqrt{1^2-4(1)(1)}}{2(1)}$$

$$x=\frac{-1\pm\sqrt{1-4}}{2}$$

$$x=\frac{-1\pm\sqrt{-3}}{2}$$

No real solution.

2. a.

$$s=\frac{-(-8)\pm\sqrt{(-8)^2-4(5)(-3)}}{2(5)}$$

$$s=\frac{8\pm\sqrt{64+60}}{10}$$

$$s=\frac{8\pm\sqrt{124}}{10}$$

$$s=\frac{4+\sqrt{31}}{5}\quad s=\frac{4-\sqrt{31}}{5}$$

check :

$$5\left(\frac{4+\sqrt{31}}{5}\right)^2-8\left(\frac{4+\sqrt{31}}{5}\right)\stackrel{?}{=}3$$

$$\frac{15}{5}\stackrel{?}{=}3$$

$$3=3$$

$$5\left(\frac{4-\sqrt{31}}{5}\right)^2-8\left(\frac{4-\sqrt{31}}{5}\right)\stackrel{?}{=}3$$

$$\frac{15}{5}\stackrel{?}{=}3$$

$$3=3$$

b.

$$m=\frac{-3\pm\sqrt{(3)^2-4(3)(-7)}}{2(3)}$$

$$m=\frac{-3\pm\sqrt{9+84}}{6}$$

$$m=\frac{-3+\sqrt{93}}{6}\quad m=\frac{-3-\sqrt{93}}{6}$$

check :

$$3\left(\frac{-3+\sqrt{93}}{6}\right)^2\stackrel{?}{=}7-3\left(\frac{-3+\sqrt{93}}{6}\right)$$

$$\frac{17-\sqrt{93}}{2}\stackrel{?}{=}7-3\left(\frac{-3+\sqrt{93}}{6}\right)$$

$$\frac{17-\sqrt{93}}{2}=\frac{17-\sqrt{93}}{2}$$

$$3\left(\frac{-3-\sqrt{93}}{6}\right)^2\stackrel{?}{=}7-3\left(\frac{-3-\sqrt{93}}{6}\right)$$

$$\frac{17+\sqrt{93}}{2}\stackrel{?}{=}7-3\left(\frac{-3-\sqrt{93}}{6}\right)$$

$$\frac{17+\sqrt{93}}{2}=\frac{17+\sqrt{93}}{2}$$

3.

$$x=\frac{-(-6)\pm\sqrt{(-6)^2-4(3)(-4)}}{2(3)}$$

$$x=\frac{6\pm\sqrt{36+48}}{6}=\frac{6\pm\sqrt{84}}{6}$$

$$x=\frac{6+\sqrt{84}}{6}\approx 2.5$$

$$x=\frac{6-\sqrt{84}}{6}\approx -0.5$$

check :

$$3(2.5)^2\overset{?}{=}6(2.5)+4 \quad 3(-0.5)^2\overset{?}{=}6(-0.5)+4$$

$$18.75\approx 19 \qquad 0.75\approx 1$$

4.

$$\frac{3}{x}+\frac{3}{x+5}=1$$

$$3(x+5)+3x=1(x)(x+5)$$

$$3x+15+3x=x^2+5x$$

$$x^2-x-15=0$$

$$x=\frac{1\pm\sqrt{(-1)^2-4(1)(-15)}}{2(1)}$$

$$x=\frac{1+\sqrt{61}}{2}\approx 4.4$$

$$x+5\approx 9.4$$

Working alone, it takes one pump about 4.4 hours and the other about 9.4 hours to empty the tank.

Exercises 9.3

	Standard form	a =	b =	c=
1.	$2x^2+9x-1=0$	2	9	-1
3.	$-x^2+3x-8=0$	-1	3	-8
5.	$x^2-x+8=0$	1	-1	8
7.	$\frac{1}{3}y^2-\frac{1}{2}y+\frac{1}{4}=0$	$\frac{1}{3}$	$-\frac{1}{2}$	$\frac{1}{4}$
9.	$2x^2-7x-5=0$	2	-7	-5

11.

$$t=\frac{-2\pm\sqrt{2^2-4(1)(-3)}}{2(1)}$$

$$t=\frac{-2\pm\sqrt{4+12}}{2}$$

$$t=\frac{-2\pm\sqrt{16}}{2}$$

$$t=\frac{-2+4}{2} \quad t=\frac{-2-4}{2}$$

$$t=\frac{2}{2} \quad t=\frac{-6}{2}$$

$$t=1 \quad t=-3$$

check :

$$(1)^2+2(1)-3\overset{?}{=}0 \quad (-3)^2+2(-3)-3\overset{?}{=}0$$

$$1+2-3\overset{?}{=}0 \qquad 9-6-3\overset{?}{=}0$$

$$0=0 \qquad 0=0$$

13.

$$x=\frac{3\pm\sqrt{3^2\quad 4(1)(6)}}{2(1)}$$

$$x=\frac{-3\pm\sqrt{9-24}}{2}$$

$$x=\frac{-3\pm\sqrt{-15}}{2}$$

No real solution.

15.

$$x=\frac{-(-6)\pm\sqrt{(-6)^2-4(1)(-7)}}{2(1)}$$

$$x=\frac{6\pm\sqrt{36+28}}{2}$$

$$x=\frac{6\pm\sqrt{64}}{2}$$

$$x=\frac{6+8}{2} \quad x=\frac{6-8}{2}=$$

$$x=7 \qquad x=-1$$

check :

$$7+6(7)-7^2\overset{?}{=}0$$

$$7+42-49\overset{?}{=}0$$

$$0=0$$

$$7+6(-1)-(-1)^2\overset{?}{=}0$$

$$7-6-1\overset{?}{=}0$$

$$0=0$$

17.

$$y=\frac{-(-4)\pm\sqrt{(-4)^2-4(1)(-1)}}{2(1)}$$

$$y=\frac{4\pm\sqrt{20}}{2}$$

$$y=\frac{4\pm2\sqrt{5}}{2}$$

$$y=2+\sqrt{5} \quad y=2-\sqrt{5}$$

check :

$$\left(2+\sqrt{5}\right)^2-4\left(2+\sqrt{5}\right)\overset{?}{=}1 \quad \left(2-\sqrt{5}\right)^2-4\left(2-\sqrt{5}\right)\overset{?}{=}1$$

$$9+4\sqrt{5}-8-4\sqrt{5}\overset{?}{=}1 \qquad 9-4\sqrt{5}-8+4\sqrt{5}\overset{?}{=}1$$

$$1=1 \qquad 1=1$$

19.

$$3p^2-4p+1=0$$

$$p=\frac{4\pm\sqrt{(-4)^2-4(3)(1)}}{2(3)}$$

$$p=\frac{4\pm\sqrt{4}}{6}$$

$$p=\frac{4\pm2}{6}$$

$$p=\frac{6}{6} \qquad p=\frac{2}{6}$$

$$p=1 \qquad p=\frac{1}{3}$$

check :

$$3(1)^2-4(1)+1\overset{?}{=}0 \quad 3\left(\frac{1}{3}\right)^2-4\left(\frac{1}{3}\right)+1\overset{?}{=}0$$

$$3-4+1\overset{?}{=}0 \qquad \frac{1}{3}-\frac{4}{3}+1\overset{?}{=}0$$

$$0=0 \qquad 0=0$$

21.

$$x=\frac{-(-6)\pm\sqrt{(-6)^2-4(4)(1)}}{2(4)}$$

$$x=\frac{6\pm\sqrt{20}}{8}$$

$$x=\frac{6\pm2\sqrt{5}}{8}$$

$$x=\frac{3+\sqrt{5}}{4} \quad x=\frac{3-\sqrt{5}}{4}$$

check :

$$4\left(\frac{3+\sqrt{5}}{4}\right)^2-6\left(\frac{3+\sqrt{5}}{4}\right)\overset{?}{=}-1$$

$$-1=-1$$

$$4\left(\frac{3-\sqrt{5}}{4}\right)^2-6\left(\frac{3-\sqrt{5}}{4}\right)\overset{?}{=}-1$$

$$-1=-1$$

23.

$$n=\frac{-(-7)\pm\sqrt{(-7)^2-4(2)(-30)}}{2(2)}$$

$$n=\frac{7\pm\sqrt{49+240}}{4}$$

$$n=\frac{7\pm\sqrt{289}}{4}$$

$$n=\frac{7+17}{4} \quad n=\frac{7-17}{4}$$

$$n=\frac{24}{4}=6 \quad n=\frac{-10}{4}=-\frac{5}{2}$$

check :

$$2(6)^2\overset{?}{=}7(6)+30 \quad 2\left(-\frac{5}{2}\right)^2\overset{?}{=}7\left(-\frac{5}{2}\right)+30$$

$$72=72 \quad \frac{25}{2}=\frac{25}{2}$$

25.

$$x=\frac{-3\pm\sqrt{3^2-4(1)(2)}}{2(1)}$$

$$x=\frac{-3\pm\sqrt{9-8}}{2}$$

$$x=\frac{-3\pm\sqrt{1}}{2}$$

$$x=\frac{-3+1}{2} \quad x=\frac{-3-1}{2}$$

$$x=-1 \quad x=-2$$

check :

$$(-1)^2+3(-1)+2\overset{?}{=}0 \quad (-2)^2+3(-2)+2\overset{?}{=}0$$

$$1-3+2\overset{?}{=}0 \quad 4-6+2\overset{?}{=}0$$

$$0=0 \quad 0=0$$

27.

$$x=\frac{-4\pm\sqrt{4^2-4(1)(3)}}{2(1)}$$

$$x=\frac{-4\pm\sqrt{16-12}}{2}$$

$$x=\frac{-4\pm\sqrt{4}}{2}$$

$$x=\frac{-4+2}{2} \quad x=\frac{-4-2}{2}$$

$$x=-1 \quad x=-3$$

check :

$$3+(-1)^2+4(-1)\overset{?}{=}0 \quad 3+(-3)^2+4(-3)\overset{?}{=}0$$

$$3+1-4\overset{?}{=}0 \quad 3+9-12\overset{?}{=}0$$

$$0=0 \quad 0=0$$

29.

$$p=\frac{-(-2)\pm\sqrt{(-2)^2-4(3)(0)}}{2(3)}$$

$$p=\frac{2\pm\sqrt{4-0}}{6}$$

$$p=\frac{2\pm\sqrt{4}}{6}$$

$$p=\frac{2+2}{6} \quad p=\frac{2-2}{6}$$

$$p=\frac{2}{3} \quad p=0$$

check :

$$3\left(\frac{2}{3}\right)^2\overset{?}{=}2\left(\frac{2}{3}\right) \quad 3(0)^2\overset{?}{=}2(0)$$

$$\frac{4}{3}\overset{?}{=}\frac{4}{3} \quad 0\overset{?}{=}0$$

31.

$$n=\frac{-3\pm\sqrt{3^2-4(4)(-1)}}{2(4)}$$

$$n=\frac{-3\pm\sqrt{9+16}}{8}$$

$$n=\frac{-3\pm\sqrt{25}}{8}$$

$$n=\frac{-3+5}{8} \quad n=\frac{-3-5}{8}$$

$$n=\frac{1}{4} \qquad n=-1$$

check :

$$4\left(\frac{1}{4}\right)^2+3\left(\frac{1}{4}\right)-1\overset{?}{=}0 \quad 4(-1)^2+3(-1)-1\overset{?}{=}0$$

$$\frac{1}{4}+\frac{3}{4}-1\overset{?}{=}0 \qquad 4-3-1\overset{?}{=}0$$

$$0=0 \qquad 0=0$$

33.

$$t=\frac{-(-7)\pm\sqrt{(-7)^2-4(1)(5)}}{2(1)}$$

$$t=\frac{7\pm\sqrt{49-20}}{2}$$

$$t=\frac{7+\sqrt{29}}{2} \quad t=\frac{7-\sqrt{29}}{2}$$

check :

$$\left(\frac{7+\sqrt{29}}{2}\right)^2+5\overset{?}{=}7\left(\frac{7+\sqrt{29}}{2}\right)$$

$$\frac{39+7\sqrt{29}}{2}+5\overset{?}{=}\frac{49+7\sqrt{29}}{2}$$

$$\frac{49+7\sqrt{29}}{2}=\frac{49+7\sqrt{29}}{2}$$

$$\left(\frac{7-\sqrt{29}}{2}\right)^2+5\overset{?}{=}7\left(\frac{7-\sqrt{29}}{2}\right)$$

$$\frac{39-7\sqrt{29}}{2}+5\overset{?}{=}\frac{49-7\sqrt{29}}{2}$$

$$\frac{49-7\sqrt{29}}{2}=\frac{49-7\sqrt{29}}{2}$$

35.

$$(n+7)(n-8)=5$$

$$n^2-n-56=5$$

$$n^2-n-61=0$$

$$n=\frac{1\pm\sqrt{(-1)^2-4(1)(-61)}}{2(1)}$$

$$n=\frac{1\pm\sqrt{1+244}}{2}$$

$$n=\frac{1\pm\sqrt{245}}{2}$$

$$n=\frac{1+7\sqrt{5}}{2} \quad n=\frac{1-7\sqrt{5}}{2}$$

check :

$$\left(\frac{1+7\sqrt{5}}{2}+7\right)\left(\frac{1+7\sqrt{5}}{2}-8\right)\overset{?}{=}5$$

$$\left(\frac{15+7\sqrt{5}}{2}\right)\left(\frac{-15+7\sqrt{5}}{2}\right)\overset{?}{=}5$$

$$\frac{20}{4}\overset{?}{=}5$$

$$5=5$$

$$\left(\frac{1-7\sqrt{5}}{2}+7\right)\left(\frac{1-7\sqrt{5}}{2}-8\right)\overset{?}{=}5$$

$$\left(\frac{1-7\sqrt{5}}{2}+7\right)\left(\frac{1-7\sqrt{5}}{2}-8\right)\overset{?}{=}5$$

$$\left(\frac{15-7\sqrt{5}}{2}+7\right)\left(\frac{-15-7\sqrt{5}}{2}-8\right)\overset{?}{=}5$$

$$\frac{20}{4}\overset{?}{=}12$$

$$5=5$$

37.

$5(x-1)=x(x+2)$

$5x-5=x^2+2x$

$x^2-3x+5=0$

$x=\dfrac{3\pm\sqrt{(-3)^2-4(1)(5)}}{2(1)}$

$x=\dfrac{3\pm\sqrt{9-20}}{2}$

$x=\dfrac{3\pm\sqrt{-11}}{2}$

No real solution.

39.

$\dfrac{x^2}{3}-x=-\dfrac{1}{2}$

$2x^2-6x=-3$

$2x^2-6x+3=0$

$x=\dfrac{6\pm\sqrt{(-6)^2-4(2)(3)}}{2(2)}$

$x=\dfrac{6\pm\sqrt{36-24}}{4}$

$x=\dfrac{6+\sqrt{12}}{4}=\dfrac{6+2\sqrt{3}}{4}$

$x=\dfrac{3+\sqrt{3}}{2} \quad x=\dfrac{3-\sqrt{3}}{2}$

check .

$\dfrac{\left(\dfrac{3+\sqrt{3}}{2}\right)^2}{3}-\left(\dfrac{3+\sqrt{3}}{2}\right)\overset{?}{=}-\dfrac{1}{2} \qquad \dfrac{\left(\dfrac{3-\sqrt{3}}{2}\right)^2}{3}-\left(\dfrac{3-\sqrt{3}}{2}\right)=-\dfrac{1}{2}$

$\dfrac{2+\sqrt{3}}{2}-\dfrac{3+\sqrt{3}}{2}\overset{?}{=}-\dfrac{1}{2} \qquad \dfrac{2-\sqrt{3}}{2}-\dfrac{3-\sqrt{3}}{2}\overset{?}{=}-\dfrac{1}{2}$

$-\dfrac{1}{2}=-\dfrac{1}{2} \qquad -\dfrac{1}{2}=-\dfrac{1}{2}$

Applications

41.

$91=\dfrac{n(n+1)}{2}$

$182=n^2+n$

$n^2+n-182=0$

$n=\dfrac{-1\pm\sqrt{1^2-4(1)(-182)}}{2(1)}$

$n=\dfrac{-1\pm\sqrt{1+728}}{2(1)}$

$n=\dfrac{-1\pm\sqrt{729}}{2}$

$n=\dfrac{-1+27}{2} \quad n=\dfrac{-1-27}{2}$

$n=13 \qquad n=-14$

n is 13.

43.

$\dfrac{40}{x}+\dfrac{40}{x+20}=1$

$40x+800+40x=x^2+20x$

$x^2-60x-800=0$

$x=\dfrac{60\pm\sqrt{(-60)^2-4(1)(-800)}}{2(1)}$

$x=\dfrac{60\pm\sqrt{6800}}{2}$

$x\approx 71 \qquad x\approx -11$

$x+20\approx 91$

Working alone it will take the lab coordinator about 91 minutes to set up the lab and it takes the lab technician about 71 minutes to set up the lab.

45.

$$\frac{40}{x+20}+4=\frac{40}{x}$$

$$40x+4x^2+80x=40x+800$$

$$4x^2+80x-800=0$$

$$x^2+20x-200=0$$

$$x=\frac{-20\pm\sqrt{20^2-4(1)(-200)}}{2(1)}$$

$$x=\frac{-20\pm\sqrt{1200}}{2}$$

$$x=\frac{-20\pm20\sqrt{3}}{2}=-10\pm10\sqrt{3}$$

$$x\approx7$$

$$x+20\approx27$$

The speed of the car was approximately 27 miles per hour.

47.

$$1000=11.8n^2+83.2n+432.9$$

$$11.8n^2+83.2n-567.1=0$$

$$n=\frac{-83.2\pm\sqrt{(83.2)^2-4(11.8)(-567.1)}}{2(11.8)}$$

$$n=\frac{-83.2+\sqrt{33689.36}}{23.6}$$

$$n\approx4$$

The value of the index was approximately 1000 in 1998.

9.4 Graphing Quadratic Equations in Two Variables

Practice 9.4

1.

x	$y=-2x^2$	(x, y)
-2	$y=-2(-2)^2=-2(4)=-8$	$(-2,-8)$
-1	$y=-2(-1)^2=-2(1)=-2$	$(-1,-2)$
0	$y=-2(0)^2=-2(0)=0$	$(0,0)$
1	$y=-2(1)^2=-2(1)=-2$	$(1,-2)$
2	$y=-2(2)^2=-2(4)=-8$	$(2,-8)$

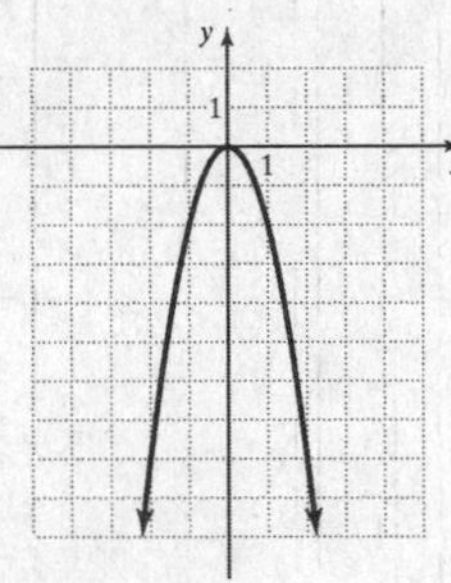

2.

x	$y=-3x^2+1$	(x, y)
-2	$y=-3(-2)^2+1=-3(4)+1=-11$	$(-2,-11)$
-1	$y=-3(-1)^2+1=-3(1)+1=-2$	$(-1,-2)$
0	$y=-3(0)^2+1=-3(0)+1=1$	$(0,1)$
1	$y=-3(1)^2+1=-3(1)+1=-2$	$(1,-2)$
2	$y=-3(2)^2+1=-3(4)+1=-11$	$(2,-11)$

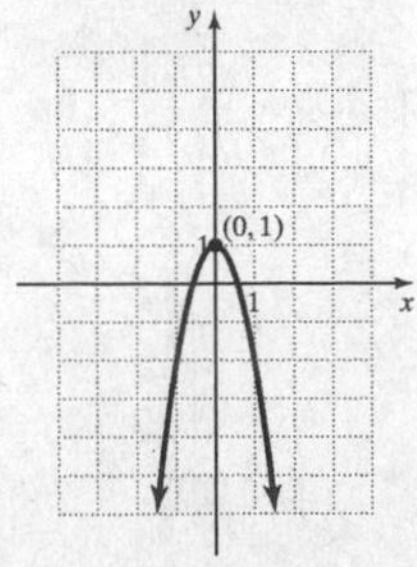

3.

a. vertex: $x = \frac{-b}{2a} = \frac{-2}{2(-1)} = \frac{-2}{-2} = 1$

$y = -1^2 + 2(1) + 3 = 4 \qquad (1,4)$

axis of symmetry: x = 1
x-intercepts:

$-x^2 + 2x + 3 = 0$

$x^2 - 2x - 3 = 0$

$(x-3)(x+1) = 0$

$x - 3 = 0 \quad x + 1 = 0$

$x = 3 \qquad x = -1$

(1,0) (3,0)
y-intercept:

$-(0)^2 + 2(0) + 3 = y$

$y = 3 \qquad (0,3)$

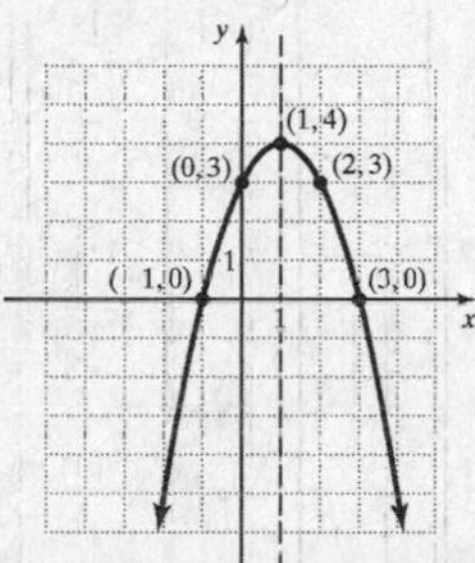

b. The graph of the parabola opens downward since a = -1 is negative. The curve turns at the point (1,4), the highest point of the graph. The equation of the axis of symmetry is x = 1.

4.

vertex: $x = \frac{-b}{2a} = \frac{-(-10)}{2(1)} = \frac{10}{2} = 5$

$y = 5^2 - 10(5) = -25 \qquad (5,-25)$

axis of symmetry: x = 5
x-intercepts:

$x^2 - 10x = 0$

$x(x-10) = 0$

$x - 10 = 0 \quad x = 0$

$x = 10$

(0,0) (10,0)

y-intercept:

$(0)^2 - 10(0) = y$

$y = 0 \qquad (0,0)$

Selling 5 items results in the most money lost ($25).

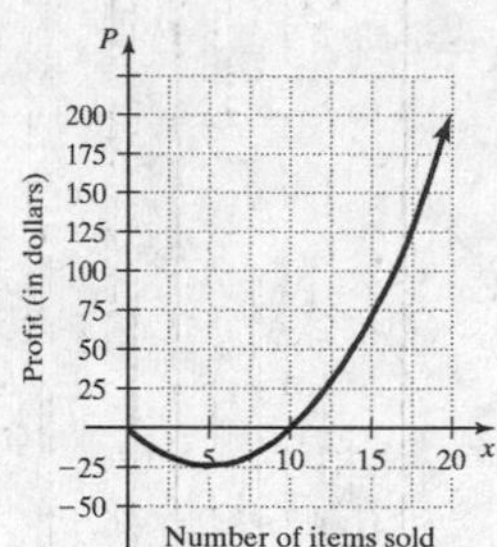

Exercises 9.4

1.

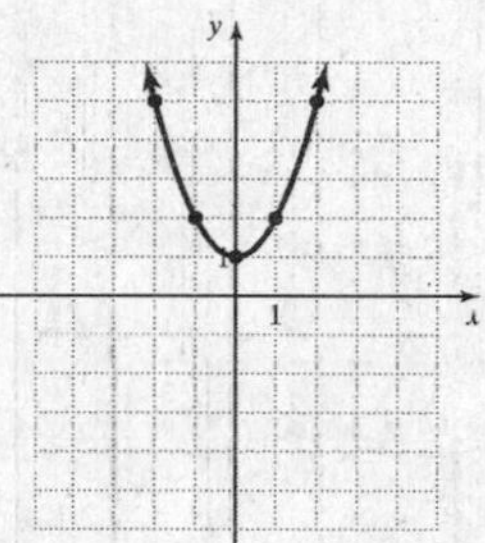

3.

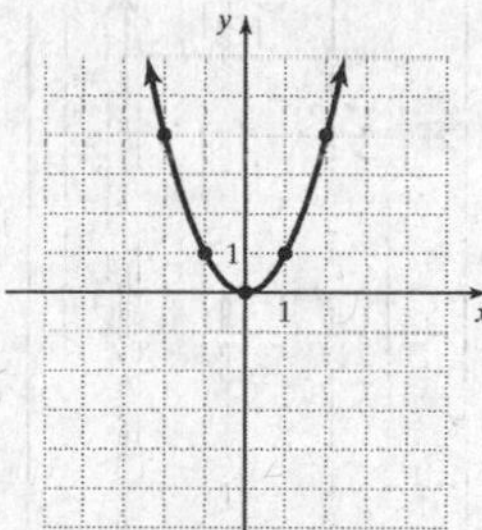

5.

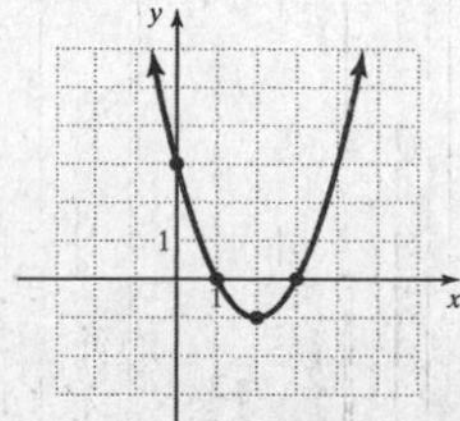

7.

$$y=\frac{1}{2}x^2$$

$$x=\frac{-0}{2\left(\frac{1}{2}\right)}=0 \quad y=\frac{1}{2}\left(0^2\right)=0$$

axis of symmetry: x= 0; vertex: (0,0)

9.

$$y=5-4x+x^2$$

$$x=\frac{-(-4)}{2(1)}=2 \quad y=5+-4(2)+(2)^2=1$$

axis of symmetry: x= 2; vertex: (2,1)

11.

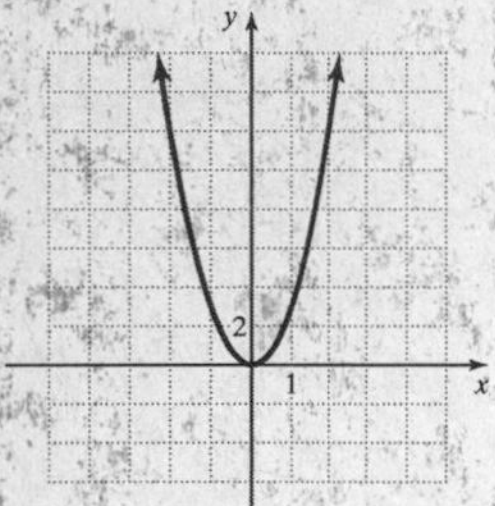

13.

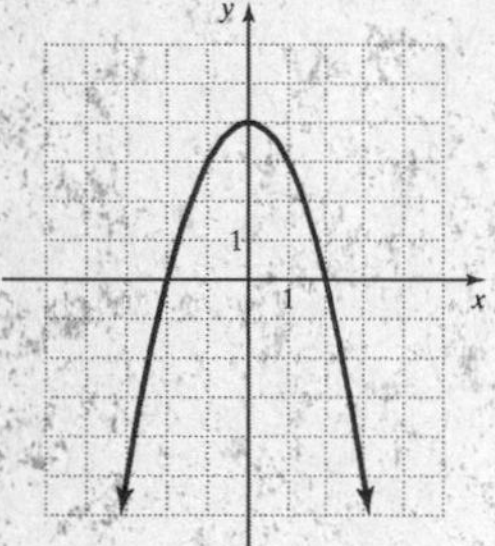

15.

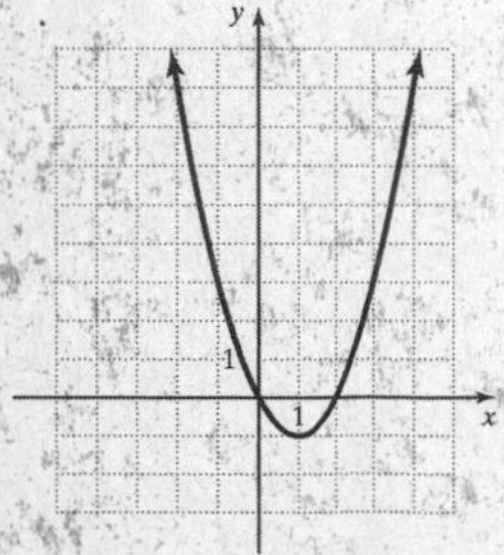

17.

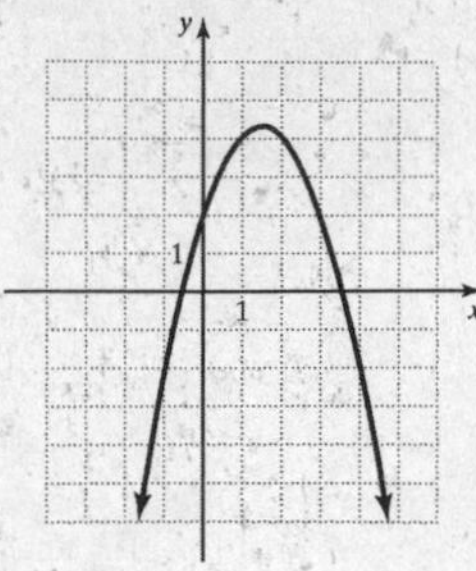

19.

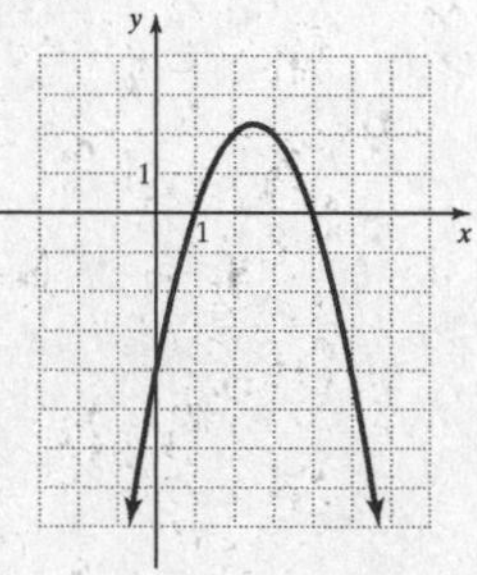

21.

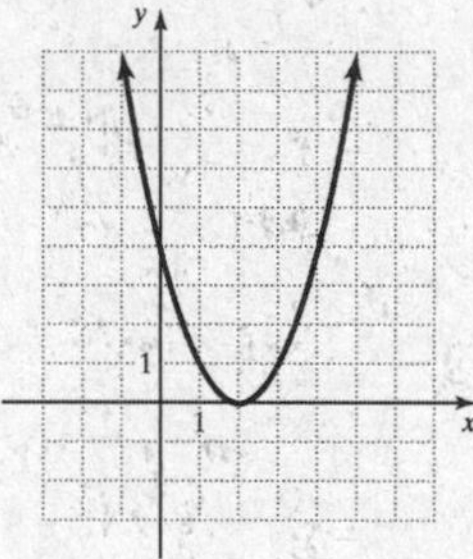

23.

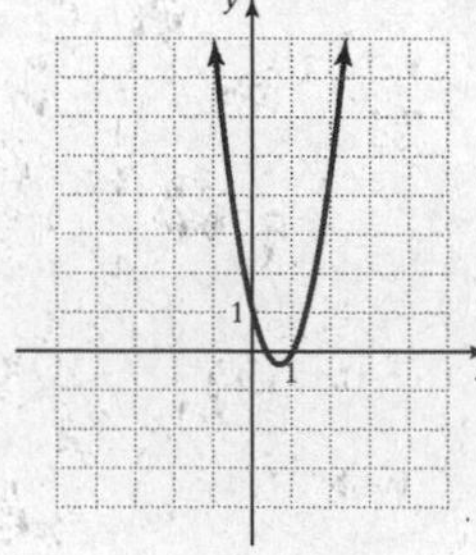

25.

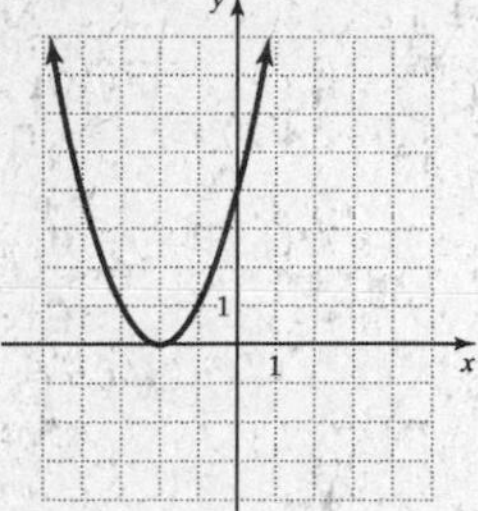

27.

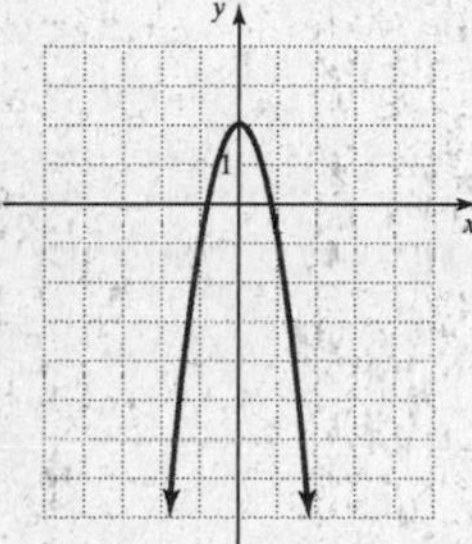

29.

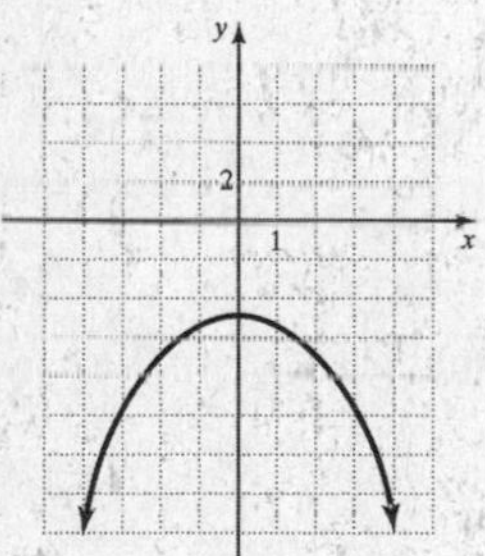

Applications

31. a.

$P = 2x + 2w$

$2x + 2w = 80$

$2w = 80 - 2x$

$w = 40 - x$

area = length x width

area = $x(40 - x)$

The perimeter of the rectangle is 80 and the length is x. The width is found by solving $2x + 2w = 80$. So the width is 40 – x. The area then is length times width, which is x(40-x).

b.

c.

$y = -x^2 + 40x$

$x = \frac{-b}{2a} = \frac{-40}{2(-1)} = 20$

$y = -(20)^2 + 40(20) = 400$

The are
a is a maximum(400 square feet) when the length is 20 feet.

33.

a.

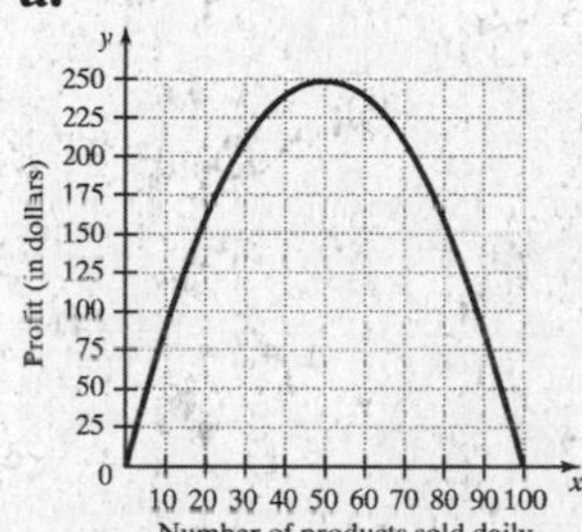

b. The company must sell 50 products to maximize the profit.

c. The height of the graph increases when 0 to 50 products are sold, but then decreases for sales of more than 50 products.

35.a.

$$R = x\left(100 - \frac{1}{4}x\right) = 100x - \frac{1}{4}x^2$$

b.

x	0	100	200	300	400
R	0	7500	10,000	7500	0

c.

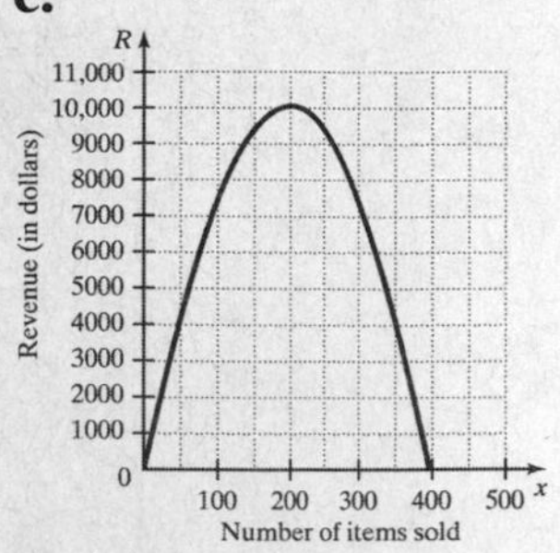

d.

$$x = \frac{-b}{2a} = \frac{-100}{2\left(-\frac{1}{4}\right)} = 200$$

The store should sell 200 items to maximize the revenue .

e.

$$R = 100(200) - \frac{1}{4}(200)^2$$

$$R = \$10,000$$

The maximum revenue is $10,000.

Chapter 9 Review Exercises

1.

$$y^2 = 24$$

$$\sqrt{y^2} = \pm\sqrt{24}$$

$$y = \pm 2\sqrt{6}$$

check :

$$\left(2\sqrt{6}\right)^2 \stackrel{?}{=} 24 \quad \left(-2\sqrt{6}\right)^2 \stackrel{?}{=} 24$$

$$24 = 24 \qquad 24 = 24$$

2.

$$4x^2 = 12$$

$$x^2 = 3$$

$$\sqrt{x^2} = \pm\sqrt{3}$$

$$x = \pm\sqrt{3}$$

check :

$$4\left(\sqrt{3}\right)^2 \stackrel{?}{=} 12 \quad 4\left(-\sqrt{3}\right)^2 \stackrel{?}{=} 12$$

$$12 = 12 \qquad 12 = 12$$

3.

$$x^2 + 5 = 0$$

$$x^2 = -5$$

$$\sqrt{x^2} = \pm\sqrt{-5}$$

No real solution.

4.

$$(2n-5)^2 = 18$$

$$2n - 5 = \pm\sqrt{18}$$

$$2n = 5 \pm 3\sqrt{2}$$

$$n = \frac{5 \pm 3\sqrt{2}}{2}$$

check :

$$\left(2\left(\frac{5+3\sqrt{2}}{2}\right) - 5\right)^2 \stackrel{?}{=} 18$$

$$\left(3\sqrt{2}\right)^2 \stackrel{?}{=} 18$$

$$18 = 18$$

$$\left(2\left(\frac{5-3\sqrt{2}}{2}\right) - 5\right)^2 \stackrel{?}{=} 18$$

$$\left(-3\sqrt{2}\right)^2 \stackrel{?}{=} 18$$

$$18 = 18$$

5.

$$A = \pi r^2$$

$$\frac{A}{\pi} = r^2$$

$$r = \sqrt{\frac{A}{\pi}}$$

$$r = \frac{\sqrt{\pi A}}{\pi}$$

6.

$$E = mc^2$$

$$\frac{E}{m} = c^2$$

$$c = \sqrt{\frac{E}{m}}$$

$$c = \frac{\sqrt{mE}}{m}$$

7. $\left(\frac{1}{2}\cdot(-10)\right)^2 = 25$

8. $\left(\frac{1}{2}\cdot 7\right)^2 = \left(\frac{7}{2}\right)^2 = \frac{49}{4}$

9.

$$n^2 - 6n = 27$$

$$n^2 - 6n + 9 = 27 + 9$$

$$(n-3)^2 = 36$$

$$n - 3 = \pm\sqrt{36}$$

$$n - 3 = 6 \quad n - 3 = -6$$

$$n = 9 \qquad n = -3$$

check :

$$9^2 - 6(9) \stackrel{?}{=} 27 \quad (-3)^2 - 6(-3) \stackrel{?}{=} 27$$

$$81 - 54 \stackrel{?}{=} 27 \qquad 9 + 18 \stackrel{?}{=} 27$$

$$27 = 27 \qquad 27 = 27$$

10.

$$y^2 + 3y = 4$$

$$y^2 + 3y + \frac{9}{4} = 4 + \frac{9}{4}$$

$$\left(y + \frac{3}{2}\right)^2 = \frac{25}{4}$$

$$y + \frac{3}{2} = \pm\sqrt{\frac{25}{4}}$$

$$y + \frac{3}{2} = \frac{5}{2} \quad y + \frac{3}{2} = -\frac{5}{2}$$

$$y = 1 \qquad y = -4$$

check :

$$1^2 + 3(1) \stackrel{?}{=} 4 \quad (-4)^2 + 3(-4) \stackrel{?}{=} 4$$

$$1 + 3 \stackrel{?}{=} 4 \qquad 16 - 12 \stackrel{?}{=} 4$$

$$4 = 4 \qquad 4 = 4$$

11.

$$2x^2 + 8x = 4$$

$$x^2 + 4x = 2$$

$$x^2 + 4x + 4 = 2 + 4$$

$$(x+2)^2 = 6$$

$$x + 2 = \pm\sqrt{6}$$

$$x = -2 + \sqrt{6} \quad x = -2 - \sqrt{6}$$

check :

$$2\left(-2+\sqrt{6}\right)^2 + 8\left(-2+\sqrt{6}\right) \stackrel{?}{=} 4$$

$$20 - 8\sqrt{6} - 16 + 8\sqrt{6} \stackrel{?}{=} 4$$

$$4 = 4$$

$$2\left(-2-\sqrt{6}\right)^2 + 8\left(-2-\sqrt{6}\right) \stackrel{?}{=} 4$$

$$20 + 8\sqrt{6} - 16 - 8\sqrt{6} \stackrel{?}{=} 4$$

$$4 = 4$$

12.

$$4y^2 = 4y+1$$

$$y^2 - y = \frac{1}{4}$$

$$y^2 - y + \frac{1}{4} = \frac{1}{4} + \frac{1}{4}$$

$$\left(y - \frac{1}{2}\right)^2 = \frac{1}{2}$$

$$y - \frac{1}{2} = \pm\sqrt{\frac{1}{2}}$$

$$y - \frac{1}{2} = \pm\frac{\sqrt{2}}{2}$$

$$y = \frac{1+\sqrt{2}}{2} \quad y = \frac{1-\sqrt{2}}{2}$$

$$4\left(\frac{1+\sqrt{2}}{2}\right)^2 \stackrel{?}{=} 4\left(\frac{1+\sqrt{2}}{2}\right)+1$$

$$3+2\sqrt{2} = 3+2\sqrt{2}$$

$$4\left(\frac{1-\sqrt{2}}{2}\right)^2 \stackrel{?}{=} 4\left(\frac{1-\sqrt{2}}{2}\right)+1$$

$$3-2\sqrt{2} = 3-2\sqrt{2}$$

13.

$$y = \frac{-(-2) \pm \sqrt{(-2)^2 - 4(1)(-1)}}{2(1)}$$

$$y = \frac{2 \pm \sqrt{8}}{2} = \frac{2 \pm 2\sqrt{2}}{2}$$

$$y = 1+\sqrt{2} \quad y = 1-\sqrt{2}$$

check :

$$\left(1+\sqrt{2}\right)^2 - 2\left(1+\sqrt{2}\right) - 1 \stackrel{?}{=} 0$$

$$3+2\sqrt{2}-2-2\sqrt{2}-1 \stackrel{?}{=} 0$$

$$0=0$$

$$\left(1-\sqrt{2}\right)^2 - 2\left(1-\sqrt{2}\right) - 1 \stackrel{?}{=} 0$$

$$3-2\sqrt{2}-2+2\sqrt{2}-1 \stackrel{?}{=} 0$$

$$0=0$$

14.

$$x = \frac{-(-8) \pm \sqrt{(-8)^2 - 4(-1)(-1)}}{2(-1)}$$

$$x = \frac{8 \pm \sqrt{60}}{-2} = \frac{8 \pm 2\sqrt{15}}{-2}$$

$$x = -4+\sqrt{15} \quad x = -4-\sqrt{15}$$

check :

$$-\left(-4+\sqrt{15}\right)^2 \stackrel{?}{=} 8\left(-4+\sqrt{15}\right)+1$$

$$-31+8\sqrt{15} \stackrel{?}{=} -32+8\sqrt{15}+1$$

$$-31=-31$$

$$-\left(-4-\sqrt{15}\right)^2 \stackrel{?}{=} 8\left(-4-\sqrt{15}\right)+1$$

$$-31+8\sqrt{15} \stackrel{?}{=} -32-8\sqrt{15}+1$$

$$-31=-31$$

15.

$$n = \frac{-1 \pm \sqrt{1^2 - 4(2)(-5)}}{2(2)}$$

$$n = \frac{-1 \pm \sqrt{41}}{4}$$

$$n = \frac{-1+\sqrt{41}}{4} \quad n = \frac{-1-\sqrt{41}}{4}$$

check :

$$2\left(\frac{-1+\sqrt{41}}{4}\right)^2 + \frac{-1+\sqrt{41}}{4} \stackrel{?}{=} 5$$

$$\frac{20}{4} \stackrel{?}{=} 5$$

$$5=5$$

$$2\left(\frac{-1-\sqrt{41}}{4}\right)^2 + \frac{-1-\sqrt{41}}{4} \stackrel{?}{=} 5$$

$$\frac{20}{4} \stackrel{?}{=} 5$$

$$5=5$$

16.

$$(x+3)(x-2)=-10$$

$$x^2+x+4=0$$

$$x=\frac{-1\pm\sqrt{(1)^2-4(1)(4)}}{2(1)}$$

$$x=\frac{-1\pm\sqrt{1-16}}{2}$$

$$x=\frac{-1\pm\sqrt{-15}}{2}$$

No real solution.

17.

$$y=\frac{-5\pm\sqrt{5^2-4(1)(0)}}{2(1)}$$

$$y=\frac{-5\pm\sqrt{25}}{2}=\frac{-5\pm 5}{2}$$

$$y=0 \quad y=-5$$

check :

$$(0)^2\overset{?}{=}-5(0) \quad (-5)^2\overset{?}{=}-5(-5)$$

$$0=0 \qquad 25=25$$

18.

$$\frac{2}{x}+\frac{1}{x+3}-1$$

$$2(x+3)+x=x(x+3)$$

$$2x+6+x=x^2+3x$$

$$x^2-6=0$$

$$x=\frac{0\pm\sqrt{0^2-4(1)(-6)}}{2(1)}$$

$$x=\frac{\pm\sqrt{24}}{2}=\pm\sqrt{6}$$

check :

$$\frac{2}{\sqrt{6}}+\frac{1}{\sqrt{6}+3}\overset{?}{=}1 \qquad \frac{2}{-\sqrt{6}}+\frac{1}{-\sqrt{6}+3}\overset{?}{=}1$$

$$\frac{6+3\sqrt{6}}{6+3\sqrt{6}}\overset{?}{=}1 \qquad \frac{6-3\sqrt{6}}{6-3\sqrt{6}}\overset{?}{=}1$$

$$1=1 \qquad 1=1$$

$$1=1$$

19.

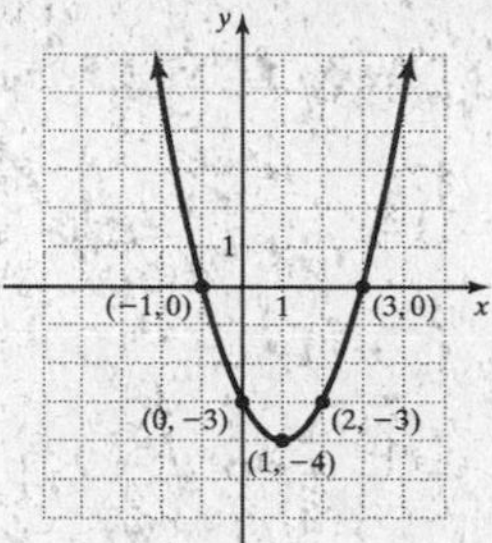

20.

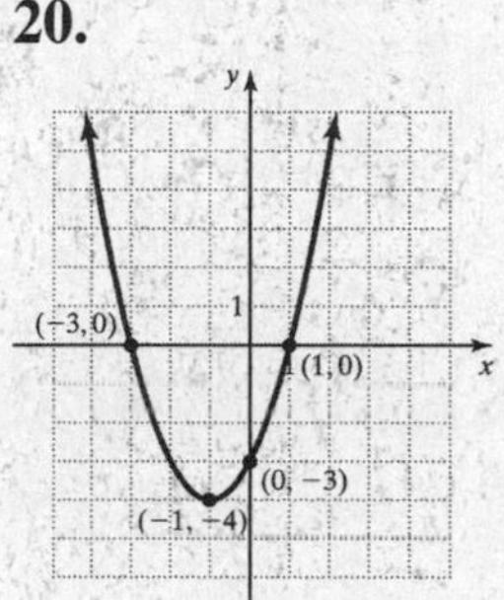

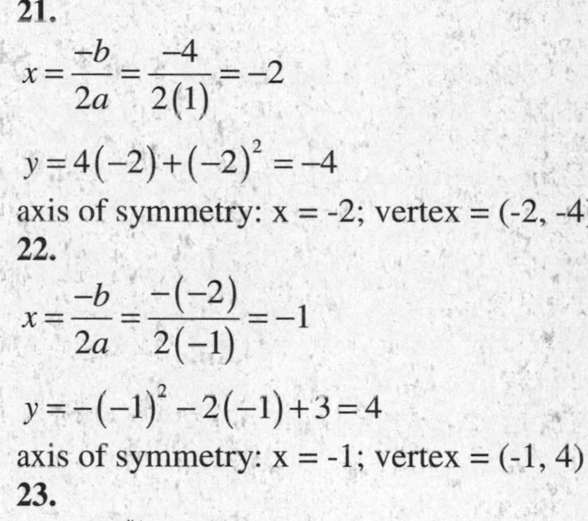

21.

$$x=\frac{-b}{2a}=\frac{-4}{2(1)}=-2$$

$$y=4(-2)+(-2)^2=-4$$

axis of symmetry: x = -2; vertex = (-2, -4)

22.

$$x=\frac{-b}{2a}=\frac{-(-2)}{2(-1)}=-1$$

$$y=-(-1)^2-2(-1)+3=4$$

axis of symmetry: x = -1; vertex = (-1, 4)

23.

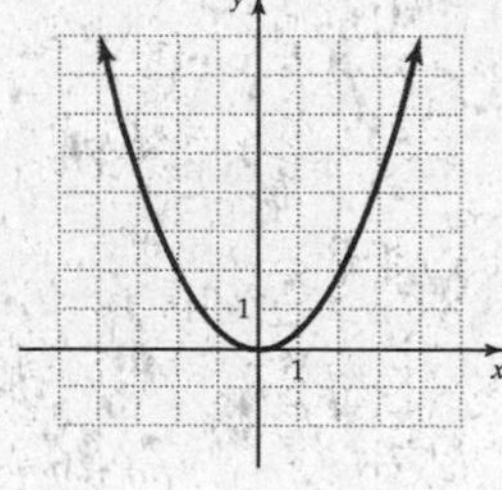

24.

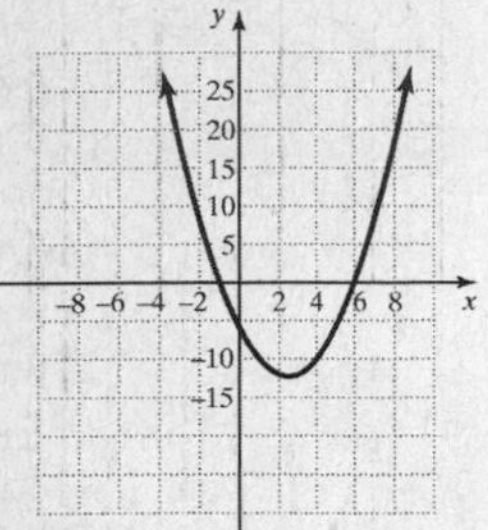

25.

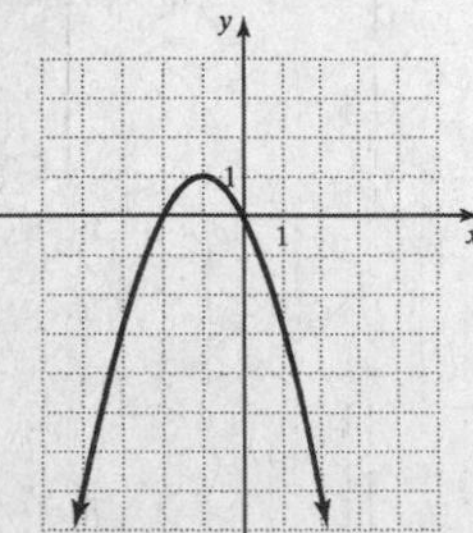

26.

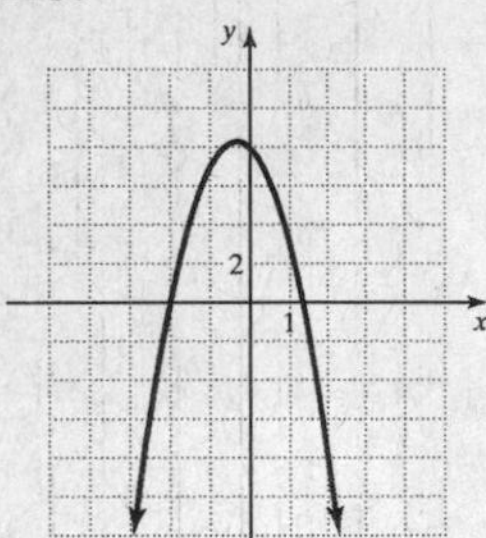

27.

$4(3.14)r^2 = 615$

$12.56r^2 = 615$

$r^2 = \dfrac{615}{12.56}$

$r = \sqrt{\dfrac{615}{12.56}}$

$r \approx 7.0$

The radius of the sphere is approximately 7.0 inches.

28.

$2000 = 90x - x^2$

$x^2 - 90x + 2000 = 0$

$(x-40)(x-50) = 0$

$x - 40 = 0 \quad x - 50$

$x = 40 \qquad x = 50$

Since the profit is \$2000 for both, the company had sold either 40 rugs or 50 rugs.

29.

$w(2w+5) = 20$

$2w^2 + 5w - 20 = 0$

$w = \dfrac{-5 \pm \sqrt{5^2 - 4(2)(-20)}}{2(2)}$

$w = \dfrac{-5 \pm \sqrt{25+160}}{4} = \dfrac{-5 \pm \sqrt{185}}{4}$

$w \approx 2.2 \qquad w \neq -4.7$

$2w + 5 \approx 9.3$

The banner is approximately 2.2 feet by 9.4 feet.

30.

$2(14) = n(n-3)$

$n^2 - 3n - 28 = 0$

$(n-7)(n+4) = 0$

$n - 7 = 0 \quad n + 4 = 0$

$n = 7 \qquad n \neq -4$

The building has 7 sides.

31.

a.

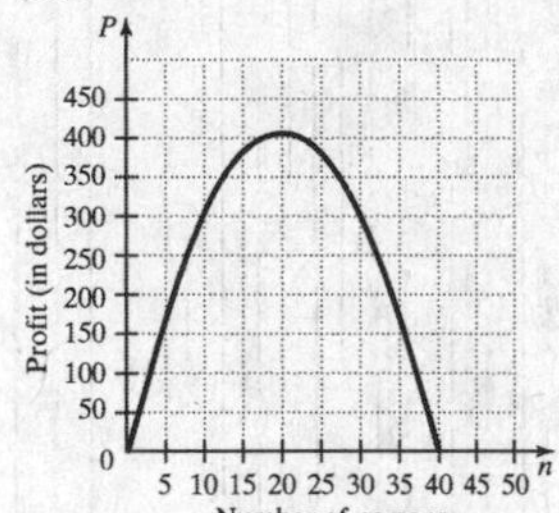

b. The profit is \$384 per person if 16 people go.

c. Since the profit is \$351 for both, either 13 people or 27 people went to the Bahamas.

d.

The maximum profit per person that the agency can make is \$400 (when 20 people go to the Bahamas).

32.

a.

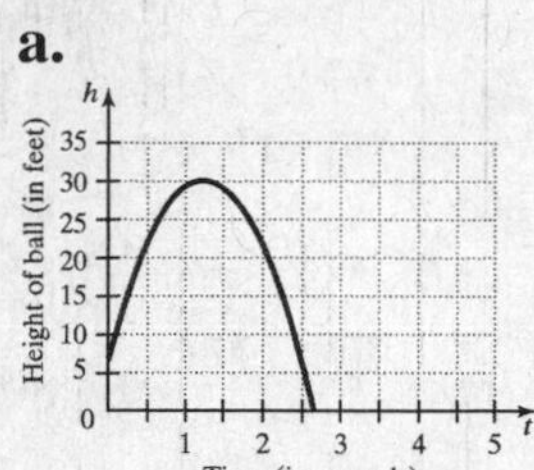

b. The ball reaches its maximum height in $\frac{5}{4}$ or 1.25 seconds.

c.

The maximum height of the ball is 31 feet.

33.

$$\frac{3}{x}+\frac{3}{x+8}=1$$
$$3x+24+3x=x^2+8x$$
$$x^2+2x-24=0$$
$$(x-4)(x+6)=0$$
$$x-4=0 \quad x+6=0$$
$$x=4 \quad x=-6$$

Working alone, it would take the experienced graphics designer 4 days to develop the website and it would take the inexperienced graphics designer 12 days to develop the website.

34.

$$\frac{15}{x}+\frac{15}{x-10}=1$$
$$15x-150+15x=x^2-10x$$
$$x^2-40x+150=0$$
$$x=\frac{-(-40)\pm\sqrt{(-40)^2-4(1)(150)}}{2(1)}$$
$$x=\frac{40\pm\sqrt{1000}}{2}=\frac{40\pm10\sqrt{10}}{2}$$
$$x\approx 36 \quad x\approx 4$$

Her average driving speed to work is 36 miles per hour.

Chapter 9 Posttest

1.

$$4(x+1)^2=32$$
$$(x+1)^2=8$$
$$x+1=\pm\sqrt{8}$$
$$x+1=\pm2\sqrt{2}$$
$$x=-1+2\sqrt{2} \quad x=-1-2\sqrt{2}$$

check:

$$4\left(-1+2\sqrt{2}+1\right)^2\overset{?}{=}32$$
$$4\left(2\sqrt{2}\right)^2\overset{?}{=}32$$
$$32=32$$

$$4\left(-1-2\sqrt{2}+1\right)^2\overset{?}{=}32$$
$$4\left(-2\sqrt{2}\right)^2\overset{?}{=}32$$
$$32=32$$

2.

$$V=\frac{1}{3}\pi r^2h$$
$$3V=\pi r^2h$$
$$\frac{3V}{\pi h}=r^2$$
$$r=\sqrt{\frac{3V}{\pi h}}$$
$$r=\frac{\sqrt{3\pi hV}}{\pi h}$$

3.

$$\left(\frac{1}{2}\cdot(-1)\right)^2=\frac{1}{4}$$

4.

$2y^2-6y-2=0$

$y^2-3y=1$

$y^2-3y+\frac{9}{4}=1+\frac{9}{4}$

$\left(y-\frac{3}{2}\right)^2=\frac{13}{4}$

$y-\frac{3}{2}=\pm\frac{\sqrt{13}}{2}$

$y=\frac{3\pm\sqrt{13}}{2}$

check :

$2\left(\frac{3+\sqrt{13}}{2}\right)^2-6\left(\frac{3+\sqrt{13}}{2}\right)-2\overset{?}{=}0$

$11+3\sqrt{13}-9-3\sqrt{13}-2\overset{?}{=}0$

$0=0$

$2\left(\frac{3-\sqrt{13}}{2}\right)^2-6\left(\frac{3-\sqrt{13}}{2}\right)-2\overset{?}{=}0$

$11-3\sqrt{13}-9+3\sqrt{13}-2\overset{?}{=}0$

$0=0$

5.

$2y^2+3y-4=0$

$y=\frac{-3\pm\sqrt{3^2-4(2)(-4)}}{2(2)}$

$y=\frac{-3\pm\sqrt{41}}{4}$

check :

$2\left(\frac{-3+\sqrt{41}}{4}\right)^2+3\left(\frac{-3+\sqrt{41}}{4}\right)-4\overset{?}{=}0$

$\frac{16}{4}-4\overset{?}{=}0$

$0=0$

$2\left(\frac{-3-\sqrt{41}}{4}\right)^2+3\left(\frac{-3-\sqrt{41}}{4}\right)-4\overset{?}{=}0$

$\frac{16}{4}-4\overset{?}{=}0$

$0=0$

6.

$6x^2=72$

$x^2=12$

$x=\pm\sqrt{12}$

$x=2\sqrt{3}\quad x=-2\sqrt{3}$

check :

$6\left(2\sqrt{3}\right)^2=72\quad 6\left(-2\sqrt{3}\right)^2=72$

$72=72\qquad 72=72$

7.

$x^2+6x=40$

$x^2+6x-40=0$

$(x-4)(x+10)=0$

$x-4=0\quad x+10=0$

$x=4\qquad x=-10$

check :

$4^2+6(4)\overset{?}{=}40\quad (-10)^2+6(-10)\overset{?}{=}40$

$16+24\overset{?}{=}40\qquad 100-60\overset{?}{=}40$

$40=40\qquad 40=40$

8.

$$25(y-4)^2=49$$

$$(y-4)^2=\frac{49}{25}$$

$$y-4=\pm\sqrt{\frac{49}{25}}$$

$$y-4=\frac{7}{5}\quad y-4=-\frac{7}{5}$$

$$y=\frac{27}{5}\qquad y=\frac{13}{5}$$

check :

$$25\left(\frac{27}{5}-4\right)^2\stackrel{?}{=}49\quad 25\left(\frac{13}{5}-4\right)^2\stackrel{?}{=}49$$

$$49=49\qquad\qquad 49=49$$

9.

$$4x^2-8x-3=0$$

$$x=\frac{-(-8)\pm\sqrt{(-8)^2-4(4)(-3)}}{2(4)}$$

$$x=\frac{8\pm\sqrt{(-8)^2-4(4)(-3)}}{2(4)}$$

$$x=\frac{8\pm\sqrt{112}}{8}$$

$$x=\frac{2+\sqrt{7}}{2}\quad x=\frac{2-\sqrt{7}}{2}$$

check :

$$4\left(\frac{2+\sqrt{7}}{2}\right)^2-8\left(\frac{2+\sqrt{7}}{2}\right)-3=0$$

$$0=0$$

$$4x\left(\frac{2-\sqrt{7}}{2}\right)^2-8\left(\frac{2-\sqrt{7}}{2}\right)-3=0$$

$$0=0$$

10.

$$(3x-4)(x+4)=-12$$

$$3x^2+8x-16=-12$$

$$3x^2+8x-4=0$$

$$x=\frac{-8\pm\sqrt{8^2-4(3)(-4)}}{2(3)}$$

$$x=\frac{-8\pm\sqrt{112}}{6}$$

$$x=\frac{-8\pm4\sqrt{7}}{6}$$

$$x=\frac{-4+2\sqrt{7}}{3}\quad x=\frac{-4-2\sqrt{7}}{3}$$

check :

$$\left(3\left(\frac{-4+2\sqrt{7}}{3}\right)-4\right)\left(\left(\frac{-4+2\sqrt{7}}{3}\right)+4\right)\stackrel{?}{=}-12$$

$$\left(-8+2\sqrt{7}\right)\left(\frac{8+2\sqrt{7}}{3}\right)\stackrel{?}{=}-12$$

$$\frac{-64+28}{3}=-12$$

$$-12=-12$$

$$\left(3\left(\frac{-4-2\sqrt{7}}{3}\right)-4\right)\left(\left(\frac{-4-2\sqrt{7}}{3}\right)+4\right)\stackrel{?}{=}-12$$

$$\left(-8-2\sqrt{7}\right)\left(\frac{8-2\sqrt{7}}{3}\right)\stackrel{?}{=}-12$$

$$\frac{-64+28}{3}=-12$$

$$-12=-12$$

11.

$$\frac{3}{n-1}+\frac{5}{n+1}=1$$

$$3n+3+5n-5=n^2-1$$

$$n^2-8n+1=0$$

$$n=\frac{-(-8)\pm\sqrt{(-8)^2-4(1)(1)}}{2(1)}$$

$$n=\frac{8\pm\sqrt{60}}{2}=\frac{8\pm2\sqrt{15}}{2}$$

$$n=4\pm\sqrt{15}$$

check :

$$\frac{3}{4+\sqrt{15}-1}+\frac{5}{4+\sqrt{15}+1}\overset{?}{=}1$$

$$\frac{3}{3+\sqrt{15}}+\frac{5}{5+\sqrt{15}}\overset{?}{=}1$$

$$\frac{30+8\sqrt{15}}{30+8\sqrt{15}}\overset{?}{=}1$$

$$1=1$$

$$\frac{3}{4-\sqrt{15}-1}+\frac{5}{4-\sqrt{15}+1}\overset{?}{=}1$$

$$\frac{3}{3-\sqrt{15}}+\frac{5}{5-\sqrt{15}}\overset{?}{=}1$$

$$\frac{30-8\sqrt{15}}{30-8\sqrt{15}}\overset{?}{=}1$$

$$1=1$$

12.

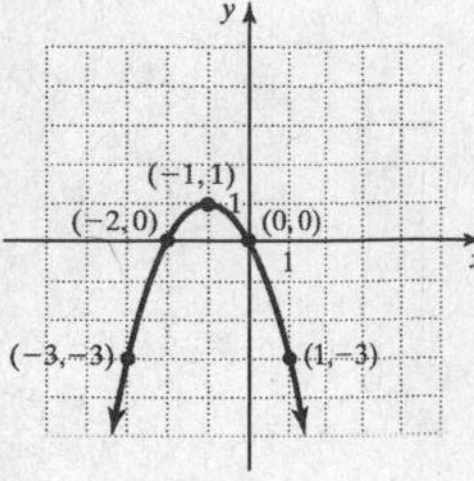

13.

$$x=\frac{-b}{2a}=\frac{-3}{2(1)}=-\frac{3}{2}$$

$$y=\left(-\frac{3}{2}\right)^2+3\left(-\frac{3}{2}\right)+4=\frac{7}{4}$$

axis of symmetry: $x = -\frac{3}{2}$;

vertex = $\left(-\frac{3}{2},\frac{7}{4}\right)$

14.

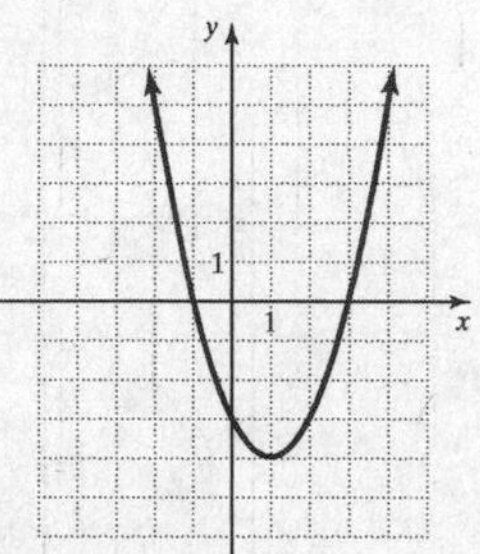

15.

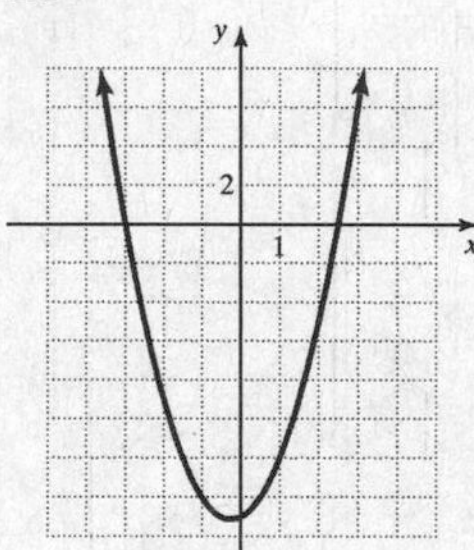

16.

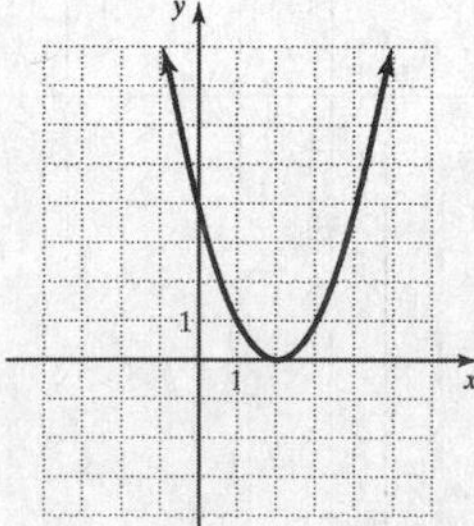

17.a.

$l = 0.81t^2$

$t^2 = \dfrac{l}{0.81}$

$t = \sqrt{\dfrac{l}{0.81}} = \dfrac{\sqrt{l}}{0.9}$

b. $t = \dfrac{\sqrt{3.6}}{0.9} \approx 2$

The period of the pendulum is approximately 2 seconds.

c. $t = \dfrac{\sqrt{90}}{0.9} \approx 10.5$

The period of the pendulum is approximately 10.5 seconds.

18.

$w(3w+4) = 24$

$3w^2 + 4w - 24 = 0$

$w = \dfrac{-4 \pm \sqrt{4^2 - 4(3)(-24)}}{2(3)}$

$w = \dfrac{-4 \pm \sqrt{304}}{6}$

$w \approx 2.2$

$3w + 4 \approx 10.6$

The length of the garden is about 2.2 yards and the length is about 10.6 yards.

19.

$4(x+2)(x+5) = 280$

$4x^2 + 28x + 40 = 280$

$x^2 + 7x + 10 = 70$

$x^2 + 7x - 60 = 0$

$(x-5)(x+12) = 0$

$x - 5 = 0 \quad x + 12 = 0$

$x = 5$

x = 5 inches

20.a.

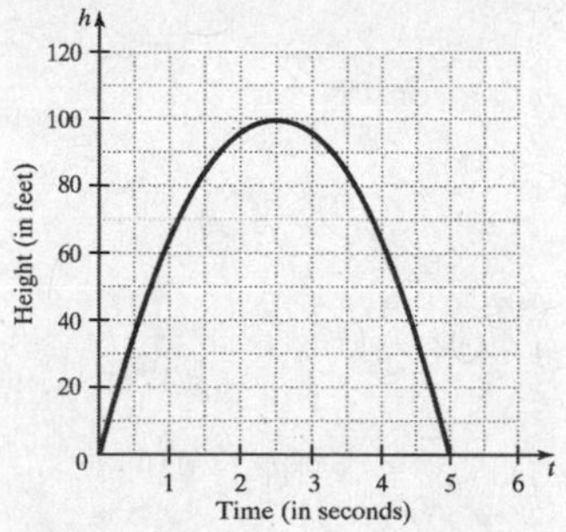

b. The object reaches a maximum height of 100 feet (when t = 2.5 seconds).

Cumulative Review

1. $(3(-2)+5)^2 + -9 = (-1)^2 + (-9) = -8$

2.

$7 + 9[11 - 4(3x + 6 - x)] =$

$7 + 9[11 - 4(2x + 6)] =$

$7 + 9[11 - 8x - 24] =$

$7 + 9[-13 - 8x] =$

$7 - 117 - 72x =$

$-72x - 110$

3.

$(3n-2)(5+6n) =$

$(3n-2)(6n+5) =$

$18n^2 + 15n - 12n - 10 =$

$18n^2 - 3n - 10$

4.

$9c^2d^2 + 6c^3d + 3cd^3 =$

$3cd\left(3cd + 2c^2 + d^2\right) =$

$3cd\left(2c^2 + 3cd + d^2\right) =$

$3cd\left(2c + d\right)\left(c + d\right)$

5.

$$\frac{4x^3}{y^2-4} \div \frac{6xy-18x}{y^2-y-6} =$$

$$\frac{\overset{2x^2}{\cancel{4x^3}}}{(y-2)\underset{1}{\cancel{(y+2)}}} \bullet \frac{\overset{1}{\cancel{(y-3)}}\,\overset{1}{\cancel{(y+2)}}}{\underset{3}{\cancel{6x}}\,\underset{1}{\cancel{(y-3)}}} =$$

$$\frac{2x^2}{3(y-2)}$$

6.

$$\frac{c}{c+6} = \frac{1}{c+2}$$

$$c^2 + 2c = c + 6$$

$$c^2 + c - 6 = 0$$

$$(c-2)(c+3) = 0$$

$$c-2=0 \quad c+3=0$$

$$c=2 \quad c=-3$$

check :

$$\frac{2}{2+6} \stackrel{?}{=} \frac{1}{2+2} \qquad \frac{-3}{-3+6} \stackrel{?}{=} \frac{1}{-3+2}$$

$$\frac{2}{8} \stackrel{?}{=} \frac{1}{4} \qquad \frac{-3}{3} \stackrel{?}{=} \frac{1}{-1}$$

$$\frac{1}{4} \stackrel{?}{=} \frac{1}{4} \qquad -1 \stackrel{?}{=} -1$$

7.

a.

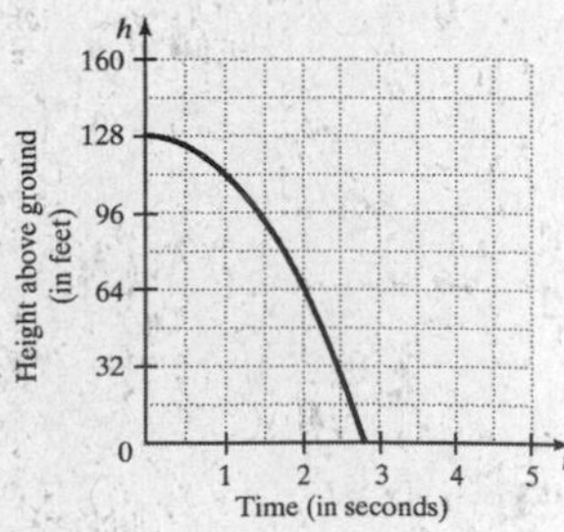

b.

$$t=0$$

$$h = -16(0)^2 + 128 = 128$$

$$t=1$$

$$h = -16(1)^2 + 128 = 112$$

$$128 - 112 = 16$$

The sandbag has fallen 16 feet. At t = 0, the sandbag is at a height of 128 feet above the ground and at 1 second the sand bag has fallen to a height of 112 feet above the ground. So it has fallen a total of 128 – 112 feet, or 16 feet.

8. a. $n = \frac{5}{6}t + 2$

b.

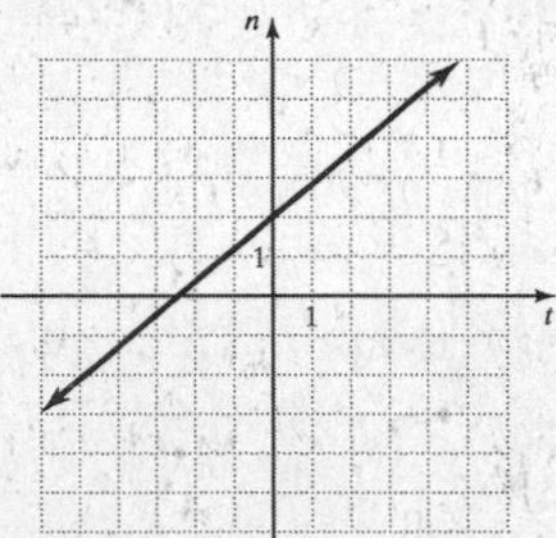

c. The slope of the graph, $\frac{5}{6}$, represents the rate of increase in the number of employees (5 employees every 6 months). In the equation it is the coefficient of t.

9.

x	$400 \bullet 3^x$	=
0	$400 \bullet 3^0$	400
1	$400 \bullet 3^1$	1200
2	$400 \bullet 3^2$	3600

For x = 0, the number of bacteria is 400; for x = 1, the number of bacteria is 1200; for x = 2, the number of bacteria is 3600.

10.

$$\frac{\sqrt{x^3y^4}}{\sqrt{xy}} = \sqrt{\frac{x^3y^4}{xy}} = \sqrt{x^2y^3} = xy\sqrt{y}$$